The Theory of Ecology

THE THEORY OF
Ecology

Edited by

Samuel M. Scheiner *and* Michael R. Willig

The University of Chicago Press

Chicago *and* London

Samuel M. Scheiner is a theoretical biologist and has been on the faculty of Northern Illinois University and Arizona State University. **Michael R. Willig** is professor of ecology and evolutionary biology and director of the Center for Environmental Sciences and Engineering, both at the University of Connecticut.

The University of Chicago Press, Chicago 60637
The University of Chicago Press, Ltd., London
© 2011 by The University of Chicago
All rights reserved. Published 2011
Printed in the United States of America

20 19 18 17 16 15 14 13 12 11   1 2 3 4 5

ISBN-13: 978-0-226-73685-3   (cloth)
ISBN-10: 0-226-73685-7   (cloth)
ISBN-13: 978-0-226-73686-0   (paper)
ISBN-10: 0-226-73686-5   (paper)

Library of Congress Cataloging-in-Publication Data

The theory of ecology / edited by Samuel M. Scheiner and Michael R. Willig.
　　p. cm
　Includes bibliographical references and index.
　ISBN-13: 978-0-226-73685-3 (cloth : alk. paper)
　ISBN-10: 0-226-73685-7 (cloth : alk. paper)   1. Ecology—Philosophy.
I. Scheiner, Samuel M., 1956–   II. Willig, Michael R.
　QH540.5.T47 2011
　577.01—dc22

2011001798

♾ The paper used in this publication meets the minimum requirements of the American National Standard for Information Sciences—Permanence of Paper for Printed Library Materials, ANSI Z39.48-1992.

SMS:

To Mark Courtney, whose many years of dedicated service have done much to advance theory in ecology.

MRW:

To Michael A. Mares, whose simultaneous skepticism and regard for theory continue to provide guidance.

CONTENTS

FOREWORD

Ecology emerged as a self-conscious discipline from the end of the 19th and into the beginning of the 20th century. In addition to describing natural systems, researchers at that time used experiments to uncover the causes of patterns that they observed in nature and speculated about how ecological systems worked. Theorizing characterized, for example, Frederic Clements's ideas on succession in plant communities. Individuals also began calling themselves ecologists, universities and colleges began offering courses in ecology, and scientific societies in ecology were formed.

In *Philosophy of Natural Science* Carl Hempel wrote: "Theories are usually introduced when previous study of a class of phenomena has revealed a system of uniformities that can be expressed in the form of empirical laws. Theories then seek to explain these regularities and, generally, to afford a deeper and more accurate understanding of the phenomena in question" (Hempel 1966, p. 70).

The papers in this volume make it clear that by Hempel's definition, theory development in ecology is alive and well. Samuel Scheiner and Michael Willig have pulled together a creative group to lead a journey from the earliest use of theory in ecology to its modern developments across a range of ecological domains.

Historically, ecological theory has had both nonquantitative and quantitative elements. Although the roots of the latter reach deep into the 19th century, the expression of ecological theory in the form of mathematical equations began to flourish in the first half of the 20th century, as nonecologists were attracted to problems in population growth, competition, and predator-prey dynamics. By the 1960s ecology had become "professionalized" as a discipline, and ecologists were building on earlier mathematical frameworks, concepts, and verbal theory to lay the broad foundations of a theoretical ecology that was both quantitative and nonquantitative.

In science, theory demarcates what we know from what we do not; it is the clearest description of the limits of our understanding. Testing or extending basic theory, therefore, is often the fastest way to advance our understanding of a scientific domain. Theory development is central to a vigorous and advancing research area: by this criterion ecology is flourishing.

There is an increasing interest in theory development across all of biology, not only ecology. By several measures, biology is the fastest growing and most rapidly changing science of the late 20th and early 21st centuries. In addition to research in what we can call the core of biology—namely, cell biology, genetics, ecology, and evolution—a rapid integration is under way between the life sciences and other disciplines, creating diverse new interdisciplinary research areas such as biophysics, bioengineering, biomathematics, geobiochemistry, chemical biology, and biology in the computer and information sciences. Theory development in biology is important, therefore, because of its capacity to accelerate advances in many interdisciplinary research areas. The latter is true for ecology in particular.

We are in a time of global change, when advances in the environmental sciences will depend on basic ecological theory. Topics covered in this volume—foraging, the ecological niche, population dynamics, predator-prey interactions, metacommunities, succession, island biogeography, ecosystems, global change, ecological and biogeographical gradients—will all contribute to the basic ecological understanding needed for adaptation and mitigation in a rapidly changing world. But challenges lie ahead.

An area such as sustainability science is an amalgam of disciplines whose scholarship draws from the natural, physical, and social sciences as well as engineering. Theory development in sustainability will draw from traditional disciplines to create an emerging research area. The process of integration affords an opportunity to test the robustness and breadth of applicability of ideas developed in a disciplinary context. But as sustainability science matures as a research area, we can predict that there will be theory developed that is unique to it. We can expect that a theory will be needed that is centered within and distinctive to sustainability and any other highly integrated areas, such as ecological economics, that emerge from a synthesis of disciplines within the environmental sciences. Accomplishing that integration and developing the required theory will not be easy.

Dr. Scheiner and Dr. Willig are known as contributors to both empirical and theoretical ecological research. Their efforts in assembling the authors of this volume reflect yet another important contribution by this accomplished pair of researchers to advancing the discipline of ecology. The authors of each chapter provide us with a look back and an assessment of where we are now. But most importantly, each chapter affords us a jumping-off point—as any good theory should—for future studies that will advance our understanding of our planet and especially the causes of global change.

James P. Collins

Introduction

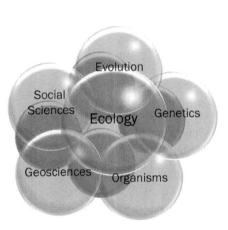

CHAPTER 1

A General Theory of Ecology

Samuel M. Scheiner and Michael R. Willig

In the absence of agreed protocols and overarching theory, Ecology with its numerous subdisciplines, can sometimes resemble an amorphous, post-modern hotel or rabbit warren with separate entrances, corridors and rooms that safely accommodate the irreconcilable.

Grime 2007

The development of theory in ecology is a lively and robust enterprise (Pickett et al. 2007). Despite claims to the contrary, the science of ecology has a long history of building theories that fruitfully guide research and deepen understanding. Our goal with this book is to reveal a selection of those theoretical structures. In doing so, our hope is that ecologists will better appreciate the theoretical frameworks within which they do research, and will more thoroughly engage those theories in designing observational, experimental, and modeling components of their research. Many theories in ecology contain unspoken or even subconscious assumptions. By bringing such assumptions to the forefront, we can understand their consequences, and discover new mechanisms, patterns, and linkages among theories. Theory sometimes seems to be distant or disconnected from everyday practice in ecology. By the end of this book, the relevance of theory to understanding in ecology and its role in advancing science should become clear.

In this chapter, we present a general theory of ecology that serves as the supporting framework—a conceptual infrastructure—for the constitutive theories that appear in subsequent chapters. Although those chapters span the disciplinary range of ecology, they are representative rather than comprehensive. We could not possibly synthesize the full richness of ecological theory in a single book without it becoming encyclopedic. We encourage others to continue the process of theory development in other venues, and to reengage theoretical discourse with ecological research (e.g., Pickett et al. 2007).

We do not claim novelty for the general theory of ecology that we put forward. Quite the contrary, the elements of the general theory have existed for at least 50 years. Many of its principles are implicit in the tables of contents of most ecology textbooks, although our previous treatise (Scheiner and Willig 2008) was their first formal explication. In this chapter, we expand our earlier discussion of the structure of theories and the framework that underlies theory in ecology, providing a foundation for the chapters that follow.

Importantly, we do not claim that the theory presented here is a final version. Rather, it should be considered provisional and ever changing, a general characteristic of theory that is often misunderstood by nonscientists. Indeed, the list of fundamental principles that we present will require additions, deletions, or refinements as ecological theory matures and is confronted by empirical evidence. Critically, this debate can occur only after explication of the theory. In the process of assembling this volume, we convened a workshop of the contributors at the Center for Environmental Sciences and Engineering of the University of Connecticut. At that workshop, a fundamental principle emerged that was not considered in our previous paper (Table 1.3, number 3 below). The theory of ecology is, in turn, embedded within an even broader theory that encompasses all of biology (Scheiner 2010). As that broader theory continues to evolve it may alter the structure of or our understanding of this theory.

The structure of theories

Before we present our general theory of ecology, we must describe the essence of theory and its structure (Tables 1.1 and 1.2). Theories are hierarchical frameworks that connect broad general principles to highly specific models. For heuristic purposes, we present this hierarchy as having three tiers (a general theory, constitutive theories, and models); however, we do not suggest that all theories fit neatly into one of these three categories. Rather, the framework will often stretch continuously from the general to the specific. The three tiers illustrate that continuum, and provide a useful way of viewing that hierarchy. The definitions and principles of the general theory are meant to encompass a wide variety of more specific constitutive theories, which in turn contain

Table 1.1 A hierarchical structure of theories including their components. A general theory creates the framework within which constitutive theories can be articulated, which in turn sets the rules for building models. Conversely, tests of models may challenge the propositions and assumptions of its constitutive theory, which in turn may result in a change in the fundamental principles of the general theory. See Table 1.2 for definitions of terms.

General Theory
Background: domain, assumptions, framework, definitions
Fundamental principles: concepts, confirmed generalizations
Outputs: constitutive theories
Constitutive Theory
Background: domain, assumptions, framework, definitions
Propositions: concepts, confirmed generalizations, laws
Outputs: models
Model
Background: domain, assumptions, framework, definitions, propositions
Construction: translation modes
Outputs: hypotheses
Tests: facts

families of models. This view of constitutive theories as families of models is consistent with how theories are treated across all of biology and in other sciences (van Fraassen 1980; Giere 1988; Beatty 1997; Longino 2002; Pickett et al. 2007; Wimsatt 2007; del Rio 2008; National Research Council 2008).

Each theory or model applies to a domain. The domain defines the universe of discourse— the scope of the theory—delimiting the boundaries within which constituent theories may be interconnected to form coherent entities. Constitutive theories are often most fruitful when they focus on one or a few phenomena in need of explanation (e.g., Hastings Chapter 6; Sax and Gaines Chapter 10). Without such boundaries, we would be faced with continually trying to create a theory of everything.

Nonetheless, we recognize that domains are somewhat arbitrary conceptual constructs and that theories or models may have overlapping domains. Changing the domain of a model can be a fruitful avenue for juxtaposing phenomena or processes that had been considered in isolation. For example,

Table 1.2 Definitions of terms for the theory components in Table 1.1 (modified from Pickett et al. 2007).

Component	Definition
Assumptions	Conditions or structures needed to build a theory or model
Concepts	Labeled regularities in phenomena
Confirmed generalizations	Condensations and abstractions from a body of facts that have been tested
Definitions	Conventions and prescriptions necessary for a theory or model to work with clarity
Domain	The scope in space, time, and phenomena addressed by a theory or model
Facts	Confirmable records of phenomena
Framework	Nested causal or logical structure of a theory or model
Fundamental principle	A concept or confirmed generalization that is a component of a general theory
Hypotheses	Testable statements derived from or representing various components of the theory or model
Laws	Conditional statements of relationship or causation, or statements of process that hold within a universe of discourse
Model	Conceptual construct that represents or simplifies the natural world
Translation modes	Procedures and concepts needed to move from the abstractions of a theory to the specifics of model, application, or test

microeconomic theory uses three concepts—utility, income, and price—to understand consumer choices (Henderson and Quandt 1971; Mansfield 1979). Choices are assumed to maximize utility, subject to income and price constraints. Behavioral ecologists study the economics of choice for nonhuman animals and have applied conceptual constructs and mathematical models from economics to understanding foraging ecology and space utilization (Stephens and Krebs 1986; see Sih Chapter 4). Recent examples of such borrowing of models across domains include the use in ecology of maximum entropy from thermodynamics theory (Harte et al. 2008; McRae et al. 2008) and connectivity models from electrical circuit theory (McRae et al. 2008).

All theories and models contain simplifying assumptions so as to focus other characteristics of a system. The problem with many assumptions is that they are unstated, even subconscious. Making such assumptions explicit sometimes may change the focus of the theory. For example, a fundamental principle of ecology is that ecological traits arise through evolution, but nearly always this is an unstated and ignored assumption. Models of community assembly usually ignore phylogenetic relationships among species. Recently, models that incorporate phylogenetic relationships have added substantially to our understanding of community assembly (e.g., Kraft et al. 2007).

Sometimes, such unstated assumptions can turn around and bite us. Most models of life history evolution assume that organisms can always adopt the optimal phenotype, instantaneously reallocating resources from growth to reproduction, and so ignoring evolutionary and developmental constraints. Ignoring this assumption led to predictions that were biologically improbable, e.g., an organism should allocate 100% of its resources to reproduction one day after devoting 100% of its resources to growth (Schaffer 1983), or an annual plant should switch multiple times between growth and reproduction (King and Roughgarden 1982).

Principles and propositions

When asked to describe a theory, we often think of a set of broad statements about empirical patterns and the processes that operate within a domain. For the sake of clarity, we use different terms to refer to those broad statements when we speak of general theories (fundamental principles) versus when we speak of constitutive theories (propositions). In part, fundamental principles are similar to propositions. Each can be a concept (labeled regularities) or a confirmed generalization (condensations of facts). They differ in that fundamental principles are broader in scope, often encompassing multiple interrelated patterns and mechanisms. Because constitutive theories are meant to guide the building of specific models, their propositions should be more precise statements that represent the potential individual components of those models.

Propositions can be laws: statements of relationship or causation. The propositions are where the fundamental principles of the general theory are integrated. For the general theory of ecology, some of the principles involve patterns, others involve processes, many involve both (see below). Thus, the causal linking of process and pattern, the lawlike behavior that we look for in theories, occurs through the propositions of the constitutive theories.

Laws reside within constitutive theories, and not as part of the general theory, because no single law is required for the construction of the models in all of ecology's subdomains. Several chapters show, however, that ecology is rich in laws that hold within more limited domains (see discussion in Willig and Scheiner Chapter 15). A brisk debate has occurred over whether ecology has any laws at the level of its general theory (e.g., Lawton 1999; Murray 2000; Turchin 2001; Berryman 2003; Simberloff 2004; O'Hara 2005; Pickett et al. 2007; Lockwood 2008), which is related to the debate about laws across all of biology (e.g., Beatty 1997; Brandon 1997; Mitchell 1997; Sober 1997). The continuing search for such laws is an important aspect of a theory's evolution.

The reaction of many to confirmed generalizations is, "Well, isn't that obvious?" In reality, the answer is no. Often such generalizations are obvious only after their explication. Generalizations serve as reminders about assumptions contained in lower-level theories or models. For example, a fundamental principle in ecology is that ecological processes depend on contingencies (see below). Yet many ecological theories and models are deterministic and ignore the role of contingency or stochasticity in molding patterns and processes in nature. Deterministic models are not wrong, just potentially incomplete. Sometimes ignoring contingencies has no effect on model predictions. At other times, the consequences can be profound. As the statistician George E. P. Box is reputed to have said, "Essentially, all models are wrong, but some are useful."

Fundamental principles keep prodding us to test assumptions. For example, one fundamental principle tells us that species are made up of individuals that differ in phenotype. Nonetheless, many ecological theories assume that species consist of identical individuals. Although this is a useful simplification in many instances, it is important to be reminded continually about this assumption and its consequences to predictive understanding. Similarly, many of the fundamental principles consider variation in the environment or species interactions, yet many constitutive theories or models average over that variation (Clark 2010).

Not all assumptions within a constitutive theory derive from the fundamental principles of its general theory. Some assumptions derive from other domains. If an assumption is taken unchanged from another domain it may be unspecified within a theory. For example, all constitutive theories in ecology take as given the conservation of matter and energy, fundamental principles from the domain of physics. We take as given the fundamental principles of any other general theory. As such, we recognize the general tenet of consilience: the entire set of scientific theories must be consistent with each other (Whewell 1858). The decision to explicitly include such assumptions as fundamental principles within the theory under consideration depends on whether those assumptions are subject to test within that theory. Since no theory in ecology would ever test the conservation of matter, it lies outside those theories.

Theories may clash, but such clashes indicate foci of research that advance understanding. In general, theories inhabiting different domains will not clash directly, although results from one domain can point to problems with theories in other domains. For example, studies of geographical distributions of clades of organisms within the domain of historical biogeography became important evidence for the theory of continental drift, a part of the domain of geology. In that instance, the need for a causal mechanism to explain distribution patterns was a factor that led to the development of new fundamental principles in another domain.

Models

At the lowest level of our theory hierarchy are models. Models are where the theoretical rubber meets the empirical road. Many ecological theories are just such models. Although scientific theories encompass a wide variety of types of models, including physical models (e.g., Watson and Crick's ball and wire model of a DNA molecule), in ecology we generally deal with abstract or conceptual models. These models may be analytic, statistical, or computational.

Models are where predictions are made and hypotheses are tested. Those predictions can run the gamut from general qualitative predictions (e.g., increases in primary productivity will lead to increases in species richness) to very specific quantitative predictions (e.g., an increase in soil nitrogen of 5 ppm will result in an increase in average species richness of 4.3 species). The prediction can be a point estimate if the model is deterministic, or it can be a distribution of values if the model is stochastic. The models that make those predictions can be very simple (e.g., equation 7.1 in Holt Chapter 7) or highly complex (e.g., figure 12.4 in Peters et al. Chapter 12). A particular constitutive theory can encompass many different types of models. Because general theories consist of families of models, they very rarely rise or fall based on tests of any one model. Alternative or competing models exist within most theoretical constructs in ecology (e.g., Pickett et al. Chapter 9) allowing a single theory to encompass a diversity of phenomena.

Recognizing that what is often labeled as a theory is but one model within a larger theory can help to clarify our thinking. For example, Scheiner and Willig (2005) assembled an apparently bewildering array of 17 models about species richness gradients into a framework built on just four propositions. A similar process of clarification can be found in Chapter 8, where Leibold shows that all metacommunity theories can be captured within a single framework of just two characteristics: amount of interpatch heterogeneity and dispersal rate. Other chapters in this book provide further examples of model unification. This process of model unification has begun to take hold in other areas of ecology (e.g., McGill 2010). We disagree, however, with McGill's claim that to be unified a theory can contain just a single model. Rather, a strength of our approach to theory unification is the ability of a theory to embrace model diversity.

Because theories often consist of families of models, it is possible for models to be inconsistent or even contradictory. Sometimes, such inconsistencies point to areas that require additional empirical evaluation or model development. But sometimes contradictory models can be maintained side-by-side because they serve different functions or are useful under different conditions. For example, in some physics models, light is treated as a particle and in others as a wave. There is no need to insist that contradictory models always be reconciled or that one always prevail. Instead, this apparent contradiction is resolved at a higher level in the theory hierarchy by a more general theory, for example one that allows for both wave-like and particle-like behavior of light. The apparently contradictory models are built from differing sets of propositions arising from different assumptions and thus refer to different domains. In a similar fashion, constitutive theories can be contradictory if they are built with different assumptions.

The domain of ecology

The domain of ecology is the spatial and temporal patterns of the distribution and abundance of organisms, including causes and consequences. Although our definition of the domain spans the definitions found in most textbooks (Pickett et al. 2007; Scheiner and Willig 2008), it differs in two respects. First, our definition includes the phenomena to be understood (i.e., spatial and temporal patterns of the abundance of organisms) and the causes of those phenomena. Some definitions include only the latter (i.e., interactions of organisms and environments). Second, and most strikingly, our definition explicitly includes the study of the consequences of those phenomena, such as the flux of matter and energy.

In general, the domain of a theory defines the objects of interest and their characteristics. Ecological theories make predictions about three types of objects: species, individuals, and traits or consequences of individuals. Parts of ecology (e.g., ecosystem theory) also make predictions about fluxes and pools of elements and energy. However, what makes these theories part of the domain of ecology is that those fluxes and pools are controlled or affected by the activities, abundances, and distribution of organisms. Thus, they are aggregate consequences of species, individuals, or the traits of individuals. Otherwise, ecosystem theory would reside firmly in the domain of the geosciences.

All three types of objects share an important property, variability (see below). This collection of objects distinguishes ecology from other related and overlapping domains. The theory of evolution makes predictions about species and the traits of individuals. Its domain differs from that of ecology in that predictions are always about collections of individuals (e.g., gene frequencies), never about a single individual. In contrast, theories in ecology may make predictions about either collections of individuals or a single individual (e.g., Sih Chapter 4). Because a given object may be part of multiple domains, understanding of that object and its characteristics depends on examining it within the context of all of those domains.

Just as a general theory has a domain, each constitutive theory or particular model has a domain. Explicitly defining each such domain is important for two reasons. First, a domain defines the most central or general topics under investigation. Second, a clear definition indicates which objects or phenomena are excluded from consideration. Many protracted debates in ecology have occurred when proponents or opponents of particular theories or models have attempted to make claims that fall outside a theory's domain. For example, the extensive debates over the causes of large-scale patterns of plant diversity (e.g., Huston 1994; Waide et al. 1999; Mittelbach et al. 2001; Mittelbach et al. 2003; Whittaker and Heegaard 2003) are based on extrapolating to continental and global scales, models that are valid only at a regional scale (Fox et al. Chapter 13).

Overlapping domains

The domain of the theory of ecology overlaps substantially with several other domains (Scheiner 2010). Of course, all scientific domains overlap in some fashion, but we speak here of those domains that make predictions about some of the same objects of study as does the theory of ecology, or constitutive theories that use fundamental principles from other domains. A constitutive theory can straddle two or more general theories if some of its models ultimately address a central question of each general theory. One way to decide whether a constitutive theory straddles two general theories is to consider the assumptions of those general theories. If the constitutive theory simply accepts all of the assumptions in a particular general theory and never questions or tests them, it likely is not a member of that general theory.

A corollary of the previous statement is that any given model of necessity explores or tests one or more of the assumptions, fundamental principles or propositions of a theory. For example, a continuing issue in ecology concerns the identity of parameters that can be treated as constants and those that need to be treated as variables in a particular theory or model. If a parameter is treated as a constant, the average value of that parameter is assumed to be sufficient because either the variation has no effect or acts in a strictly additive fashion relative to the causative mechanisms under examination.

In some instances, ecologists make assumptions without ever testing them. For example, it is reasonable to assume that we can average over quantum fluctuations (from the domain of physics) in ecological processes. On the other hand, the physiological variations that occur in a mammal so as to maintain body temperature (from the domain of the theory of organisms) (Scheiner 2010; Zamer and Scheiner in prep.) may matter for ecological processes and should not be averaged in some instances. For example, basal metabolic rates in large mammals can vary substantially between winter and summer. Failure to account for this variation can seriously overestimate winter energy expenditures and underestimate summer energy expenditures and the concomitant consequences for food intake requirements (Arnold et al. 2006).

A subdomain can overlap two domains. For example, ecosystem science has some constituent theories that are part of ecology and some that are part of the geosciences. Such overlaps can extend to the level of individual models. For example, foraging theory (Sih Chapter 4) contains some models that are ecological, others that are evolutionary, and others that are both. This sharing of subdomains shows that the boundaries of domains are not distinct and can be somewhat arbitrary.

A domain as defined by a general theory, constitutive theory, or model should be a coherent entity. Some named areas are not domains, but collections of domains. For example, evolutionary ecology consists of a set of constituent theories, some of which are within the domain of the theory of ecology and others that are within the domain of the theory of evolution.

The fundamental principles of ecology

The general theory of ecology consists of eight fundamental principles (Table 1.3). The roots of these principles can be traced to the origins of ecology in the 19th century. They were in place and widely accepted by the 1950s, were recently codified as the components of a general theory (Scheiner and Willig 2008), and continue to evolve (compare this treatment with somewhat different versions in Scheiner and Willig 2008, and Scheiner 2010). In par-

Table 1.3 Eight fundamental principles of the general theory of ecology (modified from Scheiner and Willig 2008; Scheiner 2010)

1. Organisms are distributed in space and time in a heterogeneous manner.

2. Organisms interact with their abiotic and biotic environments.

3. Variation in the characteristics of organisms results in heterogeneity of ecological patterns and processes.

4. The distributions of organisms and their interactions depend on contingencies.

5. Environmental conditions as perceived by organisms are heterogeneous in space and time.

6. Resources as perceived by organisms are finite and heterogeneous in space and time.

7. Birth rates and death rates are a consequence of interactions with the abiotic and biotic environment.

8. The ecological properties of species are the result of evolution.

ticular, we have added an eighth fundamental principle (number 3), so that the numbering of this set differs somewhat from our previous list, and revised the wording of several others.

Heterogeneous distributions

The first fundamental principle—the heterogeneous distribution of organisms—is a refinement of the domain of the theory of ecology. The heterogeneity of distributions is one of the most striking features of nature: all species have a heterogeneous distribution at some if not most spatial scales. Thus, this principle encompasses a basic object of interest, is its most important property, and serves to guide the rest of the theory. All of the other parts of the theory of ecology serve to either explain this central observation or to explore its consequences. Arguably, the origins of ecology as a discipline and the first ecological theories can be traced to its recognition (Forster 1778; von Humboldt 1808). This heterogeneous distribution is both caused by and a cause of other ecological patterns and processes.

Environmental interactions

The second fundamental principle—interactions of organisms—includes within it the vast majority of ecological processes responsible for heterogeneity in time and space. They include both intraspecific and interspecific interactions such as competition, predation, and mutualism, as well as feedbacks between biotic and abiotic components. Within this principle, particular interactions that are part of constituent theories act to elaborate the general theory (see later chapters). Many definitions of ecology are restatements of this principle (Scheiner and Willig 2008).

Variation of organisms

The third principle—the variation of organisms—is the result of processes that derive from the theory of organisms (Scheiner 2010; Zamer and Scheiner in prep.). Ecological theories make predictions about the characteristics or aggregate properties of species, individuals, or traits. The majority of ecological theories make predictions about species or collections of species (e.g., species richness of communities; see Chapters 8–10, 13, 14). Some theories, such as population ecology and behavioral ecology, concern themselves with predictions about individuals or collections of individuals (e.g., numbers of individuals in a population; see Chapters 4–8). Some theories make predictions about the properties of individuals or species (e.g., body size distributions; see Chapters 4, 5, 8, 10, 13, 14). Finally, some theories make predictions about the aggregate properties of individuals or species (e.g., ecosystem standing biomass; see Chapter 11).

Groups of species or individuals share the property that the members of those groups differ in their characteristics, even though many theories and models assume invariance. For example, one of the most common hidden assumption in models of species richness is that all individuals within a species are identical (e.g., Fox et al. Chapter 13). Such assumptions may be reasonable for the purposes of simplifying models. Violations of this assumption may not substantially change predictions. However, in some cases relaxing this assumption has led to substantial changes in predictions. For example, when the chances of survival are allowed to vary among individuals within a population, treating all individuals as identical turns out to substantially misestimate the risk of local extinction from demographic stochasticity; depending on the model used for reproduction, treating all individuals as identical can over- or underestimate that risk (Kendall and Fox 2003).

Contingency

The fourth fundamental principle—contingency—has grown in importance in ecological theory and now appears in a wide variety of constituent theories and models. By contingency we mean the combined effects of two processes—randomness and sensitivity to initial conditions. Contingency is an important cause of the heterogeneous distribution of organisms, both at very large and very small extents of time and space (e.g., a seed lands in one spot and not another; a particular species arises on a particular continent). This principle exemplifies the dynamic nature of a theory. A theory is constantly evolving, although substantive change typically occurs over decades. One hallmark of that dynamic is the emergence of new principles, such as this one, which arose during the 1960s to 1980s.

Heterogeneity of environmental conditions

The fifth fundamental principle—environmental heterogeneity—is a consequence of the interaction of processes from the theory of organisms and the theories of earth and space sciences when the environmental factors are abiotic, as well as the consequences of the second principle when those factors are biotic. For example, seasonal variation in temperature is the result of orbital properties of the Earth, whereas a variety of geophysical processes create heterogeneity in environmental stressors like salt (e.g., wave action near shores) or heavy metals (e.g., geologic processes that create differences in bedrocks). This principle is part of many constituent theories and contains a broad class of underlying mechanisms for the heterogeneous distribution of organisms, as seen in many of the constitutive theories presented in this book. As with the second principle, particular mechanisms pertain to particular constituent theories.

Finite and heterogeneous resources

The sixth principle—finite and heterogeneous resources—is again a consequence of processes from the theory of organisms, and the theories of earth and space sciences or the second principle. Although variation in resources is similar to variation in environmental conditions, a fundamental distinction is the finite, and thus limiting, nature of these resources. Unlike an environmental condition, a resource is subject to competition. For example, seasonal variation in light and temperature are caused by the same orbital mechanisms, but light is subject to competition (e.g., one plant shades another) whereas temperature is a condition and not subject to competition. This distinction in the nature of environmental factors with regard to competitive processes can result in different ecological outcomes. For example, β-diversity in plant communities is high in warm deserts and low in arctic tundra because diversity in warm deserts is controlled by water, a limiting resource, whereas diversity in arctic tundra is controlled by temperature, an environmental condition (Scheiner and Rey-Benayas 1994). Whether a particular environmental factor is a condition or a resource can be context dependent. For example, water is sometimes a resource subject to competition (e.g., plants in a desert) and sometimes a condition (e.g., fish in the ocean). Some heavy metals (e.g., manganese) can be limiting to plants if at low levels, so acting as a resource, and be toxic at high levels, so acting as a condition.

Birth and death

The seventh fundamental principle—the birth and death of organisms—is the result of processes that come from the domain of the theory of organisms (Scheiner 2010; Zamer and Scheiner in prep.). One of the fundamental characteristics of life is reproduction. While birth comes about through cellular and organismal processes, such as fertilization and development, the rate that it occurs depends on interactions of an organism with its environment, such as the uptake of nutrients or mating.

Similarly, a defining characteristic of life is that all organisms are mortal. By "mortal" we mean that no organism is invulnerable, i.e., any organism might die as the result of predation, stress, trauma, or starvation. Thus, the rate of death depends on environmental interactions. We do not mean that all organisms senesce. The senescence of organisms, a decrease in function or fitness with age, is a more narrow version of this principle that would apply to particular constituent theories. This fifth principle forms the basis of a large number of constituent theories concerning phenomena as wide ranging as life histories, behavior, demography, and succession (e.g., Chapters 4, 6, and 9).

Evolution

The eighth principle—the evolutionary cause of ecological properties—is the result of processes that derive from the theory of evolution. The inclusion of evolution within ecological thinking was an important outcome of the Modern Synthesis. Although evolutionary thinking about ecological processes goes back at least to Darwin (1859), evolutionary thinking had been influencing ecology widely since at least the 1920s (Collins 1986; Mitman 1992), and its widespread acceptance occurred primarily in the latter half of the 20th century. The acceptance of this principle led to such disciplines as behavioral ecology (Sih Chapter 4) and population biology, and contributed to the demise of the Clementsian superorganism theory (Clements 1916, 1937).

This principle illustrates how theories in overlapping domains can interact with each other. One of the fundamental principles of the theory of evolution is that evolutionary change is caused primarily by natural selection (Mayr 1982; Scheiner 2010). Fitness differences among individuals, a key component of the process of natural selection, are caused in large part by ecological processes. So ecology drives evolution, which in turn determines ecological properties.

Overview

This chapter only begins to delve into the many issues that relate to theory structure and development in ecology. For a much more comprehensive discussion, we recommend Pickett et al. (2007). One purpose in articulating a general theory is to clarify thinking, bringing to the fore aspects of science that may not be recognized consciously. For example, it is notable that five of the eight fundamental principles are about variability. Although ecologists sometimes decry the variation among the entities that they study and claim that such variation prevents the development of laws or predictions, we suggest that progress in ecology requires that ecologists embrace this variation and explicitly encompass it in theories. More important, recognizing that variation is a pervasive property of our discipline helps explain why ecologists sometimes have difficulty communicating about ecology to colleagues in other disciplines, where the focus is on the shared properties of organisms rather than on their variability.

From the general overview of the theory of ecology given here, Chapters 2 (Kolasa) and 3 (Odenbaugh) consider the role that theory has played in ecology from the perspectives of a practicing ecologist and of a philosopher of science. Then, the eleven chapters that make up the heart of the book delve into the theoretical underpinnings of a broad range of ecological subdisciplines. Each of those chapters develops a constitutive theory by identifying the domain of the theory, listing its propositions, explaining the structure of the theory, and exploring one or more models that can be derived from that theory. In doing so, they show how theory formalization enhances our understanding of the theory and improves our ability to build models. Finally, we provide a brief synthesis chapter highlighting the linkages among the constitutive theories and exploring their similarities and differences in approach to theory development and structure.

Throughout the process of developing and articulating the general theory and the constitutive theories of ecology, we have been impressed by how often the statement and full consideration of the seemingly obvious can lead to deep insights. The chapters that follow demonstrate that process. Our hope is that such insights will substantially improve how we do our science. Ecologists often despair over the seemingly endless variety of their science with no clear overarching structure. The theories discussed in this book present a critical set of steps in unifying that structure.

Acknowledgments

We thank Todd Crowl, Jay Odenbaugh, Steward Pickett, and two anonymous reviewers for thoughtful comments on an earlier draft. Many of the ideas presented in this chapter emerged from or were clarified by a workshop of the contributors to this book. We thank all of the participants for their stimulating interactions. Support to MRW was provided by the Center for Environmental Sciences and Engineering at the University of Connecticut and by a grant (DEB-0614468) from the National Science Foundation in support of long-term ecological research in the Luquillo Mountains. Support for the workshop that brought the chapter authors together was provided by the Center for Environmental Sciences and Engineering in cooperation with the Office of the Vice Provost for Research and Graduate Education at the University of Connecticut. This manuscript is based on work done by SMS while serving at the National Science Foundation. The views expressed in this paper do not necessarily reflect those of the National Science Foundation or the United States Government.

Perspectives on the Role of Theory in Ecology

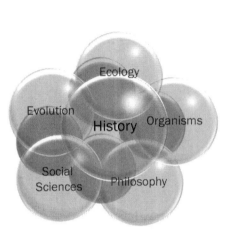

CHAPTER 2

Theory Makes Ecology Evolve

Jurek Kolasa

He who loves practice without theory is like the sailor who boards ship
without a rudder and compass and never knows where he may cast.

Leonardo da Vinci

In this chapter I comment on the importance of theoretical thought in the
development of ecology. I pursue two themes. One draws on a history of ecological theories or ideas that led to the formulation of theoretical frameworks,
and another reviews the benefits and limitations of using theory in the broad
sense. Finally, using these two perspectives, I conclude with comments on the
recent proposal to erect a general theory of ecology.

The determination of when particular theoretical ideas entered and influenced ecology relies on making judgments. However, evaluating the role of
theory in ecology is not easy, particularly because some aspects of theory are
not generally acknowledged as such. Indeed, the task is rather straightforward
when we consider quantitative models (but see Chapters 1 and 3 for the distinction between models and theories; also Pickett et al. 2007) and propositions that are recognized by most if not all ecologists as belonging to the
realm of theory. However, not all theory is recognized immediately as such,
particularly when it is not quantitative or has not been traditionally discussed
as theory in ecological literature.

Qualitative, sometimes informal theoretical ideas play an important role
in focusing, refining, and advancing ecological research as well. To acknowl-

edge and assess their role, it is necessary to present some broad criteria for distinguishing between purely empirical work, if indeed such a thing exists at all, and ideas of substantial theoretical content. Sometimes theoreticians recognize explicitly that math is preceded by crucial conceptualizations. Andrew Sih (Chapter 4) says, for example, that the key challenge for foraging theory (and for optimality theory in general) is not the math but capturing the key elements of the biology of the system. Again, that task appears much easier in retrospect, but it is much harder when theoretical ideas are in the early stages of development.

Another difficulty in evaluating the role of theory stems from the low cohesion of concepts, generalizations, and mathematical models in relation to one another. The lack of cohesion contributes to the intellectual immaturity of ecology (Hagen 1989) in spite of its spectacular growth. Many of the observations identified as the principles of Scheiner and Willig (Table 1.3) have been recognized at various stages of development of ecology, but were not collected methodically into a system of propositions. Once they are, their theoretical significance becomes unambiguous. Whenever possible, I will try to identify their first articulations.

Biologists were interested in ecological questions well before the science of ecology emerged as an identifiable discipline. Indeed, if one defined the domain of ecology as emergence and interactions of ecological entities (organisms and supraorganismal formations such as family groups, herds, or communities although emergence of organisms is an exception, as it falls into the domain of developmental biology) with each other and with their environment (but see Scheiner and Willig 2008; Scheiner and Willig Chapter 1), many biological processes would be included. Any consideration of surrounding biological phenomena is likely to involve some ecology. Greek philosophers were interested in things that today fall within the scope of ecology. As early as 4000 years ago, Idūa, a Sumerian, modeled population growth of a hypothetical heard of cows (Gelb 1967; James Drake, personal communication). Theophrastus developed the conception of an autonomous nature that interacts with man and described various interrelationships among animals and between animals and their environment as early as the 4th century BCE (Ramalay 2008). His accounts entitled "On the Causes of Plants" represent the first known efforts to organize, interpret, and expand knowledge of plant reproduction, requirements, and uses. These early interests in the relations between traits of organisms and the environment grew slowly. Initial questions were pragmatic and had limited theoretical content. Naturalists of 18th and 19th centuries frequently had interest in ecological interactions, even if such interactions were not their main focus. Darwin was acutely aware of the importance of competition and predation, and made a spectacularly successful use of these notions in developing the assumptions of evolutionary theory.

Systematic exploration of pragmatic ecological questions began in earnest in England. The Rothamsted Experimental Station, founded in 1843, initiated a series of experiments between 1843 and 1856 that were aimed at a variety of applied yet clearly ecological problems related to agriculture. Parallel to the practical concerns, ideas that we recognize as modern began to emerge as well. The ecological concept of integrated communities of organisms first appeared in the 19th century with the studies of A. Grisebach (1839), a German botanist. The importance of this perspective was reflected in the need to define the discipline. In 1866, Haeckel recognized and captured the multi-faceted complexity of interrelationships among organisms (see Fig. 2.1). He coined the term "ecology" to carve a separate scope for investigation of those relationships. Other biologists followed Grisebach's ecological approach to natural history studies: Möbius (1883) investigated Danish oyster banks whereas Forbes (1887) described a lake community as a microcosm (a conceptualization that laid the foundation for important modern views such as the ecosystem concept or habitat-based conservation). Many others subsequently explored ecology of water bodies, agricultural systems, or forests. The first professional journal in this area, *American Naturalist*, started publishing in 1867.

Yet when asked about the beginnings of ecology, we will, most likely, hear the names of Forbes, Cowles, Clements, Tansley, Lotka, Volterra, Gause, Gleason, and others, all working in the early 20th century. The question one might pose is why these names specifically? Much was known about natural phenomena before these scientists gained the recognition they deserve. So what made them famous while dozens of others like them, who investigated lakes or vegetation changes, are largely unmentioned by modern accounts?

These individuals stand out in ecology textbooks not because they are associated with explorations of the natural world, but because they proposed new and broad ideas or because they presented empirical studies within the framework of such ideas. Sometimes, they offered formal and flexible tools for developing theory (e.g., Lotka, Volterra, Gause) while sometimes they framed broad classes of phenomena as powerful and appealing concepts (e.g., Forbes, Cowles, Clements, Gleason, and Tansley).

Theoretical components

Before we examine the role of theory in the development of ecology, it is useful to note that scientific theory is a broad notion. Its borders may be fuzzy. This

Figure 2.1 Haeckel's depiction of mosses. Note his desire to place various species together as a natural system. (Haeckel 1904, plate 72.)

is because any empirical research is permeated by theoretical constructs, and any theory has an empirical content, at least in the natural sciences. A couple of extreme examples illustrate this point. Most ecologists would view measurements of temperature as a purely empirical activity. I agree that collecting temperature data is an empirical activity. However, temperature, defined as a mean kinetic energy of molecules, is a complex theoretical construct. Mean is a theoretical term. Kinetic energy is a theoretical term. A molecule is a rich theoretical conceptualization of small portions of matter (also a theoretical notion). Finally, temperature indicators, whether mercury, bimetallic sensor, or some other device, involve a whole suite of theoretical assumptions that link temperature to directly observed behavior of some other material processes.

Thus, while measuring temperature data is an empirical activity, in doing so ecologists subscribe to a plethora of theoretical concepts. At the other end of the spectrum lie concepts such as a community. Community ecologists routinely measure community metrics, which implies acceptance of a number of theoretical premises such as the existence of links between the metrics and community attributes. For example, the rationale for measuring evenness or Shannon-Wiener diversity indices relies on the belief that they convey information about processes determining the distribution of abundances, which may or may not be true. Meaningful empirical work is difficult to imagine without first accepting a set of theoretical constructs.

Indeed, because empirical and theoretical constituents combine to form various terms that ecologists use, it is difficult to clearly separate the theoretical and observational components to the development of understanding. Such a difficulty may be particularly serious when the conceptual device is not fully developed (one might think, for example, of a historical situation when heat and temperature were not yet clearly defined as different concepts). In ecology, such difficulties arose when the potential for, process of, and result of competition were used interchangeably to some degree and stimulated a debate about "the ghost of competition past" (e.g., Strong 1984; Keddy 1989). Furthermore, some situations, such as interference competition, may fail to fit neatly either the process-based view of competition or the outcome-based view because it emphasizes one-on-one interactions among individuals as opposed to the population level outcome or population level resource limitations. Different meanings may lead to different experimental tests and different interpretations of results (Pickett et al. 2007).

The critical question we thus face is when we can usefully credit theory as a driving factor for scientific advances and when credit should be given to empirical discoveries. I will not attempt to resolve this methodological question. Rather, I will restate ideas that may help in addressing the role of theory in shaping progress in ecology, without claiming that these ideas represent more than a convenience. Addressing this question requires accepting that theory and its various components exist at different levels of development (articulation, clarity, formalization). When ecologists explicitly propose a model such as a competition equation, we have no difficulty accepting it as theoretical construct. When however the theoretical facet of a proposition is less formal, some might raise their brows. So, how do we decide that a proposition has significant theoretical contents? One way of answering this question is to examine a proposition for the nature of its central notions.

Notions

Early in the development of a theory, scientists may employ imprecise or incompletely articulated concepts, metaphors, or analogies—notions. Because notions do not have explicit assumptions or form consistent conceptual structure, they are not considered parts of theory per se. Nevertheless, even at an early stage of development, they may influence the growth of theory as well as inspire and stimulate empirical research. When applicable and convenient, I will identify notions that were seminal to advancement of understanding.

Some of the early proposed notions developed into more formalized structures (e.g., niche), others continue to function informally (e.g., source-sink dynamics or even metacommunity). Furthermore, current definitions of metacommunity (e.g., Leibold Chapter 8) rely on a vague criterion of local communities being "linked by dispersal." This criterion works well at this stage of theory development in spite of being rather unruly: We do not know how much dispersal is needed, at what timescales, and whether it includes all, some, or just one species (Leibold believes one is sufficient, though; Chapter 8). Depending on the arbitrary decisions regarding such specifications, a metacommunity either comprises just one community, the whole earth, or anything in between. Clearly, notions can be helpful without being conceptually polished. Still some notions have been recognized as confusing and therefore not very useful because the confusion may hinder progress (e.g., balance of nature; cf. Wiens 1984).

Generalizations and idealizations

Two other terms are much more diagnostic of theory and thus easier to interpret in the course of the growth and development of ecology. Both have roots in the inductive process. Without delving into the fine distinctions between them, I will attempt to identify generalizations and idealizations that have been made at different times by ecologists, particularly when they led to a flurry of conceptual or empirical activity. In brief, generalization is a statement reflecting regularities extracted from a set of observations—in common parlance is it a condensation of facts. Although a generalization, by virtue of the process through which it is arrived at, has substantial empirical content, it acquires theoretical flavor during its articulation. It is then that its scope is specified (which observations qualify for inclusion), relationships are exposed (what is shared among these observations), and language (definitions of facts and relationships), sometimes formal, settled. For example, in this book (Scheiner and Willig Chapter 1), the generalization that individuals are different (i.e., share the property of being different) from each other is clearly based on numerous observations. However, it is not a sum of observations by any means. The generalization uses a relationship between two individuals of "being different." Thus, a condensation of observations is produced by introducing a new concept, a concept of (non)similarity that does not apply to a single individual (or species) but instead requires at least two entities. This generalization implicitly defines the level of detail and dimensions of comparison. The level of detail may even change along a sliding scale depending on the evolutionary affinity of the individuals being compared: comparing individuals within a species would require consideration of different characteristics than comparing different species within a genus.

Species area curves summarized as $S = CA^z$ (where S is species richness, C is a constant, A is area sampled, and z is a scaling constant) or allometric relationships such as $D_i \propto M_i^{-\frac{3}{4}}$ (where D_i is density of species i and M_i is its mass) are good examples of other empirical generalizations that commonly function in ecology (Gould 1979; Whittaker 1998), in spite of doubts as to their validity (e.g., Scheiner 2004b). One characteristic indicator of generalization is the (initial) absence of theory that is capable of reproducing the underlying pattern without exceptions.

Usually, an empirical generalization provides a strong stimulus for finding a conceptual framework that is capable of providing an explanation for the empirical pattern. This was certainly the case with $S = CA^z$ for which Preston (1962a) developed a statistical explanation and the equilibrium theory of island biogeography provided additional biological mechanisms. Also, the metabolic theory of ecology (Brown et al. 2004) currently proposes answers for the observation that $D_i \propto M_i^{-\frac{3}{4}}$.

Idealizations, on the other hand, are theoretical devices created to explore consequences of a feature or a relationship of interest as if no other interfering factors were at play. To continue with an earlier example, one could assume that species are randomly distributed in homogeneously diverse space (i.e., space in which a mix of features repeats itself over and over; cf. Hutchinson 1961). Then one could derive a formula describing the relationship between diversity and the size of the area sampled, as did Preston (1962a). If one further assumes that species abundance distribution is lognormal, this idealized relationship would produce $S = CA^z$, independent of historical events, uneven habitat availability, or the presence of dispersal barriers. In the process of idealization, habitat properties and species traits are rendered irrelevant. Such idealizations are commonly used in quantitative (unstructured) models where, for example, a population may be viewed as a collection of identical particles all with the same set of attributes (same probability of reproduction, death, resource acquisition, and others, selected as convenient). Similarly, ecosystem models employ a common currency (carbon, calories) to represent movement of materials and energy among components. Each flow, however, being different from other flows in the real world, is thus stripped of all properties save its currency value to create an idealized representation of the ecosystem. A brief trip into the past

Ecology made a much greater use of notions, and generalizations than of formal models and broader theories. Nevertheless, all were combined over time into a loose mix that insured a considerable growth of understanding and conceptual sophistication. At the beginning, ecology was strongly influenced by other sciences, from which it grew, borrowed, or was inspired (McIntosh 1985).

Although the Malthusian geometric population growth model was not developed by an ecologist, its implications influenced Darwin's ideas on the selective pressure of the environment. Darwin brought an ecological process to bear on evolutionary considerations. He may not have cared about the distinction between ecology and evolution, nor even been exposed to Haeckel's later coinage of the term "ecology"; nonetheless, he recognized that the ecological stage is essential for the evolutionary play to proceed. A mathematical model had undoubtedly contributed to the emergence of the most powerful theory of biology, evolution by natural selection, with continuous and deep implications for ecology.

A little later, Forbes (1887) published his oft cited paper, "Lake as a microcosm." Although he may not have been the first to do so, his paper makes use of a number of important concepts that soon became, and still are, at the core of many ecological paradigms. First, and most importantly, Forbes saw that interacting organisms in a habitat form a system. When he referred to predator-prey interactions as an example, he used the phrase "a close community of interests among these . . . deadly foes." Second, by emphasizing the relative autonomy of lake biotic and abiotic processes, he identified another important aspect of ecology—patchiness or partial discreteness of ecological phenomena. These ideas were later pursued in ecosystem (Golley 1993), patch dynamics (Pickett and White 1985), and succession studies (Clements 1916), among others. I note here that Scheiner and Willig's framework (Chapter 1) does not explicitly address the discreteness of ecological things and processes. Perhaps this is not necessary, or perhaps the proposed framework will find room for such aspects of nature later on.

Clements' ideas of a community of species acting as a superorganism are now deemed a failure (Hagen 1989). However, Clements' ideas had a considerable pretheoretical and theoretical content, which provoked and inspired ample empirical work that, ironically, largely intended to disprove his superorganism perspective. Ultimately, these ideas were replaced by a more individualistically oriented interpretation of species assemblages (now also known as the Gleasonian view). Nevertheless, Clements infused into ecology notions that continue to raise their head for good reasons. Although ecological systems do not behave like organisms, they share some features with organisms, even if the expression of these features is much less prominent. Forbes preceded Clements in emphasizing interdependence of components, boundedness, and the equilibrial nature of the ecological systems. Clements' and Forbes' perspectives on communities were pretheoretical. However, it is Clements who made as strong a case as was then possible, and who left ecologists thinking about these issues for a good while. Most recently, Loreau et al. (2003) returned to some of the issues (e.g., component interdependence, conceptualization of ecosystems as interacting entities) in a modern way, thus completing another cycle of refinement of theoretical thinking. Evolutionary ecologists (e.g., Wilson 1997) emphasize that coevolved species have the potential to form communities with meaningful integration (and thus a degree of entitization). Thus, while the ideas of Clements on the organismal nature of communities may not be applicable to a single trophic level such as assemblages of plants, they are far from irrelevant, as noted by Tansley (1935).

However the argument about integration may unfold, it is clear that a meaningful consideration of many ecological processes requires system identification. Often such identification remains an implicit assumption as it has been, for example, during much of the development of the theory of succession. Only recently (see Pickett et al. Chapter 9) has a systematic analysis of the structure of that theory led the authors to augment the model of succession with the concept of system boundary. Bravo!

These observations would be incomplete without commenting on the role of competition and predation models, proposed independently by Lotka (1925) and Volterra (1926). From today's perspective, these models represent more than clever formalizations. Indeed, they have been successful in advancing a field of primarily two-species interactions, in addition to inspiring food web modeling (Cohen 1978) and the examination of the effects of diversity on stability (e.g., Pimm 1980). More importantly, they led ecology by showing that complex processes can be distilled into manageable characteristics whose behavior can be examined using mathematical tools. In doing so, these models inspired many other quantitative forays into ecology. It may not be possible to make direct connections between these models, but their impact on the way ecologists think is undeniable. Every general textbook on ecology presents and explains these models; their form lends itself to analytical solutions and numerical simulations, their generality permitted adding realistic terms (self-limitations, time delays, stochasticity, patchiness) to explore specific factors, and overall, they contributed greatly to the development of quantitative ecology (Silvert 1995).

Landmark theories and concepts appeared in ecology only sporadically. The major ones are shown in Figure 2.2 and Table 2.1. Although I recognize that Figure 2.2 is incomplete in its coverage, and that different ecologists would be likely to compile different lists of theories and concepts, two tentative observations come to mind. First, there was a considerable acceleration in the number of major ideas, often well formalized, between 1930 and 1970. Second is the slowdown in the appearance of new propositions. I do not interpret this slowdown as a real problem. The period in question witnessed tremendous development of ecological theory and its penetration into all facets of ecology, from the language that ecologists use to the design of field research. This growth occurred within the broader theories established earlier, through their refinement, expansion, testing, or application in conservation or management. Chapters in this book provide numerous examples of the continuing progress along those lines. Thus, Figure 2.2 is by no means to be interpreted as evidence of limited growth of ecological theory. The growth has been impressive. However, this growth may have unintended consequences, such as the fragmentation of ecology into subdisciplines dominated by their own concepts, theoretical constructs, and conventions. Finally, it is possible that the newest theories not listed here will need time to register widely in the discipline of ecology before they can appear in an analogue to Figure 2.2.

The history of theoretical developments in ecology involves more than the timing and appearance of those constructs, or their direct effects on the discipline. In the long run, these constructs effectively define ecology, its domain, questions, and directions. They do so in a rather unsystematic manner. Nevertheless, the nature of ecology would be very different if not for the various relationships that theory has with the rest of ecology.

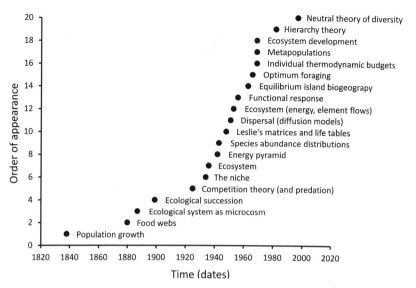

Figure 2.2 Chronology of major theories and concepts. The *y*-axis represents a particular concept. Choice of the initial dates was based on either a foundational paper for a given theory such as Lindeman's (1942) analysis of lake energy flows or Hubbell's (1997) neutral theory of diversity, an early influential paper such as Forbes' (1887) microcosm perspective, or a paper that revived or reinvigorated an already existing and accepted idea such as Odum and Odum's (1953) perspective on the ecosystem concept. The only motivation in each case was to anchor a theory or an important concept in a time period indicative of the beginning of its rise to prominence. Even if one reshuffled the order of appearance to satisfy some other, equally valid criteria, the overall temporal trend would not change. The selection of the theoretical components, with the exception of the three most recent, was guided by the compilation of Real and Brown (1991) and by characteristics of propositions as introduced in Chapter 1.

Theory supports ecology

I believe that ecology has been molded by theory to a much greater extent than commonly recognized. Although this thesis may appear trivial to those interested in theoretical progress and the unification of ecology, a great number of ecologists express some skepticism and even scorn of theoretical endeavors. Revisiting older and more recent developments and arguments shows that theory plays a central role in stimulating ecological research. Indeed, functions and relationships of theory in science are numerous. Some of these functions and relationships can be identified in ecology. Below, I comment on several common facets of the use of theory in ecology, from historical patterns to current evidence for the contribution of or need for theory. These comments

Table 2.1 Major theoretical developments, their approximate year of appearance, and comments on their influence.

Seminal paper	Concept or theory	Comments on influence
Velhurst 1838	Population growth model	Spawned substantial theoretical and empirical work
Camerano 1880	Food web models	Attempted to capture connections among species from the equilibrium perspective
Forbes 1887	The ecological system as microcosm	Articulated a persistent theme of nature's balancing act and oneness
Cowles 1899	Ecological succession	Spawned substantial empirical and conceptual work
Volterra 1926	Competition theory (predation models included)	Laid foundations for a strong and vibrant field of theoretical and empirical studies on interspecific interactions in the context of limiting resources, habitat variability, structured populations, time lags, heterogeneity, and many others
Gause 1934	The niche	Spawned a mix of substantial empirical and theoretical work, primarily but not exclusively on competition; as a nontechnical (pretheoretical) concept, niche had been used earlier (Elton 1927)
Tansley 1935	The climax	Refined climax concept (originally introduced by Cowles 1899); subsequently, the concept greatly affected debates about the nature of change in composition of ecological systems, primarily plant associations
Lindeman 1942	Energy pyramid	Led or interacted with numerous ecological ideas, from foraging theory to metabolic theory of ecology, both in theoretical and empirical realms
Fisher et al. 1943; Williams 1966	Models of species abundance	Spawned voluminous amount of theoretical and empirical work
Leslie 1948	Leslie's matrices and life tables	Provided a foundation for a tremendous amount of refinement to analysis of population dynamics

Seminal paper	Concept or theory	Comments on influence
Skellam 1951	Dispersal (diffusion models)	Boosted primarily theoretical work in the area of foraging theory
Odum and Odum 1953	Ecosystem	Spawned primarily empirical work cast in terms of physical constraints (the ecosystem concept itself had been formulated 20 years earlier)
Holling 1959*	Functional response models	Provided a major building block for optimum foraging theory; spawned a mix of empirical and theoretical studies
MacArthur and Wilson 1963	Equilibrium island biogeography	Became one of the main engines of ecological sciences, with largely empirical but also some theoretical work
MacArthur and Pianka 1966	Optimal foraging theory	Spawned abundant theoretical and empirical research, much with links to evolution
Levins 1969	Metapopulation theory	Provided formal tools for investigation of population dynamics of fragmented habitats and ultimately contributed to the rise of metacommunity framework
Porter and Gates 1969	Individual thermodynamic budgets	Provided a clear framework for ecophysiology and evolutionary ecology
Odum 1969	Theory of ecosystem development (homeostasis)	Refinement of the ecosystem concept
Allen and Starr 1982	Hierarchy theory	Brought awareness of scale to the forefront of data interpretation; suggested cross-scale analyses
Hubbell 1997	Neutral theory of diversity	Spawned considerable amount of empirical (testing) and theoretical (refinement) research; long-term impact still unclear

* Paper presented in 1956 but published in 1959.

focus on ways in which theory stimulates, guides, or assists empirical research in some other ways. Under separate subheadings, I provide examples of situations where theory (e.g., mathematical interpretation of species abundance patterns) inspired efforts to accumulate new observational cases in order to verify patterns suggested by that theory, where broad empirical patterns were generalized (allometric relationships) and tentatively explained by theory (metabolic theory), where a new theory (neutral theory of diversity) led to many tests of its assumptions and predictions, or where early ideas (succession) underwent several cycles of refinement and empirical challenge. I also comment on the practical benefits that theory provides, whether by helping with the design of experiments or by providing an intellectual reassurance for conducting research. Because not all theories that ecology uses have been initially formulated in ecology, I note some links that ecological theories have with other disciplines, and with the theory of evolution in particular.

Theory, although an indispensable vehicle for generating understanding and organizing knowledge, has occasionally hindered progress, at least in other sciences. I note at least one circumstance related to theory that may result in an unnecessary hurdle on the path of advancement of ecology: excessive emphasis on mathematical models at the expense of efforts to formulate a general and unifying theory.

Empirical efforts arise in response to theory

Historically, much empirical research was a response to natural history questions and to theoretical propositions, even if the latter represented unsuccessful attempts to capture and explain observed patterns. Species abundance curves provided a richness of examples. The observation was that most collections of species representative of a local community are characterized by a particular abundance distribution: a few species are abundant, many are rare, and some are in the middle of the abundance range. Initially, this observation was reported in the form of an empirical generalization summarized in the form of a mathematical distribution. Preston (1948) proposed that the abundance of species in a community, when plotted on the log scale, shows a hump and generally fits a lognormal distribution. Later on, the focus shifted to mathematical models with biological underpinnings. MacArthur (1957), Whittaker (1965), May (1975), Tokeshi (1993), and many others proposed a bewildering selection of models mostly based on assumptions about partitioning of resources (e.g., broken stick model), with resources being logically tied to abundance. Others invoked habitat structure (Kolasa and Pickett 1989) or population dynamics (He 2005) to explain species abundance distributions. Ecologists have been and continue collecting data to test (fit) those models.

Broad empirical generalizations beg for theory

The role of empirical generalizations is different from that of theoretical frameworks or models. Perhaps one of the more important aspects of broad empirical generalizations is that they provoke questions about causes of patterns and, as a result, give birth to theoretical explanations and subsequent tests. Examples of this are abundant in most ecology textbooks. Consider allometric relationships and their impact on theory and, more importantly, the continuous interests in those relationships because of theory. Scientists have known for nearly two centuries that larger animals have slower metabolisms than do smaller ones. A mouse must eat about half its body weight every day to avoid starvation; a human gets by on only 2% (Whitfield 2004). This coarse empirical generalization, like others, begged for explanation. The first theories to explain this trend were based on the ratio between body volume and surface area. The square-versus-cube relationship makes the area of a solid proportional to the two-third power of its mass, so metabolic rate should also be proportional to mass. But a thorough study by Kleiber (1932) found that, for mammals and birds, metabolic rate was mass to the power of 0.73 (approximately three-quarters) and not 0.67 (two-thirds as postulated based on surface to volume ratios). Other research supported this new three-quarter-power law, although consensus is still elusive (Grant 2007). It was, however, much harder to find a theoretical reason why metabolic rate should be proportional to mass raised to power of three quarters. Furthermore, it was not clear why quarter-power scaling laws should be so prevalent in biology. Nevertheless, the generalization and the difficulties with the putative causes of it led West et al. (1997) to develop a new explanation of why metabolic rate should equal the three-quarter power of body mass.

Indeed, the generalization that larger organisms have lower metabolism is such a noisy one that some question its relevance. What is important is that an early strong empirical generalization inspired development of sophisticated theoretical explanations. A considerable value of these explanations to ecology arises not from whether they are correct or accepted but rather whether they have been formulated and tested. The very process of doing so identifies further problems and questions, and directs new empirical and theoretical pursuits, which advances understanding.

Another theory formulated in response to an empirical generalization is the neutral theory of species diversity (Hubbell 1997). Most communities, irrespective of the scale of data collection, harshness of the environment, or productivity, show a general qualitative regularity of species abundances: few species are abundant while many more are rare, with species of intermediate abundance filling the range in between according to one of several abundance partitioning models. Numerous theoretical efforts have been made in the past to account for this general pattern (e.g., May 1975; Ugland and Gray 1982; Tokeshi 1993). The most familiar of those is MacArthur's (1957) broken stick model. None has gained full acceptance. Hubbell's proposition is a culmination of these efforts and a stimulus for further tests.

Admittedly, sometimes a good generalization does not lead to the development of a theory (but see Colwell Chapter 14 on gradients regarding a possible solution). Holdridge's schematic, illustrating vegetation types as a function of evapotranspiration and annual precipitation (Fig. 2.3), represents a useful generalization that does not need or inspire a complex theory. In case of vegetation types, a simple combination of nonbiological variables and a few physiological assumptions together provide an excellent explanation of the patterns. This example illustrates further that almost all efforts to systematize

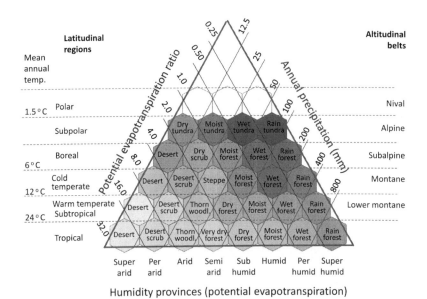

Figure 2.3 Vegetation types are defined by mean annual "biotemperature" (growing season and temperature index), annual precipitation, and a ratio of potential evapotranspiration to mean total annual precipitation. From Holdridge (1947).

ecological information, including generalizations, involve some theoretical content, a point that I presented earlier.

New propositions point toward new fruitful directions

A fairly recent metabolic theory (West et al. 1997) of ecology gave rise to an intensive activity (over 700 citations) that concentrates on two major foci: the establishment of empirical generalizations and the testing of assumptions. While attracting a fair share of praise, it has proved a magnet for criticism as well (Grant 2007). Perhaps one of the most stimulating features of this theory has been its direct predictive power. Predictions are relatively easy to derive and test. The mathematical relationships at its core seem to make it particularly amenable to testing. Few propositions with a broad scope and potential generality have this advantage. However, it is too early to make conclusions about the acceptability and ultimate place of this theory in ecology, as critics have identified various difficulties (e.g., Kozlowski and Weiner 1997; Li et al. 2004; and other comments in Ecology's Forum). Nonetheless, its influence on current research remains high, as demonstrated by the number of papers addressing various aspects of the theory, from empirical patterns to nuances of assumptions.

Old propositions continue to mature and be refined

Succession theory is a prime example of the importance of theory to ecological research and its progress. It started with the simple observation that vegetation changes over time. First attempts to record and explain the patterns of change are due to Cowles (1899) and Clements (1916). These attempts led to more empirical research aimed at verifying the patterns or contesting early generalizations (e.g., succession leads to a climax; Connell 1972). The debates and new data culminated in a sequence of models of succession, of which Connell and Slatyer's (1977) tolerance-inhibition-facilitation model has become common textbook knowledge. Once these more logically precise models were available, it was possible to expose their weakness and to attempt "a professional" grade formulation of theory (see Pickett et al. Chapter 9 for an in-depth account).

Experiments invoke theory

In addition to the impact of well-defined, broad-scope propositions such as succession theory, island biogeography theory, or metabolic theory of ecology that I have already mentioned, theory affects ecology in many indirect ways. As ecology expanded its use of experiments, theory was one of the most common motivations for experimentation. While attributing credit to theory may be difficult for empirical generalizations, particularly those accumulated from numerous cases over long periods of time, experiments almost invariably find roots and rationale in theory. Whether this is a general theory, a specific model, or even a vague model depends on the preferences or awareness of the authors of the experiment. It would be difficult to find a contemporary paper that presents experiments but makes no reference to theory.

Theories provide comfort

Most ecologists are familiar with, or have at least heard of, population growth models, foraging theory, competition models, metapopulations, niche theory, succession theory, or island biogeography theory. This is because much of ecological research is either motivated by these theories and models, or their specialized offshoots, or that theories are used to interpret the results of empirical studies. Theory often provides a compass to studies whose primary purpose may be gathering or compilation of data. In this manner, theory also provides reassurance as to the validity and significance of efforts by offering a context within which to make strategic research decisions. Leonardo da Vinci was right. Without these and some newer or more specialized theories, ecological research would not make much sense beyond some practical cataloging of observations.

General science theories enrich ecological frameworks

Examples of general science theories that enrich ecological frameworks include systems theory and biogeography. In 1982 Allen and Starr (1982) published a book entitled *Hierarchy*. It was meant largely as a methodological book in the broadest sense. The book refined and recast general scientific understanding of how the world is constructed and perceived by scientists, with a special focus on ecology. Although the book presented no quantitative formulations or theory per se, it emphasized the inevitability of changing interpretations of the observed data sets as a function of scale. Its impact (enhanced by a couple of companion books with Allen as coauthor) was impressive. Although few studies directly tested premises that form the constellation of ideas jointly called hierarchy theory, Allen's writings and conference presentations made ecologists aware of the effects different observational scales on the interpretation of results. As of 2008, more than 500 papers appeared in searches with keywords "ecology" and "scale"; this provides evidence for the extensive impact made by this theory (Schneider 2001). In this book, Peters et al. (Chapter 12) draw substantially on the ideas presented by Allen and Starr not so long ago. MacArthur and Wilson's (1967) theory of island biogeography had a similarly stimulating impact. Not only did it encourage numerous tests that incidentally illuminated the importance of a number of additional factors such as island heterogeneity, differential colonization leading to compositional disharmony, or incomplete saturation, but it also became a major conceptual ingredient of conservation strategies (see Burke and Lauenroth Chapter 11).

Evolutionary theory led the way

The theory of evolution has been tapped by ecologists from the beginning, and has provided a sound and rich framework for the studies of habitat selection, parental investment, tradeoffs, and behavioral ecology, just to give a sample of the range of addressed topics. However, the main and readily identifiable signs of the link between the theory of evolution and ecology appeared relatively late (e.g., Lack 1954; Brown and Wilson 1956; Hutchinson 1965). The existence, caliber, and continuous impact of important journals largely devoted to the exploration of the interface between evolution and ecology (*American Naturalist, Evolutionary Ecology Research*) testify to the vitality of evolutionary theory in ecology. They further add to the weight of the argument that theory plays a fundamental role in advancing ecology, even when ecology cannot claim the primary ownership of that theory.

Theory is a vehicle for sharing knowledge across ecological subdisciplines

I offer an informal observation that ecologists share important information across subdisciplines by using theory. Although this sharing may involve simplified versions of theory, it is nevertheless the main material that glues the science of ecology together. For example, whether one conducts research in plant physiology, ant behavior, nitrogen pathways in soil, reproduction of deer, squirrel allocation of time to foraging versus caching, clines of diversity, or global carbon cycle, one is likely to be familiar with most theories and concepts listed in Figure 2.2. Furthermore, theory is often seen as an agent for dissemination of ideas among subdisciplines. Ecologists working on streams, soils, or other habitats reach to general theories for inspiration and guidance concerning questions and problems, and adopt the concepts and definitions associated with such theories. For example, Lake et al. (2007) explicitly call for linking ecological theory with stream restoration. Similarly, Barot et al. (2007)

identify the need for soil ecologists to make greater use of evolutionary theory and modeling to shift emphasis of soil ecology from particularities of empirical observation to generalities that they associate with ecological theories.

Ecologists often think of theory as mathematical models

A strong debate developed in the 1950s and 1960s among the proponents of theory in ecology based on the response of organisms to resources and habitat conditions (Andrewartha and Birch 1954) and those who saw the need for inclusion of evolutionary and community processes (Lack 1954; Orians 1962). Battles for a conceptual vision of ecology, for its domain, main assumptions, and structure of theory are important. Many research programs were undoubtedly influenced by arguments arising in the course of such debates, whether a specific debate was about the inclusion of evolutionary processes, energetic principles, or nonequilibrium perspectives. However, books devoted to theoretical ecology (e.g., Yodzis 1989; Roughgarden 1998; Case 2000) make no mention of this or similar conceptual debates. These texts present theory as collections of mathematical models. Although such models are powerful and illuminating theoretical constructs, their very dominance of the theoretical landscape of ecology may have unintended consequences because it may detract from or undervalue the significance of efforts to reorganize the conceptual framework of ecology. Fortunately, the project initiated by Scheiner and Willig may mitigate this potentially negative effect.

Indeed, there were earlier attempts to refine this framework. Schoener (1986) attempted to accommodate the diversity of models and perspectives by calling for "pluralistic ecology." Restricting his scope to community ecology, Schoener believed that the best approach was to develop separate mathematical models using six primitive (e.g., body size, motility), six environmental (e.g., spatial fragmentation, severity of physical factors), and six derived (e.g., relative importance of competition and predation) axes to classify collections of species into different community types. Different types of communities would then be approached as separate theoretical problems that require separate treatment leading to separate solutions. Others (Reiners and Lockwood 2009) extend this view to all of ecology. I am not convinced, however, that the best course of action in the face of conflicting theoretical positions is a compromise of this kind (see also McIntosh 1987). A plurality of approaches may represent a necessary stage, but little precedent supports the idea that will offer an effective solution to the conceptual mix that ecology offers today. Also, it is likely that pluralism or fragmentation of the domain and methodology will impede discovery of and consensus about empirical generalizations.

Ecological data spurs theory

In contrast to the previous section, empirical research may also be motivated by theory or condensed to form generalizations. Generalizations call for explanations, which are provided by theory, and so the cycle continues. However, empirical discoveries may sometimes open new avenues of inquiry, independently of the influence of any particular theory. The discovery of hydrothermal vents and their associated communities (Lonsdale 1977) had significant implications on ideas about the origin of life (Wächtershäuser 1990), on perspectives concerning food chains, and on thinking in the new field of exobiology. Records of pest outbreaks pose a continuing challenge for theoreticians and stimulate further empirical work and theoretical approaches (Dwyer et al. 2004; Stone 2004). Cases where new observations and discoveries spurred a flurry of research activity are numerous (interspecific carbon exchanges among plants, coral bleaching, acid rain, morphological changes in zooplankton in the presence of predators, appearance of zebra mussels in North America, and many others). However, the meaning and value of such discoveries were quickly enhanced when wedded to a theoretical construct (generalization, working explanatory hypothesis, or broader theory). On their own, empirical findings, unlike theory, rarely give impetus to a robust program and steady research direction.

Difficulties

In spite of numerous examples of theory being important to ecology, the direct use of theory is still modest and perhaps limited by tradition, training, or skepticism. In 2008, the November and December issues of *Ecology* contained a total of 57 articles and comments, of which only 15 were clearly motivated by theory or formal models. Many papers did not invoke theory to any extent. However, even the most empirically oriented papers were firmly seated within modern paradigms of ecology, which itself is strongly shaped by theoretical views and constructs. *The American Naturalist* favors theoretically justified papers to a much greater extent, although many papers in that category involve evolutionary models. For example, the November and December issues in 2008 contained 32 papers, with 25 primarily motivated or driven by theory. The 1929 issues of these two journals had almost no papers involving ecological or evolutionary theory; only 3 of 28 in *Ecology* and 1 of 21 in *American Naturalist* were expressly motivated by theory. A search of JStor (Ecology and Environment section) shows that the number of papers with keywords (see legend to Fig. 2.4) that indicate theoretical motivation or links increases at a faster

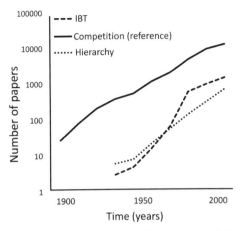

Figure 2.4 Number of papers found after searching for "competition" (deemed as a reference term with a balance of empirical and theoretical connotations), "hierarchy theory + scale" and "island biogeography" by JStor search engine between 1900 (for competition) and 1930 (for Island Biogeography Theory (IBT) and Hierarchy) and the year 2000. The number of all three categories of papers increased exponentially but their slopes were higher for the theory-based papers (0.85 and 1.18 for Hierarchy and IBT, respectively) as compared to competition (0.67).

rate than does the number of papers without such a designation (Fig. 2.4). Whatever the particular trends, the influence of theory on ecology grows quickly and, very likely, faster than that of the observational component.

The growth of theory (together with an associated increase in the depth of analyses), improved expertise in translating theory into lab and field research. Additionally, the development of many specific theories increases the overall influence of theoretical approaches. This growth, however, continues to contribute to the splitting and fragmentation of ecology into areas with rather independent existences. Growth is good, but fragmentation introduces and nurtures conceptual incongruities that, in my view, hamper ecology. In support of this idea, I cite two examples: one explicitly identifies the deficiencies due to conceptual isolation whereas the other shows how unification and progress could be achieved if cross-fertilization of ideas took place. For example, a considerable debate developed about the nature of competition among plants. Specifically, the relationship between nutrient availability as affected by potential competitors, allocation to root growth, and consequences of these factors for plant growth and reproduction developed as major concerns among plant ecologists (Craine 2007). Yet specific answers are unlikely to interest animal ecologists, and the conceptual refinements associated with them are unlikely to apply beyond plant ecology. The second example concerns two separate research traditions of parasite-host and predator-prey ecology. Raffel et al. (2008) argued that developments in predator-prey ecology, such as temporal risk allocation and associational resistance, can contribute to development of new hypotheses for parasite-host systems. Conversely, concepts developed in parasite-host ecology, such as threshold host densities and phylodynamics, might enrich predator-prey ecology (e.g., Holt Chapter 7). Propositions such as trait-mediated indirect effects and enemy-mediated facilitation provide opportunities for the two fields to forged a shared theoretical perspective that would foster advances in both fields (Raffel et al. 2008).

This troubling and somewhat perennial state of affairs motivated several attempts at unification. At various times, theories that we now recognize as of limited domain were advanced to fulfill that role. In the 1960s, ecosystem theory attempted to explain all by ignoring detail, and by focusing on physical and chemical fundamentals or constraints. At the same time, population ecologists thought of extending the mathematical framework of population ecology to community ecology, macroevolution, and biogeography (Odenbaugh Chapter 3). More recent examples are the neutral theory of diversity and the metabolic theory of ecology. There were others with a lesser bang but still offering evidence of interest (e.g., Belgrano and Brown 2002) in bringing more cohesion to the fragmented science. Clearly, ecology itches for a general theory. Unfortunately, so far ecology lacks a well-articulated general theory and even a general framework for relating available theories (but see Scheiner and Willig Chapter 1, which aims to lay a foundation for the erection of such a general theory by a thorough examination of the domain and foundational principles).

Perhaps this last obstacle—the lack of a general theory—will come as no surprise. Ecology is focused on the natural world and is a science that draws on the passion of those who love to study it. As such, it does a limited job at preparing young scientists in the use of theory. This tradition results in ecologists being more familiar with the requirement to formulate testable hypotheses but less familiar with identifying the domain, assumptions, central relationships, or the developmental status of various components. Consequently, ecology suffers from a level of resistance to new theoretical propositions (entrenched paradigms?), particularly those derived from traditions other than its own (e.g., the skeptical reception that hierarchy theory received initially) (e.g., Keitt 1999). Fortunately, the eloquence and persistence of its main proponents (T. F. H. Allen and R. O'Neill) led to a wider acceptance of some of the theory's premises (e.g., pattern shifts with scale) and benefited research in subsequent years, if the benefit is measured by the number of times the term "scale" appears in the literature (Schneider 2001). The resistance to theory may come also from apparent success of descriptive or quasi-descriptive work. As many studies uncover and describe important facts of nature by simply asking "why" or "how," they ease the urgency of using theory and of building a synthetic theory of ecology. The attitude that "it's all very well in practice, but it will never work in theory" (a French proverb) may be instantaneously satisfying but a serious hindrance in the long run.

Conclusions: pros and cons of this reality

Pros—clarity of focus and economy of effort

The benefits of organizing research within a theoretical framework or by directing research towards a prospective framework are fairly well understood (see Pickett et al. 2007). Theory assists researchers in identifying appropriate questions, provides guidance as to what approach will be suitable to test the hypotheses, and offers criteria to judge the progress in answering these questions. At each of these tasks, theory helps in making specific decisions or assessments. By streamlining and directing research, theory is the only guarantee, although not infallible, that the journey from observation to question and to the best available answer is as short as possible. Of course, wandering minds may by chance find shortcuts and do better than will a researcher who systematically uses theory. However, wandering minds offer no guarantee of success and, more often than not, travel in unproductive directions.

Cons—exclusion and barriers to ideas not sanctioned by current theory

Any established framework resists change and may suppress diversity of thought (Roughgarden 2009). Although the current body of ecological theory is an aggregate of loosely related and unrelated constructs, it still offers substantial resistance. This resistance may come in the form of marginalization of some research avenues by disproportionate emphasis on others (e.g., International Biological Program has been dominated by trophic or energetic concerns couched in ecosystem theory) (Golley 1993), misinterpretation of the promise new propositions may offer, or falling back on a minimalistic agenda. For example, a search for general laws of ecology may be discouraged because of skepticism as to whether they can be found (Lawton 1999; Knapp et al. 2004).

It is possible that in being very successful, Lotka and Volterra delayed other theoretical developments by shifting many ecologists' attention to the application of differential equations at the expense of other approaches to formalizing ecological processes. I am not arguing that this happened, but rather that our praise of their contribution may be unbalanced, and that the matter needs careful consideration in the future.

Misunderstandings may stimulate empirical research, which ultimately may lead to an alignment of theory and observations. When Robert May (1973) published his analysis of the effects of diversity on the stability of ecological systems, many were puzzled, but appeared to accept the result that diversity is destabilizing (McCann 2000). Specifically, by using linear stability analysis on models constructed from a statistical universe (i.e., randomly constructed communities with randomly assigned interaction strengths), May found that diversity generally destabilized community dynamics. However, May's results did not imply at all that diverse communities should be unstable; rather, he showed that communities comprising random species should be. This crucial difference was overlooked for a couple of decades. Furthermore, May included no modularity in connecting species and treated the interactive network as a single unit. More recent research indicates that this makes a major difference, because modularity insulates unstable species groups from each other (e.g., Kolasa 2005; McCann 2005). Nonetheless, May's approach and results stayed at the center of discussions on the role of diversity in maintenance of ecosystems and their properties. Importantly, May's articulation of the problem led to further theoretical and empirical work that continues to provide valuable insights regarding the role of species diversity in ecological systems. In this case, misunderstanding put ecology on a wrong path from which it recovered gracefully and with the benefit of stimulating an important area of inquiry.

Future: need to integrate and use theory more thoroughly

Charting the course for the future may be made easier by looking at types of theoretical work and their relationships to the different goals ecologists choose to pursue, whether it is generality, precision, or realism (Fig. 2.5). Although these goals are not mutually exclusive in principle, in practice, theoretical constructs show more strength in addressing one of the goals than others. Without pretending any analytical rigor, I would like to register some impressions. Theoretical constructs in ecology (models, low-level theories, or even generalizations) fall somewhere between realism and generality (Fig. 2.5). I suggest this is so as a compromise between their features: many models such as those based on population dynamics (predator-prey, competition, metacommunity) are general in formulation. This generality contrasts, however, with

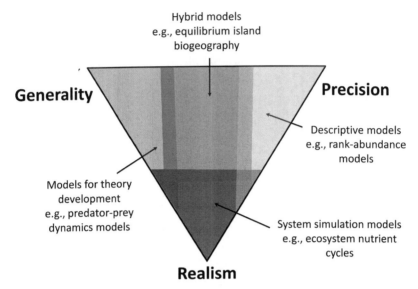

Figure 2.5 A simplified way of arranging ecological models relative to each other in order to assess the strengths and gaps. It is possible to place models along three axes of typical tradeoffs of realism, generality, and precision, together with some examples. Partial overlaps of areas suggest possible models of an intermediate or mix nature.

their limited scopes. Take a logistic population growth model. The model can be applied to any population—it is thus general. However, a population is only a tiny fraction of the ecological universe. The model does not necessarily apply to the whole species, as its carrying capacity and genetic constitution may vary in space and time. Moreover, the model does not apply to family groups, colonies, or symbiotic systems, or to coevolved multispecies systems. From this perspective, the model is not general at all. It is just one in a collection of complementary, and sometimes interacting or overlapping, models. By having a narrow scope, the model can be reasonably realistic if parameters are estimated for a limited natural universe. Still, models of populations, or interactions among populations in different contexts, are a major stronghold of theory in ecology.

In contrast, ecosystem models are often developed for particular natural systems (Fitz et al. 1996). Such models rely on general physical principles to account for the energy and material budgets, and on biological relationships to provide system-specific and local contents. They can be quite realistic—at least this is the intention of the investigators—but attempts to make them general would require stripping biological contents and so removing these models from the domain of ecology. In any case, I believe that ecology has benefitted from developing and accumulating ecosystem models in several ways such as building bridges to cybernetics and engineering, providing management tools (e.g., adaptive management; Holling 1978), and explicitly incorporating abiotic environment into the dynamics of ecological systems.

Still another category of models appears to be driven by the need for precision. In an attempt to explain species rank-abundance distributions, ecologists developed a number of statistical models. They are judged by the precision with which they describe actual collections of data. In the absence of a good theory about underlying mechanisms, these models cannot be seen as general. In fact, these models seem to work well in particular situations only; for example, large data sets from undisturbed habitats generally follow a lognormal model of abundance distribution. Different models usually apply to disturbed habitats or small data sets. As these models do not make good explicit links to underlying processes, they can hardly be seen as realistic, even when they provide a precise fit to the data.

It is surprising, and it holds some promise, that some of the early, potentially important, concepts have not advanced theoretically for over 100 years. Forbes' microcosm embodied the idea of wholeness and relative autonomy. He articulated, early on, a central notion that the ecological world exists in the form of aggregates of interdependent parts. In his view, this aggregation would be responsible for the emergence of partially bounded systems whose components have an appreciable ability to adjust internally to each other. It is rather surprising that this idea, while often invoked in the context of conservation, has not developed theoretically to any major extent. Although ecosystems ecology attempted to deal with properties of entity-like aggregations of organisms, it now appears to be stuck in narrow mechanistic representations of flows and budgets [e.g., Potter et al. (2001) compare boreal ecosystem models with respect to annual carbon and water fluxes to evaluate carbon budget]. One of the biggest and most interesting ideas of early ecologists is still a promising area for theoretical development. Calls to remove some of the past interdisciplinary barriers such as those between community ecology and ecosystem perspective are worth listening to, especially when they come from experienced theoreticians such as Roughgarden (2009).

The recent proposition by Scheiner and Willig (Chapter 1) that is at the heart of this book will undoubtedly stimulate and provoke further work. It will mature, evolve, lose unnecessary components, or relate them more efficiently. From the perspective of a practicing ecologist, the eight principles they propose are difficult to employ because the relationships among them are not yet established succinctly. Also, an important element may be still missing.

Future: we need to define the domain of ecology and the level(s) of organization

Ecologists have traditionally been rather reluctant (or oblivious) to the problem of domain. However, clarity as to the domain of a proposition is a sine qua non for evaluating a proposition. Scheiner and Willig do a good service (Chapter 1) to ecology in the way they tackle the question of domain—they comment on and assess domains of various constituent theories of ecology and their overlaps or exclusions with other sciences. In the process they show that currently available theories cover portions of the entire domain of ecology. This is a work in progress, but its completion may help to consolidate the theory of ecology.

A unified theory of ecology must, satisfactorily, deal with several levels of organization. Some believe that a successful theory of ecology must be based on the distribution of organisms; including its causes and consequences (e.g., Gleason 1917; Andrewartha and Birch 1954; Wiens 1984; Scheiner and Willig Chapter 1). Others take an even more radical position and claim that natural selection at an individual level is a sufficient explanation, and thus a basis for a general theory, of all ecological phenomena (Williams 1966). Still others, with Clements (1916), Odum (1969), Wynne-Edwards (1962), Patten (1981), Ulanowicz (1997), and Jørgensen (2007), emphasize entities arising from interdependence of species (and environment), such as communities and ecosystems, to be the central objects of interest to a general theory. All of these ecologists make good arguments; the troubling problem is that these arguments appear incompatible at the moment.

Two general strategies may hold promise in solving the century-old split. One is to gather existing theories in a unifying superstructure. This is the approach taken in this book. While it does not specifically address the problem of multi-individual or multispecies entities, in principle the hierarchical structure of a unified general theory has the capability of accommodating many well articulated lower level theories. Alternatively, a unified general theory that does not center on one or on narrow range of organizational scales may be capable of resolving the conflict as well. But such a theory may need another book.

In any case, it may be useful to add one more principle to the eight identified by Scheiner and Willig (Chapter 1) that reflects a generalization that all theories of ecology, and thus the prospective general theory of ecology, deal with ecological entities. Ecological entities appear at different levels of organization: they may be individuals, kin groups, local populations, metapopulations, symbiotic systems, local communities or ecosystems (e.g., Wilson 1988), and many others. Scheiner and Willig are clearly aware of this when they mention variation among individuals—a label that includes individual organisms and individual species. However, different degrees of entitization constitute a universal feature of living nature. An explicit recognition of it as a ninth principle might aid in advancing the general theory of ecology.

In a nutshell

1. Theory has had strong direct and indirect influences on the paradigm (mindset, culture, standards, and directions) of ecology.
2. Ecology without theory would be a science of accumulating cases, without the ability to develop and evolve, and without the ability to make sense out of the multitude of cases, except in the light of theory of evolution.
3. Even though the systematic pursuit of theory has been modest, the successes have largely been idiosyncratic, addressing particular questions without an overarching framework.
4. Awareness and use of theory increase with time across the discipline, but weaknesses such as poor links among subdisciplines of ecology persist.
5. In spite of efforts to remedy the situation, ecological theory has until recently comprised theories and models of narrow scope.
6. A larger integrating theory is lacking, but promising efforts towards its formulation are under way.

Acknowledgments

I thank Sam Scheiner and Mike Willig for creating an opportunity to reflect on historical and contemporary issues of general ecological theory. Two anonymous reviewers generously helped to clarify my ideas and their presentation. Jim Drake provided information on the first known attempt to model population growth by a Sumerian scribe.

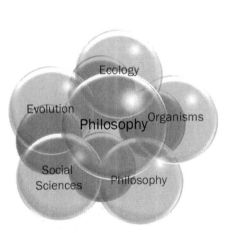

CHAPTER 3

A General, Unifying Theory of Ecology?

Jay Odenbaugh

Samuel Scheiner and Michael Willig (Chapter 1) have provided a philosophical framework for understanding ecological principles, theories, and models. Fundamentally, they contend that contrary to many ecologists' views about their own discipline, ecology already possesses a general, unified theory. In this essay, I first present their framework. Second, by way of comparison, I consider the work of the population and community ecologist Robert MacArthur. MacArthur's own work was thought of as providing unifying theories. In contrast, I argue it focused more specifically on integrative theories and models. Finally, I expand on several points in the Scheiner and Willig framework.

The Scheiner-Willig framework

According to Scheiner and Willig (SW), a theory in the sciences consists of two elements, a set of principles and a domain (Chapter 1). What is the domain of ecology? According to SW, it is the spatial and temporal patterns of the distribution and abundance of organisms, and includes the causes and consequences of those ecological processes (Chapter 1). Lest one think this is too "organism-focused," it includes biotic and abiotic factors that affect organisms along with groups of organisms at a variety of levels including populations, communities, and ecosystems. The principles describing this domain are selected on the basis of two criteria (Scheiner and Willig 2008): an inclusionary rule (for something to be a principle of a given domain, the principle must be shared by many constituent theories of that domain) and an exclusionary rule (for something to be a principle of a given domain, it must distinguish this domain from some other distinct domain).

Put simply then, something is a principle of a given domain if, and only if, it is shared by the constituent theories of that domain and is not shared with constituent theories of a distinct domain.

So what are the basic principles of this general and unified theory of ecology? They are principles that ecologists already accept given the domains they study. In a way, SW are simply making those principles explicit. More importantly, they are presenting them together, which highlights their role in structuring ecological thought collectively. Although ecology may appear to be a fairly disunified discipline, there is actually much unity undergirding various theories and models. See Table 1.3 for their list of these unifying principles.

SW recognize that theories come in degrees of generality or abstractness. They distinguish between general theory, constituent theory, and models (Chapter 1). A general theory consists of confirmed generalizations that are abstracted from facts that have been systematically tested. Using general theory in addition to more concrete considerations, we can arrive at constituent theories. A constituent theory consists of confirmed generalizations or laws, and from them models may be derived. Finally, there are models that are propositions by which hypotheses can be articulated and evaluated. Thus, there is a hierarchy of theoretical structures that become less abstract as one moves down the hierarchy. The principles of the general theory are listed above and the constituent theories include succession theory, foraging theory, metapopulation theory, and many others.

As a particular example of the SW framework, consider work on predator-prey theory. There are several general principles that can be used to derive constituent theory concerning predator-prey relationships that include the claim that organisms interact with their abiotic and biotic environments, resources are finite and heterogeneous in space and time, and birth rates and death rates are a consequence of interactions with the abiotic and biotic environment. Let's assume that the growth rate of the prey is determined by the growth rate of the prey population independent of the predator minus the capture rate of prey per predator multiplied by the number of predators, and that the growth rate of the predator population is determined by the rate at which each predator converts captured prey into predator births minus the rate at which predators die in the absence of prey multiplied by the number of predators. Here we have a classical theory of predator-prey interactions. From this we can derive more concrete models that may be tested by data. For example, let's further assume as Lotka-Volterra models do that the prey grows exponentially in the absence of the prey, predator and prey encounter one another randomly in proportion to their abundance, the predators have a linear functional response, the numerical response of the predator is a constant multiplied by the functional response, and the predator declines exponentially in the absence of the prey. Here we start with general principles, add detail that results in a constituent theory, and then arrive at a model only once we have added, in this case, quantitative detail. It is crucial to note that from general principles one can derive many different constituent theories and from them many different models. We could have devised a different model by assuming a different type of functional response for example.

Each of the principles above can be found in different areas of ecology, though articulated in different ways. For example, consider principle one—organisms are distributed in space and time in a heterogeneous manner. In population ecology, organisms are distributed unevenly as a population over habitat. For example, they may be distributed unevenly vertically in a lake or in a forest (Begon et al. 1996b). Likewise, in metapopulation theory, organisms are grouped into a population of populations, and their dynamics are largely controlled by local extinction and migration amongst distinct patches. This may be the case with forest patches or oceanic islands (Hanski and Gilpin 1997). Similarly, in landscape ecology, we see organisms distributed in different ecosystems or biomes (Turner et al. 2001). Depending on the subdiscipline of interest, each of these principles can be made concrete in different ways. As another example, consider the third general principle—variation in the characteristics of organisms results in heterogeneity of ecological patterns and processes. In behavioral ecology, the characteristics of interest involve different foraging strategies such as being generalist or specialist; in life history theory it may involve the characteristics of being an annual or perennial plant. Finally, consider the principle that the distributions of organisms and their interactions depend on contingencies. We can see how this principle is made concrete in different ways by considering the introduction of stochastic growth rates in population ecology and in community ecology through the notion of "ecological drift" in neutral theories of biodiversity.

The fact that ecological theory is composed of this hierarchy of general principles, constituent theories, and models has fundamentally important implications for evaluating general principles, constituent theories, and models. Specifically, it challenges an excessively narrow Popperian view of theory testing in ecology (Peters 1991). Suppose a model fails some particular test or tests. This does not necessarily impugn the constituent model from which it was derived nor the general principles on which it is based. The general principles and constituent theories are made concrete in models based on the particular domains, assumptions, backgrounds, and definitions that are considered. Thus, one falsifies a general principle only after many different constituent theories and models have been evaluated in light of the relevant facts.

SW have focused on the notion of generality and unification as the relations that exist between ecological theories and data, about which more will be said below. However, I now will introduce what I believe is a different conception of how theories relate, namely that of integration. As an example, consider the work offered by the eminent ecologist Robert MacArthur.

MacArthur's integrative approach

Robert MacArthur stands as one of the most influential and controversial ecologists ever to work in the discipline (Fretwell 1975; Pianka and Horn 2005). He is recognized for having done exceptionally original theoretical and empirical work. However, many believe that he took ecology down the wrong path both theoretically and methodologically. As an example, MacArthur and his colleagues' work on limiting similarity is often seen in this light (MacArthur and Wilson 1967; May and MacArthur 1972). The project was to understand why species are spaced along resource spectra given their niche breadths and widths respectively. The MacArthurites argued that there is some maximum degree of similarity in resource use such that a set of species can coexist. However, it was argued that this theoretical claim was argued to be very fragile given the assumptions made in the models and did not hold up with respect to the data (Abrams 1983). Independent of one's opinion on this matter, MacArthur's work serves as an interesting case study. He too attempted to provide a framework for understanding how ecological theories relate to each other. However, in my mind, it was not a unificationist but an integrationist approach. More on this later; first, a bit of history (for more details, see Odenbaugh 2006).

In July 1964, Robert MacArthur, Edward. O. Wilson, Egbert Leigh, Richard Levins, Leigh van Valen, and Richard Lewontin met at MacArthur's lakeside home in Marlboro, Vermont. The subject of their conversation was their own research in evolutionary genetics, ecology, and biogeography, and the overall future of what is termed "population biology." Ironically, the subject matter of these conversations was not simply population biology understood as population genetics, population ecology, and possibly ethology. It clearly included disciplines like community ecology, macroevolution, and biogeography, given the sorts of models formulated and questions asked. More importantly, there was a general tendency to approach these areas with mathematical theory as represented in theoretical population biology. For two days, each participant discussed their work and how a "central theory" could be achieved (Wilson 1993, pp. 252–253). The work that resulted from these collaborations was important and changed much of evolutionary and ecological theory. In 1972, collaborator E. O. Wilson and mentor G. E. Hutchinson wrote the following of MacArthur after his death.

> [He] will be remembered as one of the founders of evolutionary ecology. It is his distinction to have brought population and community ecology within the reach of genetics. By reformulating many of the parameters of ecology, biogeography, and genetics into a common framework of fundamental theory, MacArthur—more than any other person who worked during the decisive decade of the 1960s—set the stage for the unification of population biology. (Wilson and Hutchinson 1982, p. 319)

Did MacArthur and his coworkers "unify" evolution, ecology, and biogeography? I will argue contrary to Wilson and Hutchinson that he did not do so.

To assess MacArthur's accomplishment, we must understand the components of the program he and others articulated. Here are some of the elements. First, MacArthur typically formulated general, simple deterministic models that lacked precision. In the terms of Richard Levins' account of model building, precision was sacrificed for generality and realism. This is not to say that MacArthur modeled ecological systems realistically; rather, the desiderata of interest were generality and realism, and precision less so. As an example of MacArthur's "realism," he devised a mechanistic consumer-resource model with two consumers and two resources, and showed how the more phenomenological Lotka-Volterra interspecific competition model could be derived from it (MacArthur 1972). Second, MacArthur also emphasized the ecological process of interspecific competition as a mechanism structuring ecological communities. This is evident is his work on limiting similarity and species distributions (i.e., the broken stick model). This is not to say that he did not work on other types of processes like predation (MacArthur 1955); rather, interspecific competition played a predominant role in his thinking. Third, MacArthur rarely evaluated model predictions statistically. There are of course exceptions to this rule, but mostly he and his colleagues evaluated models by looking for corresponding dynamical patterns such as stable equilibria and various types of cycles. Finally, he was a master at presenting complex mathematical results with graphical representations (MacArthur and Levins 1967; MacArthur 1970a). Specifically, MacArthur used isocline analysis to not only present theory in pedagogically useful ways but also to draw interesting and unobvious implications (Rosenzweig and MacArthur 1963), a hallmark of good theory.

MacArthur and his colleagues produced a variety of different models involving environmental heterogeneity, density-dependent selection, optimal foraging, limiting similarity, and equilibrial island biogeography. As an example of MacArthur's theoretical work, let us consider his modeling of density-dependent selection. This is a case where MacArthur attempts to integrate ecological and evolutionary concepts that connects to several of SW's principles (specifically principles two, five, and eight) .

In most evolutionary models according to MacArthur, population geneticists use r, the intrinsic rate of increase of a population, as a measure of fitness. He writes, "For populations expanding with constant birth and death rates, r, or some equivalent measure (Fisher used r; Haldane and Wright used e^r which Wright called W) is then an appropriate definition of fitness" (MacArthur 1962).

However, as MacArthur notes, present values of r may not be reliable predictors of the number of descendants a group of individuals will have because r is an accurate measure of fitness only if the environment is relatively stable. One way in which the environment may be unstable is if population density affects fitness. In fact, MacArthur writes, "to the ecologist, the most natural way to define fitness in a crowded population is by the carrying capacity of the environment, K." (MacArthur 1962, p. 146). MacArthur offers the following mathematical model. Let n_1 and n_2 represent populations of alleles 1 and 2, respectively, and let them be governed by the following equations:

$$dn_1/dt = f(n_1, n_2) \tag{3.1}$$
$$dn_2/dt = g(n_1, n_2) \tag{3.2}$$

To understand this model, it is simplest to examine it graphically (Fig. 3.1).

Suppose we have a phase space where the x-axis represents the population of allele 1 n_1 and the other y-axis represents the population of allele 2 n_2. Thus, a point in the space represents the joint abundances of population n_1 and n_2. Suppose there is a set of values of n_1 and n_2 such that there is a solution $f(n_1, n_2) = 0$, or equivalently, $dn_1/dt = 0$ for those values of n_1 and n_2. If the population of n_1 is to the left of the f-isocline, then it will increase. Likewise, if the population of n_1 is to the right of the f-isocline, then it will decrease. Let us further suppose that there are a set of values of n_1 and n_2 such that there is a solution $g(n_1, n_2) = 0$, or equivalently, $dn_2/dt = 0$ for those values of n_1 and n_2. If the population of n_2 is below the g-isocline, then it will increase. Likewise, if the n_2 population is above the g-isocline, then it will decrease.

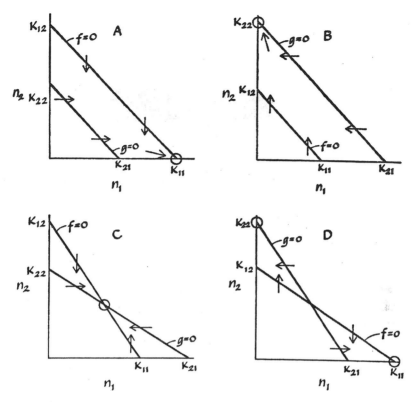

Figure 3.1 Density-dependent selection and competitive exclusion of alleles. From Mac-Arthur and Wilson 1967, p. 147.

There are four different ways the two isoclines can relate to one another. In part A of Fig. 3.1, we can see from the vector arrows that allele 1 will outcompete allele 2. Likewise, in part B, allele 2 will outcompete allele 1. In part C, the vector arrows show that there is a stable equilibrium between allele 1 and 2. Finally, in part D, whichever allele is more frequent at the outset will outcompete the other.

We can now explain how this model represents both ecological and evolutionary features. The f-isocline intersects the axis at the value K_{11}. In this circumstance, the population consists only of allele 1, and K_{11} represents the number of allele 1 homozygotes that can maintain themselves in this environment. In other words, K_{11} is the carrying capacity of the allele 1 homozygotes. Likewise, the f-isocline intersects the axis at the value K_{12}. K_{12} is the number of allele 2 that can keep allele 1 from increasing and represents the carrying capacity of the environment for heterozygotes expressed in units of allele 1. We can similarly designate the end of points of the g-isocline as K_{22} and K_{21}. MacArthur concludes, "We have now replaced the classical population genetics of expanding populations, where fitness was r, as measured in an uncrowded environment, by an analogous population genetics of crowded populations where fitness is K" (MacArthur 1962, p. 149).

Let us now consider what in fact MacArthur accomplished theoretically in this and other examples. First, let me define the notion of a unifying theory. A unifying theory applies a single theoretical framework (for example, common state variables and parameters) that applies to a variety of different phenomena. Often philosophers of science consider a theory to be a *unifying theory* in just the sense (for discussion and debate see Friedman 1974; Kitcher 1989; Morrison 2000). As Margaret Morrison writes of Newtonian mechanics and Maxwell's electrodynamics,

> The feature common to both is that each encompasses phenomena from different domains under the umbrella of a single overarching theory. Theories that do this are typically thought to have "unifying power"; they unify, under a single framework, laws, phenomena or classes of facts originally thought to be theoretically independent of one another. (Morrison 2000, p. 2)

In the case of Newtonian mechanics, Newton showed how his laws of motion and the law of universal gravitation could account for both Galileo's laws of terrestrial mechanics and Kepler's laws of planetary motion.

One might argue that the model of density-dependent selection just provided shows that MacArthur actually unified parts of population and community ecology. The equations describing density-dependent selection apply likewise to competing species in the familiar Lotka-Volterra model of interspecific competition. Put differently, MacArthur applied exactly the same framework to different genotypes within a population that he applied elsewhere to different species within a community. Just as Newton showed that the same laws apply to the sublunary domain as to the supralunary domain, MacArthur showed the same biological principles apply to populations and communities. However, I resist this conclusion because the two cases are importantly disanalogous. In the Newtonian case, sublunary and supralunary domains actually are one domain, and hence any principle that applies to the former should apply to the latter. However, in MacArthur's case, we have two distinct domains, alleles within a population and species within a community. An analogous case would be between work done using diffusion equations in physics and spatial ecology. Both use the same mathematical formalism, but apply them to quite different domains altogether. We would not say the same laws apply in both cases; rather both use similar mathematical tools to describe the nature of interactions.

If unification with respect to scientific theories or models minimally consists of a "single overarching theory" accounting for a variety of phenomena, then it appears that MacArthur's framework could not have unified population biology. If one examines the various models that MacArthur devised—models of environmental heterogeneity, density-dependent selection, limiting similarity, and equilibrium biogeography, for example—they form an extremely diverse group. The state variables and parameters are rarely the same across models; i.e., they rarely represent the same phenomena. The state variables of the limiting similarity models are population abundances and the parameters are intrinsic rates of growth, carrying capacities, and interaction coefficients, whereas in the equilibrial models of island biogeography the state variable is species richness and the parameters are rates of immigration and extinction. Likewise, in the density-dependent selection model presented above, the state variables are populations of alleles and the parameters are carrying capacities. There is no common overarching structural framework to unify population genetics, population and community ecology, and biogeography. This is not, of course, to say that there is nothing that these models have in common. However, the common ingredients are usually that the models represent equilibrial behavior, make important optimality assumptions, and are represented with deterministic equations. Nonetheless, that which is at equilibrium is sometimes population abundances, species numbers, or populations of alleles. Likewise, what is considered optimal is sometime phenotypes, genes, or the numbers of species and their abundances in a community. This methodology certainly does not generate a theory like Newtonian mechanics, which consists in a small set of schematic equations concerning the motion of objects. Hence, MacArthur provided no theoretical framework of the sort needed to unify population biology much less ecology.

Nonetheless, MacArthur did show how one could represent *both* evolutionary and ecological factors at different scales in mathematical models. These different areas of population biology had largely proceeded independently. However, if evolutionary and ecological processes are commensurate, then it was increasingly important to theoretically integrate these different processes at work in biological systems. It surely is correct that MacArthur "brought population and community ecology within the reaches of genetics" as claimed by Wilson and Hutchinson (1982). However, he did not do so by "reformulating many of the parameters of ecology, biogeography, and genetics into a common framework of fundamental theory." We can now see how MacArthur approached the relations between theories. Here is another definition. An *integrating theory* takes a variety of theories (different state variables and parameters) and combines them in their application to a variety of phenomena. He supplied a variety of models that incorporated many different evolutionary and ecological state variables and parameters, thus taking a first step toward integrating population biology.

The SW program differs from the MacArthur program in that it explicitly looks for *common* principles across ecology's subdisciplines, whereas the MacArthur program was looking for "piecemeal" connections. In fact, we can "harmonize" the SW and MacArthur programs if we recognize that both are emphasizing different parts of the theoretical hierarchy. SW have worked hard in identifying the key general principles that constituent theories and models share. MacArthur spent most his time attempting to articulate novel constituent theories and more specifically models for understanding the distribution and abundance of organisms. Thus, unification and integration are regions along a continuum. Unification is largely to be found at the most abstract level of the theoretical spectrum. Integration, on the other hand, is to be found at the level of constituent theories and models. These two features of theoretical structures are points of emphasis and are complimentary.

Having said this, MacArthur at times was clearly engaged in the same project as SW. MacArthur most famously wrote,

> Science should be general in its principles. A well-known ecologist remarked that any pattern visible in my birds but not in his *Paramecium* would not be interesting, because, I presume, he felt it would not be general. The theme running through this book is that the structure of the environment, the morphology of the species, the economics of species behavior, and the dynamics of population changes are the four essential ingredients of all interesting biogeographic patterns. Any good generalization will be likely to build in all these ingredients, and a bird pattern would only be expected to look like that of *Paramecium* if birds and *Paramecium* had the same morphology, economics, and dynamics, and found themselves in environments of the same structure. (MacArthur 1972, p. 1)

Clearly, MacArthur believed that there were general principles concerning morphology, economics, and dynamics that would be made concrete in possibly different ways in different constituent theories and models, depending on the taxonomic group under study.

Elaborating on the SW program

Let us now turn again to the SW program. First, how do we distinguish between general theory, constituent theory, and models? Are they different in kind or degree? One thing one might argue is that they are not different in kind but vary continuously along some dimension. For example, a principle is more general than another when the former's domain is a superset of the former's domain; or conversely a domain is a proper subset of another. However, one might also argue that structures differ in kind. For example, it is customary to believe theories comprise a small set of natural laws. Consider Newtonian mechanics with its three laws of motion and gravitation as familiar case in point. Models, on the other hand, are often thought of as not consisting in natural laws at all; rather, they are idealized representations of natural systems without natural laws. If this is so, then theories and models are distinct. So, there is a general question about how these different sorts of structures relate to one another.

Second, biologists make much out of the notion of *contingency*, and SW do so in principle three. However, what is "contingency"? In what sense it is a cause of the ecological patterns? There is much work to be done clarifying the role of contingency in ecological theories. Here is one way of construing contingency. An effect variable Y is *contingent* on a causal variable X to the degree that slight changes in values of X greatly change values of Y. Of course, this is just sensitivity to initial conditions—a species of nonlinearity—and there are various quantitative measures of it. Moreover, we could generalize with regard to a multivariable system where small changes in a set of causal variables $X_1, X_2, X_3, \ldots, X_n$ lead to a large change in the effect variable Y. In the way that I have characterized contingency, it is not a cause of anything; rather, it is a pattern concerning causes and their effects.

Third, SW claim that evolution causes the ecological properties of species. As the eminent ecologist G. E. Hutchinson (1965) argued, ecology is the theater of the evolutionary play. Put less metaphorically and only in terms of natural selection, ecological processes create selective regimes. These ecological processes cause or determine mechanistic or proximate differences in reproductive success. One way of construing SW's insight is that they are insisting that *current* ecological processes are in play because of *past* evolutionary processes. Hutchinson's idea can then be coupled to this proposition with the claim that *current* evolutionary processes are in play because of *past* ecological processes. Thus, properly understood, ecological and evolutionary processes are spatiotemporally interdependent. Thus, there is a crucial interaction between ecological and evolutionary processes.

Finally, where does ecosystem ecology fit in the prescribed domain of the abundance and distribution of organisms? The domain of ecosystem ecology is roughly the cycling of nutrients and flow of energy. For example, ecosystem ecologists focus on the nitrogen and carbon cycles or gross and net primary production. One could and some do argue that ecosystem ecology really just is biogeochemistry and not ecology per se since organisms—the currency of ecology and other biological sciences—have disappeared from the science (Cooper 2003). However, in my view, this would inject a bias in favor of population and community ecology and the history of ecology has been ensconced with ecosystem ecology just as much as these other disciplines. In fact, historians of ecology have spent more time writing about ecosystem ecology than about population or community ecology. I am unsure of why this is, but it is an interesting fact about the history, or historians, of science.

If ecosystem ecology is a genuine branch of ecology as I have suggested that it is, then this is where an integrative framework is important since it can couple energy flows and nutrient cycles with food web dynamics for example. Of course, there may be even more general principles one can provide that bring ecosystem, community, and population ecology together.

Conclusion

In this essay, I have presented the SW unification framework and have also presented a similar though importantly different integrative framework through the work of Robert MacArthur. Importantly, unification concerns finding the most general principles of a domain, and integration consists in bringing together different constituent theories and models. However, unification and integration are complementary because they concern different regions of the theoretical hierarchy. I also considered some specific elements of the SW framework including the notion of contingency, the relationship between ecology and evolution, and the place of ecosystem ecology in their general principles. Whether SW have provided a complete account of the unifying principles of ecology or not, they have certainly made an excellent and productive start.

Acknowledgments

I thank Samuel Scheiner and Michael Willig for their support. Many of the ideas in this chapter arose from interactions among the chapter authors who were participants in a Workshop on the Theory of Ecology supported by the University of Connecticut through the Center for Environmental Sciences & Engineering and the Office of the Vice Provost for Research & Graduate Education.

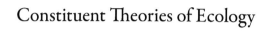

Constituent Theories of Ecology

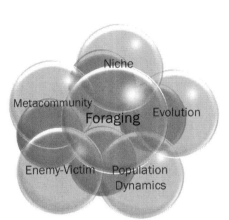

CHAPTER 4

Foraging theory

Andrew Sih

This chapter is about the predator-prey interaction viewed at the individual level. It discusses ecological theory designed to explain or even predict how predators and prey adjust their behavior in response to changes in their external and internal environments. The resulting patterns of predation often then have major impacts on population dynamics, species interactions (e.g., competition and predator-prey interactions), community dynamics, and ecosystem patterns. Thus, foraging theory links with and potentially provides a mechanistic underpinning for many of the other constitutive theories in ecology.

The underlying framework for foraging theory, as well as for evolutionary ecology in general, is the optimality approach. The key premise is that natural selection has shaped organisms to exhibit foraging behaviors that enhance lifetime fitness. Behaving optimally is a complex endeavor. It requires the organism to account for its own relatively fixed traits (e.g., its morphology, physiology), its current state (e.g., energy reserves), its biotic environment (e.g., competitors, predators, diseases) and abiotic environment. Often, organisms have only imprecise information about many, if not all of these factors. Although we do not expect organisms to necessarily be capable of coming up with the optimal behavior that balances this complicated mix of factors, the notion is that the optimality framework can predict qualitative behavior well enough to be useful.

Here, I will (1) define the domain of foraging theory—the fundamental questions and research goals; (2) outline basic propositions that guide the theory; (3) describe propositions on classes of factors that influence foraging behavior; (4) outline several propositions on general predictions (outcomes) from foraging theory; (5) provide a brief overview of major types of models including early simple models and more complex models that include more aspects of reality; (6) discuss links to other fields of ecology; and (7) note promising future directions.

Domain

In keeping with the notion that "the domain of ecology and its general theory is the spatial and temporal patterns of the distribution and abundance of organisms" (Scheiner and Willig 2005), the domain of foraging theory is the spatial and temporal patterns of behavior of organisms foraging for resources. Predation—"tooth and claw"—might conjure up an image of carnivores chasing down animal prey (e.g., lions chasing ungulates, big fish chasing smaller fish). However foraging theory can be applied to a broader range of consumers including herbivores foraging on plants, parasites foraging for hosts, pollinators foraging for nectar or pollen, frugivores searching for fruit, breeding adults searching for oviposition sites, and even plants sending out roots or shoots in search of resources. Although this may be taking some liberty with semantic fine distinctions, I will interchangeably use the terms "predators," "consumers," or "foragers" foraging on "prey" or searching for "resources" to refer to any of these systems. When three trophic levels are involved, I will refer to them as predators consuming foragers that consume prey. Regardless of the type of forager, key goals are to explain the forager's allocation of time or energy to alternative foraging options, e.g., habitats, patches, times of day, or food types.

Foraging theory has found it useful to partition the overall foraging process into two main stages: (1) behaviors that influence encounter rates between predators and prey; and (2) behaviors that influence the probability of consumption given an encounter (Fig. 4.1). Stage 1 involves decisions by predators on where to forage (habitat or patch use), when to forage (diurnal or seasonal activity patterns), and details of foraging mode (active vs. sit-and-wait foraging, foraging speed, search paths etc.). The main focus of foraging theory for this stage has been on where to forage—optimal patch use (Stephens and Krebs 1986). Stage 2 is often split into several steps: the probability of attack, given detection of prey; the probability of capture, given an attack; and the probability of consumption, given prey capture. The main focus of foraging theory in this stage has been on the forager's decision on whether to attack

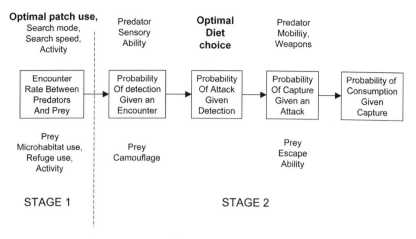

Figure 4.1 The predator-prey sequence beginning with an encounter, proceeding to consumption of prey. Predation rates depend on encounter rates and the sequence of probabilities shown in boxes. Predator and prey traits (including behaviors) that influence these rates and probabilities are shown above and below the boxes, respectively. Foraging theory focuses primarily on two predator behaviors—predator optimal patch use and optimal diets. The flowchart also shows some other predator and prey behaviors that potentially influence predation rates.

prey that are encountered—i.e., on predicting the prey types that should be part of the forager's optimal diet (Stephens and Krebs 1986; Sih and Christensen 2001).

Patch use and diets have classically been viewed as key parts of an organism's niche (Schoener 1971, 1974, 1989; MacArthur 1972; Chase Chapter 5). Thus, in a sense, the domain of foraging theory is to explain major parts of individual and species niches in functional terms relating to foraging and avoidance of predation.

Basic propositions

The basic premise of foraging theory is that foraging patterns reflect behavioral rules that have evolved under natural selection to be adaptive within constraints. This premise obviously relates to the theory of ecology's fundamental principle number 8—the evolutionary cause of ecological properties (Scheiner and Willig Chapter 1). If foraging behaviors have indeed been shaped by natural selection, then the adaptive evolutionary ecology framework, often

Table 4.1 Summary of propositions of foraging theory

A. Basic propositions that foraging theory builds on

 1. Foraging patterns maximize fitness or a correlate of fitness.

 2. Foraging patterns depend on the range of options available to the forager and on how each available option affects fitness or a correlate of fitness.

 3. Foraging behavior balances conflicting demands—tradeoffs are important in shaping foraging behavior.

B. Propositions about key factors influencing foraging behavior

 4. Foraging behavior depends on a forager's other traits.

 5. Foraging behavior responds to environmental heterogeneity (including other species).

 6. Foraging theory requires foragers to estimate parameters that influence the fitness associated with different foraging options.

C. Propositions that are major predictions of foraging theory

 7. A forager should stay in a patch as long as its current reward rate (the "marginal value" of the patch) is higher than the average reward rate for the rest of the habitat, and it should leave as soon as the current reward rate is no better than the expected rate elsewhere.

 8. When high-value prey are abundant, foragers should specialize on them (i.e., reject low-value prey), whereas when high-value prey are scarce, foragers should have generalized diets (i.e., they should attack both low- and high-value prey when they are encountered).

 9. With a sigmoid fitness function, foragers with low average feeding rates have little to lose, and should be risk prone whereas foragers with high average feeding rates have little to further gain, and should be risk averse.

 10. Animals that have low assets (e.g., low energy reserves, or low reproductive value) should be bold (accept greater predation risk) to gain more energy, whereas animals with high assets should be cautious to protect their assets from catastrophic loss (e.g. death).

 11. If foragers are free to move whenever they could do better elsewhere, then at the ideal (optimal) free distribution, all patches should yield equal reward rates.

 12. Foraging behavior often balances the conflicting demands of gaining energy and avoiding predation risk.

D. Propositions that relate foraging theory community ecology

 13. Foraging theory provides a mechanistic basis for understanding niches and patterns of competition that often underlie patterns of coexistence.

 14. Predators often have strong nonconsumptive effects on communities that are mediated by shifts in foraging behavior.

---

referred to as the optimality approach, should be insightful. The optimality approach as applied to foraging involves three basic propositions.

Proposition 1

Foraging patterns maximize fitness or a correlate of fitness. In the optimality approach, the investigator must first specify an objective function or goal. Since natural selection maximizes fitness, in evolutionary ecology theory, the ideal goal is fitness maximization. However, because fitness is often difficult to measure, much of foraging theory is based on the simpler surrogate goal of maximizing net energy intake (Stephens and Krebs 1986; Stephens et al. 2007). The idea is that having more energy available often translates into higher fitness (better long-term survival, greater reproductive success). While other considerations (e.g., mating, avoiding risk) clearly can influence fitness, perhaps during foraging bouts, it is often reasonable to assume that net energy intake rates are positively correlated with fitness. The underlying assumptions that resources are limiting and that adaptive foraging behavior is required to enhance resource intake relate to fundamental principle 6 (finite and heterogeneous resources) of Scheiner and Willig (Chapter 1).

Proposition 2

Foraging patterns depend on the range of options available to the forager and on how each available option affects fitness or a correlate of fitness. This proposition is where the biologist's expertise comes into play. Applying proposition 2 requires insight on constraints (e.g., physicochemical, genetic, physiological, or morphological constraints) that limit available options and on the mechanisms for how ecological and social factors, along with constraints, might influence the relationship between behavior and fitness. For example, if we ask how deep a duck should dive to forage and what prey should it take, a good answer probably requires a complex mix of considerations involving energy availability in different habitats with different prey, energy costs, oxygen loads, and perhaps temperature and light considerations. Most of these factors depend on the duck's physiology and energy state. Furthermore, the duck's optimal foraging behavior could also depend on competition with other ducks and with other animals foraging on the same prey, as well as on predation risk. In that case, one might want to include information about the traits of potential competitors and predators. Foraging theory, at its best, accurately includes all factors that need to be included to understand the forager's behavior.

Proposition 3

Foraging behavior balances conflicting demands. Such tradeoffs are at the heart of many ecological patterns. Indeed, one could say that in the absence of tradeoffs we would see much less organismal diversity (see Chase Chapter 5; Fox et al. Chapter 13). The one type of organism (in each trophic level) that exhibits superior performance in all situations would dominate, outperforming all other types. In reality, of course, due to tradeoffs, no one type can do best in all environments. A species that is a great competitor in one environment typically does poorly in other environments (Tilman 1982). Even in any particular environment, the best competitor often does not cope well with predators or other enemies (Sih 1987; Werner and Anholt 1993).

In many cases, the mechanism underlying performance tradeoffs involves foraging behavior. Superior foragers in one environment typically do not forage as well in other environments, and foraging behaviors that increase net energy intake rates (e.g., high activity, use of habitats with more resources) often also expose foragers to high predation risk (Sih 1980; 1987; Werner and Anholt 1993; Lima 1998; Bednekoff 2007). The flip side of this is that foraging behavior is often influenced by multiple conflicting demands (i.e., predation risk and other fitness-related factors often have important effects on foraging decisions).

With the biology of the system well represented, the final step in foraging theory involves "doing the math"—using analytical or computer-based optimization methods to find the option that yields the highest fitness. Importantly, the key challenge for foraging theory (and for optimality theory in general) is not the math. Instead, the main challenge is to capture the key elements of the biology of the system associated with propositions 2 and 3.

Propositions about key factors influencing foraging behavior

Proposition 4

Foraging behavior depends on a forager's other traits. This proposition and the following one emphasize the key, simple insight that foraging behavior is variable—among species, among individuals within a species, and within individuals. It depends both on the individual's traits (e.g., its morphology, physiology, energy state, life history state) and on the environmental context (see the next proposition). This variation in foraging behavior relates to fundamental principles 1 (the heterogeneous distribution of organisms), 3 (variation in organismal characteristics affect ecological patterns and processes),

4 (contingency), and 5 (environmental heterogeneity) of Scheiner and Willig (Chapter 1). Foraging behavior depends on the individual's morphology and physiology because the way the animal (or plant) is built clearly influences its ability to search for, detect, recognize, capture, and consume prey efficiently. In foraging theory, morphology (e.g., limbs, feeding morphology) and sensory physiology (e.g., ability to detect different kinds of prey using various types of cues) are typically considered fixed constraints that provide a mechanistic explanation for parameters and functions that determine optimal behavior. For example, a key determinant of the optimal diet is the relative value of different prey. Prey value depends on the prey's assimilable energy content, the probability of capture, and handling time. Assimilable energy content depends on the interplay of prey defenses that reduce digestibility and forager digestive physiology. Probability of capture often depends on prey versus predator mobility, which depends on their morphologies. Similarly, handling time often depends on prey and predator morphologies.

Foraging theory predicts how foraging behavior should change in response to changes in the forager's traits. Although the morphological and physiological traits discussed in the previous paragraph are usually relatively fixed as compared to behavior per se, they often show plasticity over a lifetime (DeWitt and Scheiner 2003). As a forager grows, its optimal patch use and diet should change. In many cases, the changes in morphology or physiology are induced responses to foraging-related challenges (e.g., induced responses to low food or predation risk while foraging; Tollrian and Harvell 1999; Relyea 2001). Most interestingly, some traits change in direct response to foraging behavior, and in turn, influence current and future foraging behavior. These traits (often referred to as state variables) include the forager's energy reserves, its condition or vigor, and its information state (e.g., how much it knows about the environment). In a later section, I discuss dynamic models that predict optimal foraging behavior in this dynamic situation.

Proposition 5

Foraging behavior responds to environmental heterogeneity (including other species). Proposition 4 emphasized how foraging behavior responds to the organism's traits—in essence, its internal environment. Obviously, optimal foraging behavior also depends on heterogeneity in the external environment (i.e., variation in resources, competitors, predators [enemies, in general], mutualists, and abiotic factors); these are fundamental principles 5 and 6 of Scheiner and Willig (Chapter 1). Much of foraging theory ignores species interactions other than the response of the forager to its resources. That is, most foraging models emphasize forager responses to heterogeneity in the value of the different available prey types (optimal diet theory) and in the spatial distribution of resources (optimal patch use theory). Species interactions, however, obviously also often influence foraging behavior (fundamental principle 7 of Scheiner and Willig Chapter 1). Accordingly, a branch of foraging theory accounts for how competition and predation alter foraging behavior. Species interactions often involve feedbacks that make dynamics quite complex. For example, although prey (while foraging) might shift patch use in an attempt to avoid predators, predators can then shift patch use to follow prey (Hugie and Dill 1994; Sih 1998; Luttbeg and Sih 2004; Hammond et al. 2007).

In parallel with the emphasis on scale in other chapters, environmental heterogeneity influences foraging behavior at several spatial scales. Foragers choose a general habitat type in which to forage (e.g., forest vs. meadow). Within that habitat, they choose particular patches, and within each patch, they attack some prey and reject (or ignore) others. Decisions interact across scales. Patch use depends on diet preferences (whether a bird prefers foraging in the forest or in an adjacent meadow can depend on its diet preferences) and patch use can determine diets (if a bird has chosen to forage in the forest, it will only consume forest prey).

Although it is obvious to behavioral ecologists that foraging behavior is quite plastic, and that therefore the outcomes (e.g., in terms of predator-prey or competitive dynamics) should be quite plastic, many other branches of ecological theory generally ignore this plasticity. For example, competition theory often assumes a constant competition coefficient between any two species, and predator-prey theory (Holt Chapter 7) often assumes a constant attack coefficient for predators on prey. Or, if theory allows for changes in coefficients, often it does not explicitly account for how adaptive behavior might influence key coefficients. Of course, there are exceptions; a subfield of theoretical ecology focuses explicitly on effects of adaptive behavior on population and community dynamics (Abrams 2000; Holt and Kimbrell 2007; Kotler and Brown 2007).

Proposition 6

Foraging theory requires foragers to estimate parameters that influence the fitness associated with different foraging options. Simple foraging models assume, in essence, that foragers are omniscient—that they have precise, complete information about the parameters needed to apply foraging theory. In reality, of course, foragers generally have only imprecise, incomplete estimates of the relevant parameters (Stephens and Krebs 1986; Sih 1992). Thus, foragers must either make decisions based on incomplete information, or they can sample, learn, and enhance the quality of their estimates of relevant parameters. Incomplete information and uncertainty are general problems in evolutionary ecology that shape all aspects of behavior and life histories (Dall et al. 2005), as well as physiology and morphology. In a later section, I summarize two main ways that foraging theory has addressed forager uncertainty: (1) sampling to get better information; and (2) risk sensitivity (here, risk = uncertainty) to either prefer or avoid options with more uncertain outcomes.

An overview of foraging models

Here, I provide a brief overview on major types of models in foraging theory from early simple models to more complex ones that include many aspects of reality. En route, I identify several major predictions of foraging theory that are listed as propositions about outcomes.

The basic optimal patch use and diet models

Modern foraging theory began in the 1960s with a set of models by, in particular, Robert MacArthur and colleagues on optimal diets and optimal patch use (Emlen 1966; MacArthur and Pianka 1966). These models were important in formalizing the notion of using the optimality approach to explain foraging behavior. In the 1970s, a new set of optimal foraging models generated the intuitively reasonable predictions that have become part of standard ecological dogma (Schoener 1971; Werner and Hall 1974; Charnov 1976a; 1976b). In 1986, Stephens and Krebs published a seminal book that summarized developments in foraging theory up to that point (Stephens and Krebs 1986), and in 2007, an edited volume on foraging behavior and ecology brought the state of the field up to date (Stephens et al. 2007).

The basic optimal patch use model pictures a forager moving through a series of prey-containing patches separated by areas without prey. Following the basic propositions listed earlier: (1) optimal patch use theory proposes that foragers attempt to maximize net energy intake; (2) the forager's focal behavior is T = the time spent foraging in each patch; (3) the forager's mean net energy intake is $g(T)/(T + t)$ where $g(T)$ is its cumulative net energy intake as a function of T, and t is the mean transit time between patches. Either because foragers deplete prey or because prey go into hiding, the forager's instantaneous net energy intake rate while in the patch (dg/dT) goes down with time spent in the patch (Charnov et al. 1976). In economic terminology, the *marginal value* of the patch decreases over time. At some point, the marginal value of the patch is low enough that the forager should leave; however, when a forager leaves a patch, it must go through a "transit time" without food before entering the next patch. That is, the forager faces a tradeoff. The next patch might have a higher initial net energy intake rate; however, to get to the next patch requires a "downtime" with no energy intake.

This problem is an example of a general "stay versus leave" paradigm where a forager is in a current patch and must decide, at each moment, whether to stay or move on to the next patch. In principle, the decision is simple. The forager should compare its current net energy intake rate (its reward rate) to its expected average rate if it left the patch. Its optimal patch use should then fit the following prediction (proposition 7).

Proposition 7

A forager should stay in a patch as long as its current reward rate (the marginal value of the patch) is higher than the average reward rate for the rest of the habitat, and it should leave as soon as the current reward rate is no better than the expected rate elsewhere. Charnov (1976b) called this the "marginal value theorem." It is one of the best know theoretical ideas in ecology (for a graphical depiction, see Fig. 4.2).

Corollary predictions are as follows. (1) In a given habitat, foragers should stay longer in patches that initially had higher prey availability. This prediction is obvious and did not need theory to say so. (2) In a given habitat, the marginal value of all patches (i.e., prey abundance in all patches) should be the same at the point when foragers leave the patch. If we assume that foragers do not know which patch is coming next, then in all patches foragers should stay as long as the current patch's reward rate is better than the average for the overall habitat (i.e., all patches use the same criterion for deciding when to leave a patch). After foragers have completed a bout, all patches should be reduced to the same food density. Or, as Brown and Kotler (2007) have emphasized, all patches should have the same "giving up density (GUD)"—the same prey density at which foragers give up and leave that patch. Conversely, the observed GUD in all patches should reflect the quality of the overall habitat (GUD should be lower in lower-quality habitats) and can thus be used as an indirect indicator of the forager's assessment of habitat quality (Brown and Kotler 2007). (3) Foragers that are in habitats with longer transit times (e.g., if patches are further apart) should spend more time in patches of a particular quality than should foragers in habitats with shorter transit times. Literature

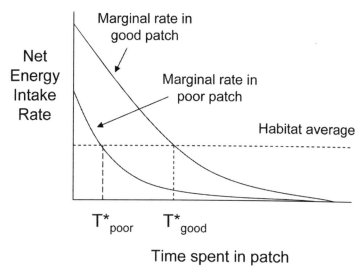

Figure 4.2 A graphical presentation of optimal patch use theory. Shown is the forager's net energy intake rate as a function of time spent in a patch. Due to prey depletion (or prey going into hiding), the forager's current net intake rate (the marginal value of the patch) should decline with time spent in a patch. The dashed horizontal line is the average net intake rate for the overall habitat. An optimal forager should stay in a current patch as long as its current net intake rate is higher than the average for the overall habitat. The optimal time to spend in a patch (T^*) is longer in better patches.

reviews suggest that in situations where reality approximates the patch use model's assumptions, optimal patch use theory does a good job of explaining observed patch use (Stephens and Krebs 1986; Nonacs 2001).

Note that the basic "stay versus leave" paradigm and the predictions of optimal patch use theory can, in principle, apply to any economic decision involving a choice among options that feature diminishing returns. For example, in our own lives, this theory and its predictions could apply for time spent persisting in a particular job, project, or partnership.

The basic optimal diet model (Schoener 1971; Werner and Hall 1974; Charnov 1976a) also, in essence, features a "stay versus leave" framework. When a particular prey item is encountered, the forager must decide whether to accept or attack it (stay with it), or reject or ignore it (leave and continue searching for other prey). For the forager, each prey type has a net energy gain associated with attacking it (which depends on its energy content, the energy cost of capturing and consuming it, and the probability of capture), and a handling time (a period of time spent on that prey item when the forager cannot search for and capture other prey). The forager should attack a prey item if its net energy intake per unit handling time is greater than the expected net energy intake if it continued searching. Put another way, when a forager attacks and handles a prey item, it suffers an opportunity cost in the sense that it loses an opportunity to possibly encounter and attack other prey. The forager should attack a prey item if the benefit of that item is larger than the opportunity cost. If we rank prey by their value, the prey type with highest value should always be taken when encountered. Since no better options exist, there is nothing to be gained from rejecting the best option to wait for something better to appear. In contrast, a lower-value prey item should be rejected if higher-value prey are abundant enough; i.e., if there is a high enough chance that during the time when the forager is handling the low-value prey, a high-value prey item might appear. The general prediction follows.

Proposition 8

When high-value prey are abundant, foragers should specialize on them (i.e., reject low-value prey), whereas when high-value prey are scarce, foragers should have broad, generalized diets (i.e., they should attack both low- and high-value prey when they are encountered). Again, this basic paradigm and predictions should also apply for other choices, such as mate choice (Crowley et al. 1991).

A literature review (Sih and Christensen 2001) found that optimal diet theory's predictions usually worked well for foragers feeding on prey that are relatively immobile (e.g., flowers, fruits, mussels, or corals), but that the theory performed poorly for foragers on mobile prey (e.g., large fish on small fish, large insects on small insects, mammalian predators on mammalian prey). With mobile prey, prey behavior (and not predator active diet choice) plays a major role in determining predator diets. Many mobile prey are hiding or difficult to capture, whereas only some mobile prey are both exposed and easy to capture. When most prey are hiding and difficult to capture, predators often attack any prey that they encounter that could conceivably be captured. The diets for predators on mobile prey then tend to be dominated not necessarily by prey with high energy per unit handling time, but by prey that are exposed and relatively easy to capture. To emphasize, this observation does not contradict basic optimal diet theory. Instead, it suggests that to understand the optimal diets of predators on mobile prey, empiricists must account for variation in prey exposure to predators, and the effect of prey escape ability in determining prey value.

Adding complexities and realities

From the 1970s on, foraging theory acknowledged and incorporated many aspects of reality that were not included in early models. That is, as a field, foraging theory matured by adding new subfields via the classic theory–empirical test feedback loop. Empirical work pointed out key missing complexities that were then incorporated into more sophisticated models, which led to new tests that revealed new complexities and so on. In terms of the principles of ecology, basic foraging theory was built on the application of principle 8 (evolutionary causes) to principles 1, 5 and 6 (heterogeneous distributions of organisms and their finite resources, as well as of the environment, in general). As the field developed, it added theory and tests associated with principles 4 (contingency and the need to account for uncertainty) and 7 (biotic interactions including competition and predation risk).

Lack of information

As stated above, early models assumed that foragers have complete and precise information about their environment—specifically, accurate estimates of all parameters needed to apply particular optimal foraging models. Presumably, this is rarely if ever true. Instead, foragers must both estimate the relevant benefit-cost parameters (proposition 6) and, given the uncertainty in parameter estimates, make decisions that account for the uncertainty. The challenge for theory has been to classify several main types of responses to lack of information and to capture their essence in simple, testable models.

One general type of theory on lack of information revolves around foragers sampling and learning to forage more efficiently in the future. Optimal sampling balances the benefits and costs of sampling. The benefit lies in the value of having more accurate parameter estimates that should result in higher foraging efficiency in the future. This benefit can depend on the interplay between environmental factors (e.g., inherent predictability and stability of the environment; one cannot learn much useful about environments that are highly unpredictable and unstable) and the organism's sensory and cognitive abilities. The cost of sampling can include temporarily reduced foraging efficiency while sampling items or patches of low value that an omniscient forager would not use, and risks taken that an omniscient forager would not take (Sih 1992). Statistical decision theory (based on Bayes' theorem), where foragers have a prior estimate of the world that is continually updated as new information becomes available, has been a useful framework for analyses of optimal sampling (De-Groot 1970; McNamara 1982; Stephens and Krebs 1986; Sih 1992).

Optimal sampling regimes have been analyzed in a range of scenarios including the following. (1) Key parameters (e.g., prey value) are fixed, but foragers need to sample items to ascertain which are good and which are poor. Signal detection theory can be used to analyze effects of variation in the ease of discriminating the relative quality of different options (Getty et al. 1987; Stephens 2007). (2) Patches exhibit diminishing returns, but foragers need to assess the rate of diminishing returns and the patches' current marginal value (Oaten 1977; McNamara 1982). (3) Options (patches or prey types) can be either good or bad, but these can change over time. The forager must sample to assess the current value of each option (Stephens and Krebs 1986). Most models have examined changes in the prey regime, but some have looked at how uncertainty about predation risk influences foraging behavior (Bouskila and Blumstein 1992; Sih 1992; Abrams 1994). Dall et al. (2005) provide a general overview on information and its use in evolutionary ecology. As with other aspects of foraging theory, optimal sampling can also be applied to human decision making (e.g., via adaptive management). We should all plan our lives to not only maximize current net gains, but also to gather the information required to make more intelligent decisions in the future.

An alternative response to uncertainty is to incorporate risk, in the sense of uncertainty, into the evaluation of an options' value (Real and Caraco 1986; Stephens and Krebs 1986; Houston and McNamara 1999). Picture, for example, a world with two options A and B, where A yields a guaranteed, moderate reward rate, and B yields the same average reward rate as A, but with high variance. B is a "boom or bust" option that could either yield a much higher or much lower reward rate. Drawing directly from economic utility theory (e.g., Keeney and Raiffa 1993), animals should prefer the uncertain, risky option B (they should be "risk prone") if the increased benefit of the higher reward rate is greater than the cost of the lower reward rate. Conversely, foragers should prefer the low-variance option A (they should be "risk averse") if the increased cost of the lower reward rate outweighs the benefit of the higher reward rate. In nature, the foraging rate versus fitness function is probably often sigmoid-shaped, where a range of low feeding rates yield little or no fitness, and high feeding rates yield diminishing returns in terms of fitness (and for humans, very high feeding rates result in obesity and reduced fitness).

Proposition 9

With a sigmoid fitness function, foragers with low average feeding rates have little to lose, and should be risk prone whereas foragers with high average feeding rates have little to further gain, and should be risk averse. In principle, these ideas could also explain risk sensitivity and gambling or investment strategies in humans.

Experimental tests of risk sensitivity theory have yielded mixed results (Kacelnik and Bateson 1997). Interestingly, foragers appear to treat uncertainty about time versus reward size differently. They tend to be risk averse about reward size, but risk prone about delay to reward (Kacelnik and Bateson 1997). Being risk prone about delays means that foragers, including many humans, tend to be impulsive—we strongly prefer immediate rewards to a degree that is difficult to explain using standard optimality theory (Stephens 2002). Extreme impulsiveness is associated with poor performance in school and addiction (Mischel et al. 1989). Explanations for impulsiveness and other aspects of risk sensitivity suggest a need to incorporate more about cognitive and perhaps neuroendocrine aspects of behavior into foraging theory.

State-dependent dynamic optimization

Although we know that animals, including ourselves, alter foraging behavior depending on hunger level or energy reserves, basic optimal foraging theory does not address this point. That is, early optimal foraging theory was not state-dependent. Seminal papers and books by McNamara and Houston (1986) and Mangel and Clark (1998) brought state-dependent dynamic models into the mainstream of foraging theory. These models address a particular key aspect of proposition 4—that foraging behavior depends on the forager's state.

In the foraging context, a state variable is a property of an individual that persists over time, with a future value that is affected by current behavior. A dynamic feedback occurs if current state affects the optimal current behavior, which affects future state and so on. Examples of ecologically relevant state variables include energy reserves, size, reproductive value, and other forms of assets, condition, vigor, and information. This framework adds several major aspects of reality into foraging theory: (1) although individual state obviously affects behavior, this was not included in earlier models; (2) state dependence connects current short-term behavior to long-term fitness, which is critical for putting risk and foraging needs into a common currency (a key issue for analyzing tradeoffs; proposition 3); and (3) unlike previous models, dynamic state-dependent models explicitly address temporal patterns of behavior—typically by incorporating a time horizon (e.g., an end of the season) and predicting changes in behavior over time.

State-dependent models help to solve what had been termed the "common currency" problem (McNamara and Houston 1986). Many empirical studies show that predation risk (or other sources of mortality) and energy needs both

influence fitness and behavior. The challenge has been to clarify how foragers (and modelers) should balance risk against energy to identify the optimal
behavior when they are in different currencies: survival versus energy intake.
How much energy should an animal give up for a unit of increased safety? This
dilemma can be solved by converting energy gain into fitness terms. The problem is that the energy gained from short-term foraging often has little or no
immediate effect on fitness. The fact that I skipped lunch to work on this paper
had no discernible effect on my fitness. Thus a corollary problem is the need
to have theory that connects short-term behavior to long-term energy budgets that affect long-term fitness. The solution is to account for state variables.
Current foraging decisions affect energy reserves. In the short-term, energy
reserves can immediately affect fitness if animals are close to starving. Over
the long-term, the accumulation of many, small, short-term foraging decisions
affects cumulative energy reserves that affect growth and future reproduction.
To date, most state-dependent models feature relatively simplistic depictions
of the relevant energetics; however, some include more sophisticated, realistic
models of the biology of energy reserves (e.g., Brodin and Clark 2007). A general result to emerge from dynamic optimization models is the "asset protection principle," which states in essence that organisms should be more cautious
when they have more to lose (Clark 1994).

Proposition 10

Animals that have low assets (e.g., low energy reserves, or low reproductive
value, and thus low prospects for the future) should be bold (accept greater
predation risk) to gain more energy, whereas animals with high assets should
be cautious to protect their assets from catastrophic loss (e.g., death). Over
time, this negative feedback process should reduce variation in assets and thus
in asset-dependent behavior. Individuals that initially had lower assets should
be more bold and active than those with initially more assets. Thus, individuals
that started with less should tend to catch up in assets compared to those that
started with more (as long as they are not killed in the process).

Simple, basic foraging theory was agnostic about how foraging patterns
might change over time. Should animals be more or less bold as they approach
the end of a growing season, or the end of their lifetime? Another improve-
ment offered by dynamic optimization theory is the fact that it generates predictions on how foraging behavior ought to change over the course of a day, a
season, or a lifetime. Opposite predictions emerge depending on whether the
time horizon represents a lethal endpoint (e.g., onset of winter for an annual
organism, or pond drying for a pond organism) or an opportunity to cash in

assets (e.g., the onset of reproduction). General predictions are that: (1) if the goal is to survive until the organism can cash in assets, then individuals should take fewer risks (i.e., be more cautious to protect assets) as they approach that horizon; and (2) if the time horizon is a lethal endpoint, then individuals should take more risks as that endpoint approaches.

Accounting for state variables also reminds us that energy is not the only benefit to be gained from food. Food also contains a myriad of other nutrients (protein, minerals, vitamins) as well as toxins. Simple foraging theory assumed, in essence, that food contains no toxins, and that all beneficial aspects of food are positively correlated; i.e., that foods that yield more energy/handling time also yield more other nutrients/handling time. A simple approach that incorporates other nutrient considerations involves using linear programming that identifies the behavior that maximizes a benefit (or minimizes a cost) under the constraint that the optimal behavior must also satisfy other minimum needs (e.g., a minimum sodium need; Belovsky 1978). In reality, effects of the mix of nutrients on forager performance can be complex (Newman 2007). In principle, multiple nutrient considerations could be incorporated into a model by including multiple fitness-related state variables. Another area of ecology that examines multiple nutrients is ecological stoichiometry (see Burke and Lauenroth Chapter 11). An exciting future step could thus involve blending principles of ecological stoichiometry with modern foraging theory.

Note that while dynamic, state-dependent models provide the significant advantage of adding many, important aspects of reality to earlier, simpler foraging models, this additional reality comes with costs. Because the models are more complex, and more specific, they are also less general. In addition, it is often difficult to get realistic estimates of the many parameters and functions that are part of these more complex models. In principle, one should do sensitivity analyses to examine effects of uncertain parameter estimates on predictions; however, the overall parameter space is often so large that this is either difficult to do or difficult to publish. Finally, with complex models with many interacting factors, it can be difficult to identify the biological mechanisms (as opposed to convenient mathematical assumptions) underlying a predicted pattern. Progress using dynamic, state-dependent models will likely depend on the balance between these benefits and costs.

Social and species interactions: competition and predation risk

Basic optimal foraging theory assumed, in essence, that predators forage alone. In reality, following fundamental principle 7 (birth and death rates depend on abiotic and biotic factors), foragers often, perhaps usually, account for competitors or predation risk (or other sources of mortality), or both. Accounting for interactions among competing foragers, and for predators or other species, adds two major types of complexity to foraging models: (1) it makes the optimization criterion more complex (as discussed above), and (2) it brings in the notion of games where the best behavior for a forager depends on decisions made by others (competitors or predators) and vice versa (Maynard Smith 1982).

Beginning almost four decades ago, the field accounted for intraspecific competition via analyses of the "ideal free distribution" (Fretwell and Lucas 1970). Picture a simple scenario with only two patches, A and B, where A initially has more resources than B. A single forager in a noncompetitive world should prefer patch A; however, if too many competitors join that forager in patch A, competition might be so intense that A is no longer better than B.

Proposition 11

If foragers are free to move whenever they could do better elsewhere, then at the ideal (optimal) free distribution, all patches should yield equal reward rates. The simple underlying logic is that if any patch is yielding higher reward rates than others, some competitors should move to that patch (thus increasing competition in that patch) until it is no better than other patches. Foragers should continue to move among patches until all patches yield equal reward rates; at that point, there is no benefit to be gained from moving. In terms of evolutionary game theory, this is an evolutionarily stable strategy (ESS; Maynard Smith 1982). In the simplest models, at the ideal free distribution (at the ESS), the ratio of consumers in two patches should "match" the ratio of resource inputs in the two patches (e.g., if patch A has a resource input rate four times as high as does patch B, then at the ESS, four times as many competitors should be in patch A than patch B). This basic framework and predictions could, in theory, apply to the distribution of competitors among any set of options—including human choice of places to live, or jobs. For example, in theory, if humans followed the ideal free distribution, the quality of life should be equal in all cities. If the quality of life, including costs of living, was higher in San Francisco than Oklahoma City, people should move to San Francisco until it is so crowded that it is no longer better than Oklahoma City.

Many studies have confirmed that, in broad agreement with ideal free distribution theory, competitors are often more abundant in areas with more resources. Although some of these studies have found that foragers indeed match their resources, most studies found "undermatching," where the consumer ratio was less than the resource ratio (e.g., if the ratio of resource inputs in two patches was 4:1, the consumer ratio was only 2:1). Subsequent theory and experiments suggested that undermatching could be explained by: (1) aggressive interactions among competitors, where dominant consumers keep less dominant individuals out of patches with more resources; (2) movement costs; and (3) imperfect knowledge (e.g., Abrahams 1986). The logic on points 2 and 3 is that consumers might stay in patches that yield lower reward rates either because it is too costly to move to a better patch or because they do not realize that other patches are better than their current patch.

Competition has also driven the evolution of alternative competitive foraging strategies within each patch (Giraldeau and Caraco 2000; Waite and Field 2007). For example, in some systems, subordinate individuals serve as "producers" who search for food, while dominant individuals are "scroungers" who steal food first found by producers. In essence, scroungers parasitize the efforts of producers. This scenario is also often seen and referred to as "piracy."

Proposition 12

Foraging behavior often balances the conflicting demands of gaining energy and avoiding predation risk. Options (patches, diet items) that yield higher energetic rewards are often also dangerous. Risk then affects foraging decisions, and feeding (energy) demands affect predator avoidance (Lima and Dill 1990; Lima 1998; Bednekoff 2007; Brown and Kotler 2007). Beginning several decades ago, both modeling and empirical work incorporated predation risk into analyses of foraging behavior (Sih 1980; Werner et al. 1983). As noted earlier, state-dependent dynamic optimization models provide insights on how foragers might balance these tradeoffs. Other notable realities about tradeoffs between foraging and risk that have been modeled and studied empirically include: (1) forager responses to multiple predators, particularly those that require different antipredator responses (Lima 1992; Matsuda et al. 1994; Sih 1998); (2) effects of temporally varying predation risk (Lima and Bednekoff 1999; Ferrari et al. 2009); (3) ideal free distributions with predation risk including models of how foragers should balance resources, competition and predation risk (Grand and Dill 1999); and (4) predator-prey space use when predator attempts to use areas with more prey can be offset by prey attempts to avoid predators (Heithaus 2001; Alonzo 2002; Lima 2002; Hammond et al. 2007).

In the four decades since optimality modeling began to be applied regularly in ecology, foraging theory has grown in a steady, organic fashion. In each stage of growth, existing theory has been criticized for being too simple, leading to new theory that incorporated important aspects of reality. Though some would dispute this assessment, my sense is that the resulting body of theory has been reasonably successful at explaining, or even predicting, major patterns in foraging behavior (Stephens et al. 2007).

Linking foraging theory to population and community ecology

Foraging theory provides a critical mechanistic bridge between organismal ecology and several other fields of ecology (Fig. 4.3). Most theory on population and community dynamics assumes that species interactions are important. The persistence and coexistence of species are thought to depend, at least in part, on competition and predator-prey interactions (MacArthur 1972; Chesson 2000a; Kotler and Brown 2007). One research goal is thus to quantify the strength of these interactions and their impacts on ecological patterns. For a mechanistic ecologist, a key additional step is to explain or even predict the strength of these interactions in terms of species traits. The strength of competition (as measured by a competition coefficient) depends on species niches (diets, habitat use; Chase Chapter 5) that, in turn, depend on species traits (e.g., morphology, physiology, life history, and behavior), and the strength of the predator-prey interaction (e.g., the predator attack coefficient) depends on

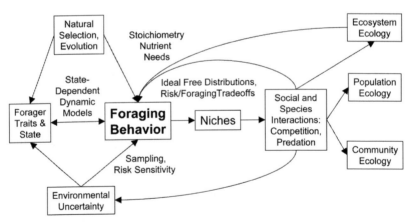

Figure 4.3 An integrative overview of interrelationships between foraging theory (which attempts to explain foraging behavior) and other major fields of ecology and evolution. Foraging behavior, shaped by natural selection, balances costs and benefits that depend on the forager's traits and state, as well as on social and species interactions and ecosystem properties. In turn, the resulting foraging behaviors are major components of resource/risk niches that underlie species interactions that explain many patterns in population, community and ecosystem ecology. All of this is influenced by environmental uncertainty.

predator and prey traits (Murdoch and Oaten 1975). Foraging theory provides an explicit quantitative basis for bridging from species traits to the interaction strengths that underlie population and community dynamics. Although most simple population or community models assume that key interaction rates are constants, in reality they invariably change with prey, competitor, or predator density in ways that depend on behavior. A central goal of foraging theory has been to explain or even predict these changes.

Proposition 13

Foraging theory provides a mechanistic basis for understanding niches and patterns of competition that underlie patterns of coexistence. Indeed, foraging theory was first created by MacArthur and colleagues (e.g., MacArthur and Pianka 1966) to explicitly address this proposition quantitatively.

A flagship example of the use of foraging theory to explain competitive co-existence is the classic work of Werner and colleagues on diets, habitat use, and competition among North American sunfish. They used optimal diet theory to predict the diets of sunfish in the laboratory (Werner and Hall 1974) and field (Mittelbach 1981). Because sunfish species differ in body shape and jaw morphology, they differ in foraging efficiency on different food types. These differences can then explain their feeding niches and how they shift in the presence of competitors (Werner 1977). Smaller sunfish, however, suffer predation risk from predatory bass. Extending theory to include size-dependent foraging-risk tradeoffs (Werner and Gilliam 1984) predicted size-dependent changes in habitat use (ontogenetic niche shifts) from within safer vegetation (feeding on vegetation-associated prey) to open water (feeding on zooplankton). These predictions were corroborated with field experiments that manipulated the presence and absence of bass (Werner et al. 1983).

Attempts to combine foraging theory and competition ran into an interesting paradox. According to foraging theory, when competition reduces food availability, foragers should broaden their diets and become generalists. Competitors should thus increase in niche overlap. In contrast, competition theory suggested that when resources are scarce, species should reduce niche overlap with competitors by specializing on preferred resources (the "compression hypothesis"). In recent years, renewed interest in the connection between foraging theory and competition has arisen around new attempts to explain this paradox (Robinson and Wilson 1998; Svanback and Bolnick 2005; 2007). One possible resolution to this paradox hypothesized that when resources become scarce, foragers should avoid competition by becoming habitat specialists while remaining a resource generalist within their preferred habitat.

Another solution suggested that all competitors prefer the same high-value prey types when they are abundant, but that under strong competition, competitors broaden their diets, but diverge to include different less-preferred prey types as shared high-value prey become scarce.

Along similar lines, foraging theory has been used to predict adaptive shifts in predator and prey behaviors that influence predator-prey dynamics (Gleeson and Wilson 1986; Abrams 2000; Brown and Kotler 2007; Holt and Kimbrell 2007). An exciting recent offshoot of this work emphasizes the ecological importance of trait-mediated nonconsumptive effects of predators on prey (Preisser et al. 2005; Orrock et al. 2008; Schmitz et al. 2008). That is, even if predators rarely consume prey, predators can have major impacts on prey fitness and prey intraspecific and interspecific interactions by causing shifts in prey habitat use, activity, and diets (Sih and Moore 1990; Lima 1998; Brown and Kotler 2007). These nonconsumptive effects can cascade down to influence the prey's prey. When carnivores cause shifts in herbivore behavior, this can have large impacts on plants (Power et al. 1985; Schmitz 2004).

Proposition 14

Predators often have strong nonconsumptive effects on communities that are mediated by shifts in foraging behavior. Strong evidence in support of this proposition comes from a meta-analysis that found that nonconsumptive effects of predators on prey (or on prey of prey) are often stronger than direct, consumptive effects of predation (Preisser et al. 2005). Foraging theory (via the risk-foraging tradeoff) can predict the nature and magnitude of trait-mediated nonconsumptive effects (Abrams 2000; Brown and Kotler 2007).

Another offshoot of foraging theory that can potentially interface with population and community ecology involves using foraging theory to predict optimal movement or dispersal among patches. In a world where many, perhaps most, habitats are distributed in a patchy manner (fragmented), a key issue for understanding species persistence and community structure is the interplay between interactions within patches and dispersal among patches. In the 1990s, a major area of study for both basic and applied ecology (e.g., conservation ecology) was metapopulation ecology (Hanski 1999), population ecology in a patchy world where local extinctions can be offset by recolonizations. In the 2000s, this field was broadened to include multispecies interactions (i.e., metacommunity ecology; Holyoak et al. 2005; Leibold Chapter 8). Theory in both metapopulation and metacommunity ecology often assumes that dispersal rates are fixed. In fact, in many animals both dispersal from one patch and settlement in another almost certainly depend on habitat quality in both patches as well as in the matrix habitat between patches. In turn, habitat quality depends on multiple factors including resource levels, competition, and predation risk. Foraging theory that predicts predator, competitor, and prey behavior as a function of these ecological factors should thus play a useful role in predicting patterns of adaptive dispersal.

Overall, the key insight from foraging theory (and more generally from evolutionary ecology) for the rest of ecology is the simple notion that behavioral plasticity (often adaptive behavior) is omnipresent and matters. Behavior generates nonlinearities (due to dynamic changes in rates) that can affect dynamics and outcomes in ways that cannot be predicted without accounting for plasticity. The general suggestion is to use foraging theory to provide a mechanistic, adaptive understanding of how plasticity in response to tradeoffs underlies ecological patterns. Behavioral plasticity is a fundamental aspect of the "theory of organisms" (Scheiner 2010); ignoring behavioral plasticity will often prove to be unwise.

Future directions

Although foraging theory logically provides a mechanistic basis for explaining behavioral responses to resources or predators that are ubiquitous and potentially important, this behavioral plasticity is often not accounted for explicitly in other ecological fields. Why not? The answer probably lies in the fact that it is an additional complexity that can be hard to incorporate in an already full research program. Ecologists often have a hard enough time getting sufficient data on their preferred level of study (i.e., the population, community, or ecosystem level) without having to also study and understand phenomena at the level of individual behavior. And yet, it is difficult to argue against the notion that consumers exhibit behavioral plasticity and that plasticity is important. A "solution" to this practical problem is to gloss over the details on foraging behavior, but to still incorporate the main patterns. Although we know that foragers often show adaptive responses, theory suggests that depending on model details, those responses and their impacts could go in diametrically opposite directions (e.g., Abrams 2000). Additional modeling and empirical work is still needed to clarify key *generalities* about flexible foraging behaviors, in particular, those that appear to be ecologically important.

Interestingly, new empirical methods hold great promise for aiding the endeavor of incorporating behavioral plasticity into ecology. To date, many of the most detailed foraging studies have been done in the laboratory, where it is relatively easy to observe animal behavior. Until recently, for many mobile animals it has been difficult, if not impossible, to get good field data on foraging behavior. New technologies (e.g., GPS collars, multicamera video systems with automated data analyses, other distributed remote sensing systems), however, are allowing us to remotely monitor more individuals than ever, with more information collected per individual, over longer periods of time, along with simultaneous monitoring of key environmental factors.

In addition, new statistical methods should help generate more accurate and nuanced insights on foraging behavior. To date, most patch use experiments have adopted the simplistic model-derived scenario of a set of equal-sized patches of food separated by areas with no food. Reality often features much more complex landscapes. New spatial statistics and landscape methods can be used to more rigorously characterize realistic patterns of spatial and temporal heterogeneity in resources, foragers, and predators, and to more rigorously identify patterns of response. Also, to date, most foraging studies have used standard p-value-driven statistical methods to compare the predictions of one or a few foraging models against null expectations. An alternative approach uses model choice methods (Burnham and Anderson 2002) to compare numerous models to identify behavioral rules that best explain observed predator and prey behavioral patterns (e.g., Luttbeg and Langen 2004; Hammond et al. 2007).

Another exciting future direction involves linking foraging theory to conceptual areas in ecology that it is not currently well linked to. These can be identified by looking at some of the other chapters in this book. Ecological stoichiometry focuses on how elemental mismatches (e.g., involving carbon:nitrogen:phosphorus ratios) between consumers and resources affect their ecology (Sterner and Elser 2002; Burke and Lauenroth Chapter 11). Although most of foraging theory uses energy (carbon) as the key measure of food value, nutrient constraints are clearly important and have been included in foraging theory (Newman 2007). Principles of ecological stoichiometry could provide a mechanistic basis for predicting how prey with different elemental ratios should differ in prey value depending on the consumer's stoichiometry. The notion that multiple chemical constituents in food should matter can be incorporated into foraging theory using dynamic state-dependent models with multiple elements included as multiple state variables. Foraging theory could then help to clarify how prey elemental content and stoichiometry (along with handling times) might balance with fitness benefits of different elements and risks of mortality to predict foraging behavior.

The metabolic theory of ecology (Brown et al. 2004) posits that basic constraints of size, temperature, and the need to move resources within organisms in vascular networks with constrained geometries might affect not just metabolic rates, but also numerous aspects of the organisms' general ecology and evolution. The basic assumption is that size and temperature play the key roles in driving metabolic rates and other outcomes. Standard foraging theory does not dispute the notion that foraging behavior likely depends on predator (and prey) size and on temperature; however, it does not typically draw on first principles to predict how size and temperature should influence foraging behavior. Thus foraging theory could gain from metabolic theory. Conversely, most of metabolic theory's models assume that resource limitation per se and foraging behaviors that govern resource intake rates do not limit metabolic rates. While metabolic theory might explain some of the major patterns in the natural world, much of the residual variation might be explained by patterns of foraging and resource limitation.

Ecosystem ecology, in general, examines fluxes of energy and nutrients (water, carbon, nitrogen, phosphorus, various elements) among major abiotic and biotic categories (e.g., among trophic levels). Ultimately, when fluxes involve consumers and resources, rates of flux should depend on individual foraging decisions. Although there are some examples of where shifts in foraging behavior fundamentally change nutrient cycles or trophic dynamics (Schmitz et al. 2008), most foraging ecologists ignore ecosystem processes, and most ecosystem ecologists ignore foraging. Clearly, developing this bridge holds great promise.

Finally, foraging theory remains poised to contribute to various fields of applied predator-prey ecology, including farm animal foraging and meat production, fish foraging and fisheries, foraging by animal disease vectors, foraging by biological control agents, and foraging by species of conservation concern. For each issue, foraging theory can provide a better understanding of predator or prey behavior that can, in turn, potentially enhance the efficacy of management.

A particularly exciting area of future study in conservation ecology and invasion ecology involves understanding variation among foragers (predators or prey) in their response to novel environments, which in the modern world often consists of human-induced rapid environmental changes. In the predator-prey context, a key issue involves prey responses to novel predators and predator responses to novel prey. Invasive, exotic predators often have major negative impacts on prey, but in other cases, prey cope well with exotic predators (Mack et al. 2000; Cox and Lima 2006; Salo et al. 2007). An important question is, why do some prey respond well to exotic predators while others do not (e.g., Rehage et al. 2005; Sih et al. 2010)? Conversely, some foragers readily utilize novel resources (e.g., some herbivores utilize crops and ornamental plants) while others do not. What explains the variation in response to novel foods (e.g., Rehage et al. 2005)? Interestingly, addressing the issue of responses to exotic invasive species will require a twist on the usual foraging theory. Foraging theory normally attempts to explain behaviors that are adaptive, presumably due to a long evolutionary history of dealing with similar situations. Now, our question is, why do some individuals respond well and others respond poorly to predators or prey in situations that they have never seen before? Traditional foraging theory might still play a useful role in explaining how responses to cues that in the past provided good indicators of prey value or predation risk might now result in mismatches that lead predators or prey to exhibit inappropriate behaviors. These mismatches have been termed "ecological traps" (Schlaepfer et al. 2002). In a confusing modern world, an important task for foraging theory might be to provide a departing point for understanding such traps.

Overall, after more than four decades, foraging theory remains a vital area of study with many exciting opportunities for bridges with other constitutive theories in ecology. Insights from foraging theory on behavioral plasticity remain to be fully integrated into other areas, and insights from other areas continue to hold promise for further refining of foraging theory.

Acknowledgments

These ideas were shaped by interactions with other chapter authors who were participants in a Workshop on the Theory of Ecology supported by the University of Connecticut through the Center for Environmental Sciences & Engineering and the Office of the Vice Provost for Research & Graduate Education. In addition, I thank Cait McGaw for many stimulating discussions that helped shape my overall views. This work was supported in part by the National Science Foundation, grant IOB-0446276.

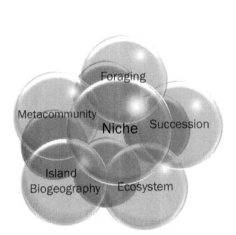

Foraging

Metacommunity

Niche   Succession

Island
Biogeography   Ecosystem

CHAPTER 5

Ecological Niche Theory

Jonathan M. Chase

Every species has a range of environmental characteristics within which it can have positive population growth, thereby circumscribing the species' niche (e.g., Hutchinson 1957). Further, we can distinguish between the fundamental niche, which consists of the range of environmental characteristics within which a species can maintain positive population in the absence of other species, and the realized niche, which consists of the range of environmental characteristics within which the species can maintain positive growth in the presence of other species (Hutchinson 1957). Thus, niche theory pervades a majority of evolutionary and ecological investigations. Evolutionary processes mold a species' niche as it interacts with its environment and other organisms, and the nature of a species' niche in turn defines its biogeography, its local and regional coexistence with other species, its strengths of interspecific interactions, its relative abundance, and its role in ecosystem functioning (Chase and Leibold 2003).

A species' ecological niche is roughly divided into two related components: (1) the suite of biotic and abiotic factors that influence a species ability to persist in a given locality, and (2) the impact of species on those factors (Leibold 1995; Chase and Leibold 2003). Importantly, I will explicitly incorporate the possible role of stochastic processes, both environmental (e.g., Chesson 2000b) and demographic (e.g., Hubbell 2001), into the propositions of niche theory. Such stochasticities are inherent to a more general niche theory, and have been investigated in concert with niche theory for some time

(MacArthur and Wilson 1967; Chesson and Warner 1981; Strong et al. 1984; Tilman 2004). Thus, even though the neutral theory espoused by Hubbell (2001) is not directly covered in this volume, my hope is to continue to move towards a more general synthesis of niche and neutral approaches as two sides of the same coin (see also Gravel et al. 2006; Leibold and McPeek 2006; Adler et al. 2007).

An abbreviated history of niche theory

Since its inception by Johnson (1910), through its more thorough development by Grinnell (1917), Elton (1927), Gause (1934), and others (reviewed in Schoener 1989, 2009; Chase and Leibold 2003), the niche concept has traversed a tumultuous road in the development of ecological enquiry. This is despite the niche consistently being a core ecological concept. Two main issues have plagued niche theory.

As a first plague, the niche concept has had multiple meanings in an ecological context, some of which are too vague or grandiose to be of use, and some of which are somewhat contradictory (e.g., Schoener 1989, 2009; Chase and Leibold 2003). The duality of niche definitions has been often discussed (Schoener 1989; Leibold 1995). One view focused more on the places that species live, including abiotic and biotic factors (e.g., Grinnell 1917; Hutchinson 1957), and the other focused on the role of species in the community (e.g., Elton 1927). Adding to this confusion are many related and overlapping terms that exist in the literature, including "habitat" (Whittaker et al. 1973), which describes the environmental features in which species live (e.g., Whittaker et al. 1973; Southwood 1977), and "guild," which describes a group of organisms with similar needs (Root 1967). Although these terms are often used, and can be useful, I avoid them and their associated concepts owing to their own inherent confusions and subjectivity.

The second plague is that niche theory has endured several rather contentious debates that interestingly seem to have arisen in ~20–30 year recurrences (see also Cooper 1993). In each case, the primary contention revolves around the predictability of community composition and its matching to environmental conditions, ranging from the historical debate between the superorganismic versus individualistic organization of communities (Clements 1916; Gleason 1926) to the contemporary debate between niche and neutral mechanisms (Hubbell 2001; Chase and Leibold 2003).

Although not formally designated in the context of niches, the debate between Clements (1916) and Gleason (1926) regarding the structure of communities revolved around how species in communities responded to environmental variation. Clements' view was more deterministic, suggesting that for any given set of environmental conditions, a particular community (the "superorganism") would develop. Gleason's view, however, considered that species associations with each other were less aligned, and that often community structure could be less predictable due to stochastic processes.

In 1957, the Cold Springs Harbor Symposium on Quantitative Biology convened a diverse group of demographers studying human and other animals, population and community ecologists, and evolutionary biologists. Though the goals of this meeting were manifold, one recurrent theme was the debate about controls of populations, and whether they were more likely to be controlled by internal processes, such as density dependence and species interactions, or controlled by external processes such as weather (Andrewartha and Birch 1954; Lack 1954). In his now classic concluding remarks to that symposium, Hutchinson (1957) presented his quantitative definition of a species' niche as an "n-dimensional hypervolume" of factors that influence the persistence of that species. He also differentiated between a species' fundamental niche, describing the factors where a species *could* possibly live in the absence of any other biotic or historical factors, and the realized niche, describing the factors that influence where a species actually does live in the presence of interacting species (and also dispersal, history, and other constraints).

The third debate where the niche concept played a central role arose during the late 1970s and early 1980s. Following Hutchinson's (1957) definition of niche, his address regarding the limits to biodiversity (Hutchinson 1959), and the works of several of his students and colleagues, most notably MacArthur (e.g., MacArthur 1958, 1960, 1964, 1970b; MacArthur and Levins 1967), a renaissance of studies that explored patterns of species coexistence and limits to niche similarity occurred (reviewed in Whittaker et al. 1973; MacArthur 1972; Vandermeer 1972; Schoener 1974). In a series of papers, Dan Simberloff and colleagues (e.g., Simberloff 1978, 1983; Connor and Simberloff 1979; Simberloff and Boecklen 1981) challenged much of the empirical evidence for this theory on statistical grounds. Specifically, these authors developed a rigorous null model approach (reviewed in Gotelli and Graves 1996) and suggested that many of the observed patterns put forth to validate niche theory based on competitive interactions and coexistence were not different from what would have been expected by chance alone. Following a rather intense, and at times venomous, debate (e.g., Strong et al. 1984), ecology became a more experimental and statistically robust science, suggesting that theory made the science of ecology evolve (see also Kolasa Chapter 2).

The ongoing debate regarding the importance of niche theory against its more stochastic alternatives intensified with the publication of Hubbell's

(2001) unified neutral theory. Neutral theory emerged from a conceptual realm related to the null model statistical approaches discussed in the preceding paragraph, along with a close allegiance with models of neutral processes in population genetic systems. Specifically, Hubbell's (2001) neutral model (see also MacArthur and Wilson 1967; Caswell 1976; Bell 2000) predicts patterns of species diversity, species composition, and relative species abundances based on stochastic processes alone, without the invocation of niche differences among species. In doing so, Hubbell's primary goal was to see how well such a neutral theory can do in predicting patterns through space and time. The answer was: often pretty well! The potential success of the neutral model begs the question of whether the premise of basing ecological models on niches is an appropriate starting place, or whether the neutral model is a better place to start (Volkov et al. 2003, 2007). As a result, a rather contentious debate has centered on the ability of the neutral model to predict patterns of community structure relative to more niche-based models (reviewed in Chave 2004; McGill et al. 2006).

In each of the above cases, proponents of niche theory emphasize determinism and predictability, whereas the opponents suggest that other forces, primarily stochastic ones, often predominate. A more complete niche theory would benefit from the inclusion of both deterministic and stochastic elements. Hubbell's neutral theory was directly borrowed from concepts of stochastic dynamics in population genetics. Ecological drift in communities is directly analogous to genetic drift in populations. Population geneticists recognize that natural selection and genetic drift act simultaneously, but differ in their relative importance as a result of several factors, including population size and dispersal (Templeton 2006). Similarly, ecological drift and niche selection act simultaneously in communities (Leibold and McPeek 2006; Adler et al. 2007; Chase 2007), and the relative importance of the two processes should vary depending on community size and dispersal (Chase 2003; Orrock and Fletcher 2005). In the following sections, I revisit niche theory from first principles, show how a variety of related theories collapse into a few general concepts, and then discuss how to integrate stochastic processes within the more traditional concepts of niche theory.

Defining the niche

Following Leibold (1995) and Chase and Leibold (2003), the niche of a species can be separated into two fundamental units: the requirement component and the impact component. Here, we define these components in the context of consumer-resource equations for simplicity of presentation and for logical progression (based on MacArthur 1972; Tilman 1982). However, this presentation is not intended to be complete or universally applicable, and instead is simply intended to serve as a backdrop for a more general formulation of the niche concept.

The *requirement* component of the niche denotes the minimum or maximum level of a particular factor that allows a species to persist in a given habitat. This can be depicted graphically by exploring how a species' birth and death rates vary with a factor that influences a species niche. For example, with increasing levels of a limiting resource, a species' birth rate should increase (Fig. 5.1a); its death rate can be constant along this range, or could decrease if higher resources allow the organism to resist mortality from other sources (either way, the qualitative conclusions are the same). If the externally supplied availability of resources is such that births exceed deaths, a species can persist, whereas when the availability of resources is such that deaths exceed birth, a species does not persist. It is important to note here that this view of the niche is orthogonal to other niche concepts that depict a species' niche (and poten-

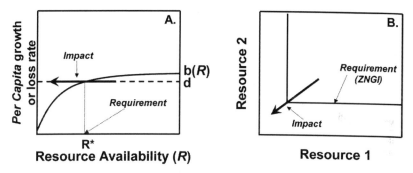

Figure 5.1 A. An example of how examining the responses of a species' birth and death rates to a factor of interest—resources in this case—can determine the requirement component of the niche. The intersection of birth and death rates indicates the level of resources necessary for the species to be at equilibrium (neither growing nor declining), and is denoted as a species' R^*. The impact of a species on this resource is denoted by the arrow, which indicates the strength of the consumption effect of the species on the resource. B. The requirement component of a species for two resources. Note, the zero net growth isocline (ZNGI) denotes the combination of resources when birth rates equal death rates (equilibrium). Here, the impact vector is the sum of the species impacts on the two resources. The R^*, ZNGI, and impact vector have underlying mathematical definitions (Chase and Leibold 2003), but have been popularized because of their graphical representation. In addition, similar graphical (and analytical) forms can be derived for any niche factor that influences a species' birth and death rates.

tial overlap with other species) as the number and types of resources on which they depend (and often the degree to which they use them) (e.g., MacArthur and Wilson 1967).

Resources will be maintained at an equilibrium when the consumer's birth and death rates are equal; this equilibrium is known as the R^* of a consumer for a resource (Tilman 1982), and defines the lower boundary of the consumer's requirement niche for that resource. As a result, when the externally supplied level of resources are such that the birth rate is equal to or greater than the death rate, the species persists. If, alternatively, the limiting factor of interest is a predator, a species can persist where the level of predation is such that births equals deaths, and is defined as a species' P^* (Holt et al. 1994). In fact, any such factor that influences births or deaths, including mutualists, environmental stressors, or spatiotemporal heterogeneities, can be determined by calculating the equilibrium where birth rates equal death rates.

The boundaries of the requirement component of the niche can be extended to any number of niche factors, though generally only two dimensions are graphically depicted. For example, the requirement component can be extended to two resources, and graphically depicted by plotting the relationship through the two-dimensional state space; the combination of factors where the species birth rate is equal to the death rate (equilibrium) is termed its zero net growth isocline (ZNGI) (Fig. 5.1b). ZNGIs such as that depicted in Fig. 5.1b can in actuality be derived for any combination of factors that influence population growth (births and deaths) of an organism, either positively or negatively. The well-studied factors include predators, stressors, and spatiotemporal heterogeneities (Holt et al. 1994; Leibold 1996; 1998; Wootton 1998; Chase and Leibold 2003), but other factors, including those that can have positive effects, such as ecosystem engineers and mutualists, can be relatively easily incorporated into this framework (Chase and Leibold 2003).

The *impact* component of a species' niche denotes the influence of the species on the niche factor of interest. For example, if the factor is a resource, the consumer will act to drive the level of that resource down towards the R^*, as long as the external supply of that resource exceeds the R^* (Fig. 5.1a). Alternatively, if the factor of interest is a predator, the species will act to drive the density of the predator higher, and thus the impact vector will point upwards along the predator axis. Finally, if the factor of interest alters a species' birth and death rates, but the species has no influence on that factor, such as externally imposed stressors (e.g., disturbance), the species will have no impact on that factor.

As with requirements, the impacts of a species can be depicted in two (or more) dimensions for any two limiting factors by taking the vector sum of the impacts on each factor. For example, if there is a consumer of two resources, the consuming species has a negative impact on both resources, and the sum of those two (horizontal and vertical) vectors represents the impact of the species on the two resource system (Fig. 5.1b). The slope of the vector indicates the relative magnitude of the impact of the species on each resource; if the species has equal effect on both resources, the slope will be 1, whereas it will be steep or shallow if the consumer of interest has a greater impact on one resource than on the other. As above, impact vectors can be drawn for any two-dimensional system, including combinations of resources (negative effects), predators and mutualists (positive effects), and environmental stressors (no effects).

Finally, in order to determine whether a species can actually persist in a given locality, we need to consider the external *supply* of the niche factor of interest. Depending on the nature of the niche factor of interest, the supply can be a rate denoted by a vector, or as a single point in the state space. If the resource has an inherent growth rate (e.g., logistic growth towards a carrying capacity), the supply vector will point towards the carrying capacity of that resource in the absence of its consumer. Likewise, the supply vector will point towards an externally supplied rate of input in the case of abiotic resources such as light or water. If, alternatively, the resource is in mass balance (sensu Holt et al. 1994), as is the case with many limiting nutrients, the supply can be depicted as a point in state space where the resource would be in the absence of any consumers. External supplies of niche factors other than resources can be depicted in a similar manner. For example, if the niche factor is some stress like a disturbance, this can be depicted as a point in the state space showing the rate or intensity of that disturbance. Finally, if the niche factor of interest is dependent on the species being modeled, for example, if the niche factor is a predator that cannot persist without prey, its external supply will be 0.

Just as in standard predator-prey theory (see Holt Chapter 7), in the simplest scenario, a species will maintain its resource at R^* regardless of the external supply of that resource. Thus, variation in the standing levels of resources does not necessarily imply variation in the supply rates of resources, but rather, can indicate differences in the R^*s of the consumers in a system. Adding a factor that increases the death rate of a species (or decreases its birth rate), such as a predator, will increase the equilibrium abundance of the resource left behind in the system (e.g., a trophic cascade) (Holt et al. 1994; Leibold 1996). Niche factors on which a species does not have an impact (e.g., a stressor such as disturbance) are not influenced by the presence of that species (Wootton 1998; Chase and Leibold 2003).

Propositions of the niche theory

Armed with these niche concepts, we can establish a series of propositions (sensu Scheiner and Willig 2008; Scheiner and Willig Chapter 1) that a synthetic niche theory can provide (Table 5.1). Importantly, this synthetic niche theory can directly incorporate deviations from the simplest predictions described above. It can do this by including complexities in species physiological, life-historical, and behavioral attributes (Tilman 1988; Vincent et al. 1996; Chase 1999; Chase and Leibold 2003), species interactions in food webs (Holt et al. 1994; Leibold 1996; 1998; Chase and Leibold 2003), a variety of spatial and temporal environmental heterogeneities (Tilman and Pacala 1993; Chesson 2000b; Chase and Leibold 2003), as well as the stochasticity inherent to the neutral theory (Chesson 2000b; Adler et al. 2007). Many of the details of this synthetic niche theory and the propositions discussed below are only superficially covered here (see also MacArthur 1972; Tilman 1982; Tilman 1988; Holt et al. 1994; Leibold 1995; Leibold 1996; Grover 1997; Chesson 2000b; Chase and Leibold 2003; Adler et al. 2007).

Proposition 1

When a single factor is limiting, the species that can maintain that factor at the lowest (or highest) level will outcompete all others. When the limiting factor is a resource, the species that exists alone will be the one that can persist on the lowest level of that resource (lowest R^*). This basic proposition is often know as Tilman's R^* rule (Tilman 1976; Tilman 1982), although its roots can be seen earlier (e.g., MacArthur 1972). The derivation of this principle is relatively straightforward. Assuming one species maintains the resource at its R^* owing to the consumer-resource dynamic, the level of that resource will be too low for any other species to invade. As alluded to above, the same principle can be derived from the standard Lotka-Volterra predator-prey model (see Holt Chapter 7), where the vertical isocline of a predator that is closest to the abscissa will outcompete any predator with an isocline further from the abscissa.

This basic principle can be applied to any limiting factor that influences a species' birth or death rates. For example, when a predator is limiting instead of a resource, the P^* rule obtains (e.g., Holt et al. 1994). In this case, through apparent competition for a shared predator, the species that is able to maintain positive population growth rate on the highest level of predators (highest P^*) will outcompete any others and exist alone. Alternatively, if a stressor influences birth or death rates, the species that will persist alone will depend on

Table 5.1 The propositions that constitute the theory of niches

1. When there is a single limiting factor, the species that can maintain that factor at the lowest (or highest) level will outcompete all others.

2. For more than one species to coexist locally, they must trade off the relative ability to compete for different limiting factors, as well as the impacts on those limiting factors.

3. Temporal variability can allow species to coexist that otherwise could not do so.

4. For two or more species to coexist regionally, they must trade off the ability to compete for different limiting resources, and those resources must be heterogeneously distributed in space.

5. Stochastic ecological drift will counterbalance deterministic niche-based processes when the local community size is small relative to the size of the regional species pool.

6. Strong niche selection will reduce the size of the realized regional species pool, and decrease the relative importance of stochastic ecological drift.

the level of the stress, since species are assumed to have no impact on stress; the more stress-tolerant species will persist alone at high stress, and the stress-intolerant species will persist alone at lower stress (Wootton 1998; Chase and Leibold 2003).

If there are multiple limiting factors, but only one species is superior on all of those factors such that it is a Hutchinsonian Demon (sensu Kneitel and Chase 2004), it will outcompete all others and exist alone. Instead, in order for more than one species to coexist locally, there must be more than one limiting factor, and those species must trade off in their ability to utilize those resources, which leads to the next proposition

Proposition 2

For more than one species to coexist locally, they must trade off the relative ability to compete for different limiting factors, as well as the impacts on those limiting factors. For two species to coexist locally in a spatiotemporally homogeneous environment when there are two limiting resources, they must trade off in their ability to utilize those resources. That is, the ZNGIs of the two species must intersect where one species has a lower R^* for one resource, and the other a lower R^* for the other resource. The same arguments can be extended to multiple resources, so long as species show perfect tradeoffs for those resources (e.g., each species has its lowest R^* on a different resource) (Levin 1970; Kneitel and Chase 2004). While such tradeoffs are necessary for local coexistence, this tradeoff alone does not guarantee local coexistence. First, there must be a second tradeoff in the impacts of those species on the resources for the equilibrium to be locally stable. Specifically, the impacts of the two species need to be such that each species has a greater relative impact on the resource that it finds most limiting (has a higher R^*). That is, if a species has traits that allow it to persist at relatively low levels of a particular resource (low R^*), those traits need to trade off with other traits that allow it to impact (e.g., consume) the resource that it finds more limiting (higher R^*). This tradeoff allows intraspecific effects of each species on itself (through their impacts on the resource) to be greater than the interspecific effects of each species on the other species, which is the standard criterion for coexistence in any multispecies community.

In addition to these coupled tradeoffs in requirements and impacts among competing species, local coexistence requires particular conditions in the environment. First, the supply rate of the resources must exceed that of the R^*s for the species. Second, the ratios of the supplies of those resources needs to be intermediate, and not skewed strongly towards any one resource (in which case, the species with the lowest R^* for that resource can outcompete the others through quorum effects). This emphasizes that a fundamental feature of niche differences and coexistence is not just differences in the traits of species, but in the characteristics of the environment in which the organisms live. Thus, environmental context is an essential component of understanding niche theory and coexistence among species, although this has not always been explicitly recognized.

These general principles of local coexistence derived when two resources are limiting are in fact true for any species that interacts with two or more factors that influence their requirement niche (e.g., birth and death rates), and on which they have impacts. For example, when two prey species share two predators, they can potentially coexist if: (1) they trade off in their P^*s, such that each prey species is least influenced by, and thus a better apparent competitor (higher P^*) for, one of the predator species; and (2) each species has a bigger positive impact (e.g., is a better food source) on the predator for which it is a worse apparent competitor (lower P^*) (Leibold 1998). Likewise, when two prey species have a common resource and a common predator, they will coexist when the species that is a superior resource competitor is simultaneously more susceptible to predation than the weaker resource competitor (i.e., a keystone predator), and the better resource competitor (lower R^*) is also better food (higher impact) for the predator (Holt et al. 1994; Leibold 1996;

Chase et al. 2000). Importantly, if the species do not impact one or more of the two limiting resources, local coexistence is not possible, but regional coexistence is (see below).

Proposition 3

Temporal variability can allow species to coexist that otherwise could not do so. An important, but sometimes overlooked mechanism by which species can coexist locally is temporal variability in the availability of resources through time. If the supply rates of resources are variable, more than one species can coexist, even on a single resource, if they trade off in their abilities to utilize the resource when it fluctuates at different availabilities. For example, rainfall and temperatures vary seasonally, and species can coexist locally by partitioning their competitive abilities across seasons (e.g., Hutchinson 1961; Grover 1997). Additionally, variability can emerge among years, for example, when precipitation varies inter-annually in arid systems. In this case, if some species are favored under wetter years, and others under drier years, they can coexist locally so long as the time frame of the variability in rainfall is not longer than the time it takes for the species to go locally extinct (i.e., the storage effect; Chesson and Warner 1981; Chesson 2000b). Similarly, species can coexist if they trade off in their relative abilities to compete for resources and in their ability to colonize habitats following temporal disturbances: the colonization-competition tradeoff (Hastings 1980; Tilman 1994). Finally, temporal variability can be internally driven by the dynamics of the consumers themselves (e.g., when responses are nonlinear), and does not have to be driven externally. If consumers create a nonequilibrial (cycling) dynamic in their resource base, they can coexist by specializing on different densities of their prey as it cycles (Armstrong and McGehee 1976; Huisman and Weissing 1999; Abrams and Holt 2002).

Proposition 4

For two or more species to coexist regionally, they must trade off the ability to compete for different limiting resources, and those resources must be heterogeneously distributed in space. Even when species show the appropriate tradeoffs for local coexistence, the supplies of resources are often skewed such that one species dominates in any given locality (i.e., no local coexistence). However, if the supply rates of those resources vary spatially (e.g., along a nutrient gradient), the species can coexist regionally if the availability of those resources varies spatially, so long as the species show perfect tradeoffs in their usage of those two resources (Tilman and Pacala 1993; Chase and Leibold 2003). As with temporal variability, spatial variability in resources can be generated by the organisms themselves, allowing species to coexist (Wilson et al. 1999; Wilson and Abrams 2005). Finally, spatial heterogeneity combined with dispersal can allow species to coexist locally and regionally as a consequence of source-sink processes, whereas higher dispersal rates can reduce local and regional diversity by allowing the better colonizers to swamp out the better competitors (Amarasekare and Nisbet 2001; Mouquet and Loreau 2003).

Importantly, the temporal and spatial heterogeneities discussed in this and the previous proposition expose a critical misrepresentation of niche theory by proponents of a purely neutral theory (Hubbell 2001; Bell 2003). Specifically, proponents of neutral theory have suggested that niche theory is predicated on the fact that the numbers of resources determine the number of species that can coexist (N resources leads to N species; Levin 1970), and that this cannot account for the very high levels of diversity often observed, particularly in tropical areas. However, simple temporal and spatial heterogeneities, even when there are only a few limiting resources, can allow very large numbers of species to coexist (Tilman 1994; 2004). Moreover, recent analyses show how individual-level variation in response to these spatiotemporal heterogeneities can promote coexistence among a diversity of species, even though the averages of species traits, without such individual-level heterogeneities, might not predict such coexistence (Clark 2010). Finally, a species can persist, and thus coexist with other species, in habitats where it experiences negative population growth (i.e., a sink habitat) if there is sufficient dispersal from adjacent habitats where the population experiences positive population growth (i.e., a source habitat) (e.g., Amarasekare and Nisbet 2001; Mouquet and Loreau 2003; Leibold Chapter 8).

Proposition 5

Stochastic ecological drift will counterbalance deterministic niche-based processes when the local community size is small relative to the size of the regional species pool. Though niche and neutral theories have been treated as dichotomous alternatives (e.g., McGill et al. 2006), the primary components of each theory, determinism and stochasticity, occur simultaneously (Gravel et al. 2006; Leibold and McPeek 2006; Adler et al. 2007; Chase 2007). Just as the stochasticity associated with genetic drift can interact with the determinism associated with natural selection in population genetics, stochastic ecological drift interacts with niche selection (e.g., the environmental filtering of the species pool) in communities (Adler et al. 2007; Chase 2007).

The relative importance of stochasticity in community assembly will be highly dependent on local and regional processes. Just as small population size increases the relative importance of genetic drift in population genetics, small local community size (individuals and species) increases the relative importance of stochastic ecological processes (Drake 1991; Chase 2003; Fukami 2004; Orrock and Fletcher 2005). One way to envision how this process could work is through simple probability; smaller communities will have fewer individuals, and thus a sampling effect would lead to a higher probability of differences among small communities than among larger communities. However, mechanistic processes, including differential effects of competitive interactions, can also lead to this pattern (e.g., Orrock and Fletcher 2005). The size of the regional species pool can also influence the relative importance of stochastic processes, but in the opposite direction. Smaller regional species pools will lead to more deterministic processes, and again this can be illustrated through probabilistic sampling; fewer species to choose from at the regional level will lead to communities that are structured more deterministically (Chase 2007; Chase et al. 2009).

Proposition 6

Strong niche selection will reduce the size of the realized regional species pool, and decrease the relative importance of stochastic ecological drift. Stochastic processes will be most important when environmental filters (niche selection) on the species pool are weak, such that the realized pool of species that can colonize a given locality is large relative to the number that can occur in any locality (Chase 2007; Chase et al. 2009). Often, a majority of species in the species pool can persist, at least in the absence of interspecific interactions, in relatively benign conditions. These conditions include relatively productive communities with low disturbances and few strong predators, among other things. Alternatively, low productivity, high disturbance and stress, or strong top predators can strongly limit the membership of the realized pool of species that can colonize those habitats. By effectively limiting the size of the realized pool of species, these harsh conditions overlay determinism on community membership, simultaneously reducing the influence of stochasticity. Indeed, stochasticity in the assembly of communities, and thus the variation in community membership, appears to be more important in environments with higher productivity (Inouye and Tilman 1995; Chase and Leibold 2002; Chase 2003; Chalcraft et al. 2008), environments with lower disturbance or stress (Booth and Larson 1999; Chase 2003, 2007; Trexler et al. 2005; Jiang and Patel 2008; Lepori and Malmqvist 2008), and environments without strong top predators (Chase et al. 2009).

Relationship to other theories in ecology

Because the niche generally describes the distribution and abundance of a particular species, as well as the factors that influence coexistence (diversity) among species, the domain of niche theory is rather inclusive of much of the domain of ecology as a whole. As a result of this broad association, it also has close ties with several of the other constituent theories in ecology.

Foraging theory (Sih Chapter 4) defines the ways in which organisms acquire the resources they need, the evolutionary causes of such acquisition, and the behavioral mechanisms by which they do so. As resource acquisition is fundamental for an organism to persist, foraging theory represents a primary mechanism by which organisms are able to achieve their requirements and have their impacts on resources. For example, an important tool in foraging ecology is measuring the density of resources in a patch on which a forager "gives up" and moves to another foraging patch (e.g., Brown 1988). This same tool can be used to understand the relative position of a species' R^* in an environment, as well as how various factors (e.g., predation risk) influence that R^* (Chase and Leibold 2003).

Population dynamics theory (Hastings Chapter 6) defines the ways in which population grow in the context of birth and death rates, and thus has obvious connections to the births and death rates that go into defining the parameters of the niche. Further, interspecific interactions, including predator-prey interactions (Holt Chapter 7), alter birth and death rates, thus directly altering the structure of the organisms' niche.

Patterns of community structure through time, as predicted by succession theory (Pickett et al. Chapter 9), and in space, as predicted by both metacommunity theory (Leibold Chapter 8) and environmental gradient theory (Fox et al. Chapter 13), depend on the interactions among species and their spatio-environmental context. As such, species' niches fundamentally underlie the mechanisms leading to these patterns.

Ecosystems theory (Burke and Lauenroth Chapter 11) examines the fluxes of energy and nutrients through organisms and their abiotic environment. In many circumstances, the traits of species, as defined by their niches, as well as the diversity and composition of species in a community, influence the importance of their roles in ecosystems.

A recent application of niche theory involves species distribution modeling (often called "niche modeling"), which correlates spatiotemporal availability of broad-scale environmental features with species occurrences (e.g., Stockwell 2007). Although species distribution modeling is based on correlations between distribution and environment, and is rather simplistic relative to the aspects of niche theory discussed here, it has potential to link small- and large-scale processes. Further, such distribution modeling is often used to try to understand how species distributions might be altered by global change (e.g., Peterson et al. 2002), and thus provides important linkages with global change theory (Peters et al. Chapter 12).

Conclusions

Despite repeated attempts over many decades (Grinnell 1917; Elton 1927; Hutchinson 1957; MacArthur 1972; Tilman 1982; Chase and Leibold 2003), niche theory has still not fully reached its potential as a central organizing theory of ecological enquiry. This is in part due to the fact that the niche is such a hard concept to define. Indeed, any definition, verbal or mathematical, of the niche can only represent a caricature of the complexity of any species' multivariate interactions with its environment and other species. Additionally, niche theory has not traditionally incorporated a variety of complexities, most notably spatiotemporal heterogeneities and stochasticity, which can often play an overriding role in determining the structure of natural communities. An emerging more general niche theory explicitly incorporates spatiotemporal heterogeneities (Tilman 1994, 2004; Chesson 2000b) as well as the stochastic processes inherent to neutral theory (e.g., Leibold and McPeek 2006; Adler et al. 2007; Chase 2007; Chase et al. 2009). Such a synthetic theory will provide the groundwork for a more general theory of ecology as a whole.

Acknowledgments

Many of the ideas in this chapter were solidified during conversations and presentations at a workshop organized by M. Willig and S. Scheiner on this topic supported by the University of Connecticut through the Center for Environmental Sciences & Engineering and the Office of the Vice Provost for Research & Graduate Education. Comments by T. Knight, S. Scheiner, M. Willig, and two anonymous reviewers greatly improved earlier versions of this chapter.

CHAPTER 6

Single Species Population Dynamics and Its Theoretical Underpinnings

Alan Hastings

A theory of single species population dynamics is one of the most basic and oldest parts of ecology (Kingsland 1995), and much of applied ecology is based on it. Management of fisheries (Clark 1990) has often been based on single species approaches. Approaches to conservation, even those embodied in the Endangered Species Act, are very much focused on single species (Lande et al. 2003). Developing appropriate approaches for these and other management questions certainly depends on the development of a theory of population dynamics. And a deep understanding of the assumptions underlying this theory is critically important in judging how and when this theory should be applied.

Not surprisingly, single species population theory has a particularly long history (Kingsland 1995), which some have argued goes back at least as far as the work of Fibonacci (who lived roughly 1170–1250), who developed his description of population dynamics in the form of the Fibonacci series. Jumping ahead, the work of Malthus (1798) over 200 years ago can be thought of as a direct precursor to some of the work that will be presented here.

Malthus presented his work verbally, but this still contained much of what was later expressed mathematically. His suppositions that populations grew geometrically while food supplies grew arithmetically is particularly relevant to the material below. This approach is one that essentially starts from the simplest description of the dynamics of a population and then proceeds to ask why this simple description does not hold.

On the other hand, there has been substantial recognition of the importance of space and heterogeneity (Levin 1992), stochasticity (Lande et al. 2003), and other factors for understanding the dynamics of populations. This begins to raise the central question of the level of appropriate detail to use in a description of the dynamics of a population. This is a difficult question, and obviously depends on the purpose for studying the dynamics in the first place. One could start from the simple approach essentially contained in the work of Malthus and build up to this more complete description. I will present this approach later. Alternatively, one could start from as complete a description of single species population dynamics as possible and consider simplifications that would potentially lead to more tractable models and descriptions. This approach, too, has its merit, and I will also present this alternate approach.

The remainder of this chapter is organized as follows. I will first present the development that starts from a relatively complete microscopic description, and continues by looking at appropriate averaging to yield more concise and useful models. Essentially this approach depends on two parts. The first part is clearly the development of as complete a description as possible. The second part is the choice of which aspects will be averaged over. Various implications and consequences of this approach will be considered. I will refer to this as the *microscopic approach*.

The alternate approach, which starts from the simplest possible descriptions of population growth, will be considered. This development, which starts with an idealized description, will be called the *macroscopic approach*. Here, the development consists of first writing down a simple model and then adding details to a model that is clearly too simple. In this approach, the biological questions and issues will be clear, but as the discussion below will indicate, it is not always clear how the models and descriptions are developed and related to underlying mechanisms. Much of the historical development has built on phenomenological rather than mechanistic models.

In some senses, there are analogues to these two approaches in the context of physical theories. One could describe the behavior of a gas by starting with idealized gas laws (expressed only in terms of bulk measures like pressure, temperature, or volume) and then subsequently add complications that recognize specific properties of either the gas or situation under examination. Alternatively, one could start with a description of the motion of each individual gas molecule based on mechanics and average appropriately to determine behavior at a larger scale. Because even single species population dynamics involves much more heterogeneity and complexity than does the dynamics of a gas, the task for the population biologist is much more difficult.

Although implications and lessons from each development will be amplified in the discussion of each development, the metapopulation approach (Hanski 1999) has been so important that it will be considered separately. The discussion of metapopulations will be brief and the main goal will be to indicate how ideas from the approaches here might inform the use and development of metapopulation theory. Connections to other levels, such as behavioral ecology or community ecology, will also be considered briefly. The subject of single species population dynamics is extremely large, so all of the discussion here will be limited. Notably, little attention will be given to specific models for populations, as the arguments developed here will not depend on the details of particular models.

Domain of the theories

The domain of the theory of single species population dynamics is the understanding and prediction of the numbers of individuals in a single species. At this level, the domain is the same for either approach. Within this context, the focus could be on total numbers of individuals, but could also be on a description of a population that is structured.

Implicit in this choice of domain is the underlying assumption that the dynamics of a single species can be understood without consideration of other species. Clearly, the extent to which this is useful depends strongly on the question asked, the purpose of the model. Single-species models are likely to be relatively poor tools for understanding biodiversity. The extent to which these models are useful for managing species, either for exploitation (fisheries) or to control undesirable species, will depend very much on the particular system.

Microscopic approach

I discuss the theory of single species population dynamics using two different but complementary approaches. The first approach is the one that is based on simplifying a description that is more or less complete, at least at the single species level. The idea here is that this top-down approach has two advantages. First, a very complete description of the domain of single species ecology will make it clear what is being left out, and at least suggest how the omitted factors might be important. Second, the specification of a relatively complete model makes clear what is being left out in a particular instance.

I begin by outlining the theory, starting with a series of propositions underlying the theory. The domain of the theory considered here is obviously single species populations. However, as will also be discussed, the approach used here has been used as well to consider issues related to interacting species.

Table 6.1 Propositions for the microscopic approach to single species population theories.

---

1. The dynamics of a population are completely determined if the timing and location and characteristics of all the offspring produced by all newborn individuals as well as all deaths are known.

2. The determination of both births and deaths are at an individual level and consequently at the level of an individual birth and death rates can be given only in terms of probabilities. Every birth or death is clearly a chance event and at the most detailed level a deterministic description is never appropriate.

Propositions

As in other chapters, the development of theory starts with a series of propositions. Here, these are very general, and in some senses phenomenological. The alternate development below in many ways is more tightly tied to the biology. The propositions are developed by considering what ingredients could provide a complete description of single species population dynamics, without reference to biological processes, and are given in Table 6.1.

Statement 1 is almost tautological—but is necessary to point out clearly what the simplifications are that need to be made when producing a description that can be effectively used in models. Statement 2 indicates some of the most dramatic problems with developing theories of single species populations. The degree to which averages can be made so that descriptions of births and deaths can be described by simple probability distributions will depend on the importance of various heterogeneities and the size of the underlying population. The description as contained in these propositions is essentially the basis for a simulation-based approach (Grimm and Railsback 2005), which has been variously called individual-based modeling or agent-based simulation. Note that by birth in these propositions any action that produces more than one individual when previously there was a single individual is included. Thus for the simplest microorganisms cell division can be thought of as either two births and a single death, or a single birth. This and other simple examples emphasize that all newborn individuals may not be identical.

Implications and consequences

As will be clear, the unfolding of the assumptions in and simplifications that can be made to these propositions provides a way to understand much of the subject of single species population dynamics. The role played by proposition 1 is central.

Specific instances of descriptions of population dynamics arise when particular assumptions are made about the population per proposition 1, and about the role of stochasticity as embodied in proposition 2. These can really be assumptions about the lack of dependence of population processes on different factors, or can be approximations. A basic approximation that is made in many cases is to describe population sizes using a continuous variable, rather than restricting attention to integer population sizes. If the population size is very large, this approximation is likely to be a good one. Conversely, for very small population sizes, descriptions based on birth-death models that focus on individual births and deaths are needed.

As another example, one can focus on a mean field (Levin and Pacala 1997) description of a model that includes the distribution of individuals in space, and ignore the role of explicit space in the dynamics of the model. With enough assumptions and approximations, the essentially complete description of population dynamics embodied in these propositions can be simplified to the most basic population models. For example, assume that all newborns are identical (and even gender can be ignored), generations are discrete and non-overlapping, the expected number of offspring is the same for every individual and does not depend on the number of individuals, spatial extent is ignored, the environment is constant, and the initial population is large enough that stochastic effects can be ignored. Then, the population description reduces to discrete time exponential (geometric growth). This is the simplest kind of description of population dynamics. Similarly, one can obtain conditions under which population growth is exponential in continuous time. Here, assume that all processes (death and birth) are independent of age, density, and location (and anything else!), and ignore stochasticity. In this case, the population grows exponentially. Of course, real populations do not behave this way.

It is rather straightforward to see how all simple descriptions of population growth in time and space (i.e., those with age structure, spatial structure, genetic structure, density dependence, or various stochastic influences) fit within this construct as a result of assumptions that simplify the model, possibly combined with approximations that simplify the model. For example, a deterministic model that includes age structure makes implicit assumptions that all individuals contribute to future growth. Also, as a deterministic model with population size as a continuous variable, the model is making the assumption (or approximation) that the population can be well described by a single number rather than by a distribution of population sizes. Thus an advantage of developing models by taking explicit averages over classes of individuals is that it helps to stress the importance of identifying these approximations and assumptions, which are important to consider when deciding if a particular model applies to a particular situation. Continuing with the steps in the derivation of the deterministic age-structured population model from the complete microscopic model, one can ask whether the population size is large enough and the environment is constant enough to ignore stochasticity. More properly, one would start with a stochastic model, and compute a full distribution of outcomes, and see whether the variance is small enough that the expected outcome is a good description. The development of essentially all single species models thus can be thought of as starting from a microscopic description, with, of necessity, averaging of various kinds (such as using expected values of a stochastic process) required to produce a macroscopic description that may be of more use in applications or in understanding biological processes.

The development from the microscopic viewpoint also makes clear the relationship of population ecology to other subfields of ecology. Fields such as behavioral ecology, physiological ecology, and evolutionary ecology all focus on ecological aspects that affect the reproduction and survival of an individual. Knowing when these aspects at the level of the individual affect the dynamics of populations is clearly a difficult, yet important, question. This idea of the state of the individual is beginning to play a more central role in some models of population dynamics (de Roos et al. 1992), going far beyond the inclusion of age or stage structure. This kind of model structure can include many further descriptions of population state, with the practical constraint that the models developed can become difficult to study or draw conclusions from.

At a higher level of organization, one needs to recognize that populations do not live in isolation. Interactions with other species, or with individuals of different species, clearly affect the dynamics of a focal species through basic processes like competition, predation, parasitism, and others. These considerations highlight the issues behind proposition 1. Proposition 1 explicitly excludes the interactions from consideration, while proposition 2 essentially says that these kinds of effects can be lumped into a category called stochasticity. The limitations of the approach described here, where interactions with other species (and physical forcing) are not explicitly included, becomes abundantly clear. Nonetheless, the approach described here can in principle be expanded to multiple species yielding interesting insights, especially in guiding simulation approaches. Yet the obvious difficulty of using the approach here to develop models of single species demonstrates the difficulties of building up from this approach to multiple species. Thus community ecology is at a level where alternate approaches may be called for, as described by Holt (Chapter 7).

Example

The stochastic version of the discrete-time Ricker model developed in Melbourne and Hastings (2008) illustrates many of the points made here. In that paper, a series of models are developed from first principles. The model is for a population that is discrete in time, where adults lay eggs, which are subject to cannibalism by adults. The adults die, and then the remaining eggs that have survived cannibalism are subject to other density-independent mortality before females lay the next generation of eggs.

Of most interest is the specification of the birth process and other processes in the model, and in particular how various stochastic processes play a role. A series of nested models are developed, where different sources of stochasticity are included, resulting in different descriptions. In particular, the birth (and death) of individuals occurs randomly through time (with a mean rate that is constant per capita), which is known as demographic stochasticity. A second source of stochasticity is the potential variation in rate due to environmental influences, which would produce fluctuations in mean rates between generations. A third source of variability that is less often included is demographic heterogeneity, differences between individuals in birth or death rates (due perhaps to different microenvironments, maternal effects, or phenotypes). A fourth source of variability is perhaps the most extreme form of demographic heterogeneity, namely sex differences.

Different combinations of stochastic influences lead to a set of eight different models with the eight different combinations of influences in addition to demographic stochasticity, which is present in all the models. Of most interest here is that these differences at the microscopic level lead to differences in behavior at the macroscopic level. All models have as their mean description the same Ricker model. However, other summary behaviors are different. Mean time to extinction depends critically on the way stochasticity is included, and these results clearly depend on starting from the kind of complete description that is outlined here. With demographic heterogeneity, not surprisingly, time to extinction can be much shorter.

Macroscopic approach

The microscopic approach has advantages in highlighting assumptions, but can be unwieldy and can potentially obscure relationships to ecological questions. An alternate way of proceeding might be to start with exponential growth and look for aspects that prevent a population from growing exponentially. Since this highlights different aspects of population biology, I present this alternative approach in detail.

Propositions

Here the propositions essentially start from the simplest description of basic population dynamics and are given in Table 6.2. This formulation can be amplified by explicitly describing the assumptions that would produce exponential growth as a consequence which further highlights the development of theory according to this approach. Thus, proposition 1 can be viewed as a consequence of the three assumptions:

1. All individuals in a population are identical.
2. The number of offspring (in discrete time) or the birth rate and death rate (in continuous time) per individual are constant through time. In particular these numbers or rates are independent of the number of individuals in the population.
3. There is no immigration.

And the second proposition also has an important amplification, namely that real populations are affected by differences among individuals, density dependence, and random events.

Although all models that are interesting ecologically and mathematically build on this amplification of proposition 2, it is informative to start without these differences. There are basically two instances of theories obeying proposition 1 that can be derived from the three assumptions. The two theories differ as to whether the focus is on discrete time or continuous time. Either formulation provides an explanation of why the assumption of no immigration enters. In the former case, the population obeys the very simple equation $N(t + 1) = RN(t)$, where $N(t)$ is the population size at time t, and R is the per capita number of individuals the following year. In continuous time, the fundamental description is $dN/dt = rN$, where N is the population size (as a

Table 6.2 Propositions for the macroscopic approach to single species population theories.

1. A population grows exponentially in the absence of other forces.
2. There are forces that can prevent a population from growing exponentially.

function of time, t) and r is the pre capita growth rate of the population, which is simply defined as the per capita birth rate minus the per capita death rate.

Both of these equations have solutions that can be written explicitly and which take the forms $N(t) = R^t N(0)$ in discrete time, and $N(t) = e^{rt} N(0)$ in continuous time. From these solutions, we can easily see the relationship $R = e^r$.

I note that both of these solutions predict only three qualitatively different kinds of behavior. In particular, if $R > 1$ ($r > 0$) the population grows exponentially, while if $R < 1$ ($r < 0$) the population declines exponentially. The third possibility is the highly unlikely one of perfect balance, $R = 1$ ($r = 0$), where the population would remain unchanged through time. These simple conclusions are very powerful, and form the basis of much of the theory underlying studies of population viability. Yet this approach is obviously too simple because it ignores many factors that affect populations, and also because it cannot explain (since the perfect balance is highly unlikely) populations that do not grow or decline exponentially.

Under this development, the study of population biology becomes one of understanding the causes of deviations from exponential growth, and then the consequences of the actions of these causes. Thus, this approach in some sense mirrors the approach described in development 1, by building up to the most complex situations, rather than by looking at what simplifications can be made to the complex system to produce a simpler one.

First, I will consider modifications that in some sense are less dramatic in that some aspects of exponential growth are still preserved. In the simplest cases just considered, the population grows exponentially from the beginning and for all time. The next simplest behavior might be for populations to grow exponentially after some initial phase of growth that is not exponential. Or perhaps a quantity other than simply population size grows exponentially.

If individuals in a population are structured by age or stage, but the other propositions all hold (with appropriate modifications to deal with structure), the conclusions of exponential growth can essentially be recovered, under some further assumptions that essentially say that all age or stage classes can be reached by all others at all sufficiently long times (see Caswell 2001 for a discussion). More specifically, one can find that asymptotically the population will grow exponentially, but that initially the growth may not be exponential since there may initially be more of an age class that contributes either more or less to immediate numbers in the following year. For example, if all individuals are immature, then it may be several years before a population grows at all. The growth rate is well defined, and can be calculated, in principle for all cases. The idea of three qualitatively different kinds of behavior is essentially recovered, with a population either growing, shrinking, or remaining constant. Yet this seemingly simple and small addition in complexity produces a theory (Caswell 2001) that is so rich and important for answering questions in population biology that despite a huge literature many questions remain unanswered, although the basic framework is completely understood.

An alternate possibility that essentially retains exponential growth, but that has very different consequences, is to allow for stochasticity (Lande et al. 2003). First, consider the case of all members of the population as identical, but allow the environment to vary from year to year, affecting the population growth rate, or allow demographic stochastic events. Demographic stochasticity occurs when the population is small enough that the chance events of either multiple births in a row or deaths in a row or, in discrete time, variability in offspring number, could have significant impact. In these cases there may still be a quantity that grows exponentially, but this exponentially growing quantity may be uninteresting from an ecological standpoint. For example, it is easy to derive models in which the mean population size grows exponentially, but the probability of extinction is 1. This is the so-called gambler's ruin paradox, which shows up even if the population has a probability of tripling each generation, which is equal to the probability of going extinct. Here, the time until extinction is a much more interesting quantity than the behavior of the mean population, since this is a more complete description of population behavior. Going further, one could consider the full distribution of population sizes at a given time. This issue was discussed earlier in the context of the Ricker model.

Before going on to density dependence with per capita birth and death rates depending on the number of individuals, it is intriguing to consider the seemingly simple combination of age structure and stochasticity. Here, even the most basic questions of the growth rate are known only under assumptions that restrict the size of the stochastic effects and the form of stochasticity (Lande et al. 2003). Nonetheless, results do indicate that growth will still be exponential (in the same sense as for stochastic models without structure). Because this is a case that is essential for applications as diverse as conservation biology and fisheries management, the difficulty of analysis highlights the need to be very careful when considering particular instances of theories in population dynamics. Assumptions and approximations used to obtain solutions need to be carefully considered. Including other forms of structure, such as space or genetic structure, would be a next natural step. This too, can be done in such ways that growth is eventually exponential. But for many applications, predictions over relatively short timescales will be important—what will the population do over a small number of years or generations? Thus, the asymptotic or long-term behavior may be less interesting. In contrast, it will be the transients (Hastings 2004) that are important, and the population dynamic behavior over shorter timescales may not be well predicted by the long-term behavior. Even a model as simple as a description of a population living in two patches, coupled by dispersal where each patch has local population dynamics with overcompensatory density dependence, has long transient behavior. Dynamics might be characterized as in phase (population sizes in both patches going up and down together) for long periods of time suddenly switching to out-of-phase dynamics or vice versa, all without any external influence.

It is clear that populations do not continue to grow exponentially. So the next step is to include factors that prevent exponential growth. Some, such as density dependence, can be considered in the context of single species. Others, such as predation, disease, or competition, require at least implicit, if not explicit, consideration of other species. Even density dependence typically implies that other species or specific resources are of interest. Thus, some of the limitations of single species population dynamics as a self-contained theory come to the fore.

Dynamics depend critically on the form of density dependence. Density dependence that is compensatory would mean that the numbers per capita the following generation are lower, but total numbers still go up as numbers go up, while with overcompensatory density dependence total numbers in the following generation go down as numbers in the current generation go up, reflecting overuse of resources. It should be noted that the simple addition of overcompensatory density dependence to population models can lead to wild and complex dynamical behavior (May 1976). (Although this arises in the simplest models only in discrete time, similar behavior is exhibited by more complex continuous time models.) Thus the domain of single species population models is large enough to produce essentially all possible kinds of dynamics.

Eventually, one can build up the macroscopic approach to the full consideration of all factors included in the microscopic approach. Neither development is primary—it is the connection between microscopic and macroscopic that is central as this allows a deeper understanding of the assumptions that are necessarily made in producing any particular model.

Metapopulation approach

The theory of metapopulations has continued to play an increasingly important role in the study of single species population dynamics. In its simplest version, metapopulation theory counts a population by looking at occupied

habitat patches and focuses on the dynamics of local extinction and colonization. The degree to which it is difficult to develop metapopulation theory
from either the macroscopic approach or microscopic approach developed
here highlights potential ways to improve on metapopulation theory. The
metapopulation approach is an example of a macroscopic approach that is
different from the ones presented above in which the averaging is taken in a
very different way. Ideas from the development of structured metapopulation
models provide one way to extend the macroscopic description. Although it
would be possible to develop a set of propositions specifically for metapopulation dynamics, or add specific ones or revise the sets given above, metapopulation models are a form of single species dynamics and therefore it should be
possible to see how metapopulation models are a specific instance of a single
population theory.

It is thus informative to think of how to "force" a description of metapopulation dynamics to fall within the structure of the microscopic approach
developed here. One needs to make specific assumptions about the effect of
individuals on each other's probability of having offspring and dying as a function of their location in space. Thinking about metapopulations this way also
points out an issue with proposition 1 of the microscopic approach. I did not
explicitly include a requirement of the knowledge of the state of an individual
(e.g., its location) throughout its life. In a metapopulation approach, the location of an individual affects population dynamics through its effect on the
production of offspring (both its own and the offspring of others).

As a way of making this discussion more concrete, begin with the simplest metapopulation model, the Levins (1969) model, which takes the form
$dp/dt = mp(1 - p) - ep$, where p is the fraction of patches or habitat occupied
by the species, m is the per occupied patch colonization rate of empty patches,
and e is the extinction rate of patches. It is obviously much easier to see how this
model fits into the microscopic approach than the macroscopic one, and there
may be more than one way to specify fully a model that looks like the Levins
model. One could specify a fixed population level for an occupied patch, and
assume that death occurs either by a single individual dying and being replaced
from the local patch, or by having all individuals die together. Reproduction
would then also be of two kinds. One would be replacing a single individual
within a patch, while the other would specify a probability of a reproduction
event (per patch or individual since with a fixed carrying capacity per patch
these are essentially equivalent) that would produce enough individuals to colonize an empty patch and immediately fill it to carrying capacity. The somewhat artificial nature of these assumptions is primarily an indication of the

kinds of averaging that go into the Levins model. In turn, acknowledging and recognizing these assumptions is an important aspect of drawing conclusions, both for basic understanding and for management, from a Levins model.

A somewhat more straightforward way to connect metapopulation theory with the content of this chapter is to start from models of structured metapopulations (Hastings and Wolin 1989; Hastings 1991; Gyllenberg and Hanski 1997). In this case, the models specify the number of individuals within a patch. Then, the connection to the single species theory becomes clearer, but the necessity for explicitly including the effect of individuals on other the birth and death probabilities of other individuals is still explicit. The specification of where offspring of newborns end up is thus a statement about the probability of remaining in the current location, or moving to a new location.

Conclusions

The study of population dynamics at the level of the single species is an area where understanding can be significantly enhanced by focusing on the fundamental assumptions and conclusions of theory. In particular, since this is a theory that clearly makes assumptions that may be unrealistic, explicit consideration of the basis for this theory can help make precise domains of applicability. This is particularly important, as many management decisions are based on single species theory, and will likely be based on single species theory into the future.

The two different approaches here (Fig. 6.1), the microscopic and the macroscopic, not only provide insights into single species population dynamics, but highlight key issues in much of modeling in ecology. How does one relate events at one level of organization, the population, to those at another level, the individual? How much simplification should be done before finalizing a model? What is the appropriate scale to focus on?

All these questions really come to the forefront when using the theory to make predictions in the face of limited data. Stochasticity clearly plays a role. The fact that model complexity is an important issue, and that simpler models may be better for making predictions (Ludwig 1999) is often not well enough appreciated in the context of management or prediction. When choosing simpler models, the more formal development of the theory that highlights the role played by assumptions and approximations can show both the possibilities and the limitations of the theory. This is particularly important in management, especially in the context of global change.

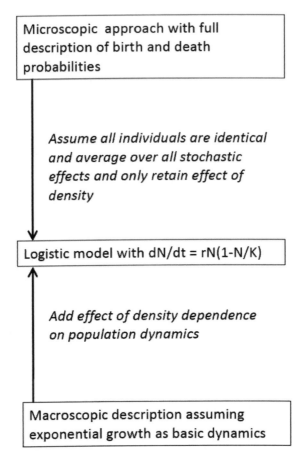

Figure 6.1 Illustration of contrast between macroscopic and microscopic approaches to single species population dynamics.

The limitations of single species theory are particularly clear when recognizing how much is not contained in it. The role played by species interactions must be an important one. Yet management is often still done within a single species context, and insights into management based on bioeconomic analyses of single species models have been very influential (Clark 1990).

Even restricted to the domain of single species theory, the microscopic approach presented here, starting from an essentially complete description of single species population dynamics, exposes the difficulties of developing a complete theory even for this case. A hope is that the formal development is useful in organizing current theory and guiding future refinements, extensions, and connections.

Acknowledgments

I thank the editors of the volume for very helpful comments. This work was supported in part by NSF Grant EF 0742674.

CHAPTER 7

Natural Enemy-Victim Interactions: Do We Have a Unified Theory Yet?

Robert D. Holt

"Nature red in tooth and claw," "eat, or be eaten": these and other popular sayings provide everyday testament to the fact that exploitation of one species by another via consumption pervades the natural world. I begin with a few words on the nature of theory in ecology, and then characterize the "domain" of natural enemy-victim ecology. I suggest that any sensible theory of interactions between natural enemies and their victims must be grounded in basic population theory, and that a necessary duality in the foundations of population biology helps identify broad principles for the theory of natural enemy-victim interactions. There is a rich existing theory for this ecological domain but also important unresolved issues and challenges for continued research.

A few words on theory in ecology

Definitions of "ecology" encompass both patterns in the distribution and abundance of organisms and the processes generating those patterns (Begon et al. 2006). Scheiner and Willig (Chapter 1) state that a general theory of ecology should contain relatively discrete "constitutive theories" comprising: (1) a *domain* (that range of the natural world encompassed in the theory); (2) a set of *propositions*, or "foundational concepts" (Turchin 2001; Berryman 2003) that help us understand the domain and are grounded in more basic principles; (3) formal, mathematical *models* that instantiate general ideas and provide a bridge between these ideas and empirical studies; and finally

(4) *connections* with other constitutive theories. In my view, these desiderata should not be viewed rigidly, but fluidly and flexibly. It may not be possible to adequately describe, understand, or even identify the basic propositions of a theory without having a formal model at hand. Or the propositions will themselves emerge from attempts to develop models, an effort that usually requires explicit or implicit linkages with other areas of theory. The boundaries of domains will often be fuzzy. And so on.

The chapter is organized as follows. First, I reflect on the domain of the body of theory that is my concern and touch on prior attempts to articulate a general theory of this domain. Second, I discuss two foundational concepts from the general theory of ecology, each of which can ground general natural enemy-victim theory. Both in my view are necessary for a well-rounded theory of this domain. Third, I develop a series of propositions (made more crisp by being cast in the language of a general, formal model) that I suggest are universal for natural enemies specialized to a single kind of victim and should thus constrain the detailed assumptions of any specific model of a natural enemy-victim interaction. I then turn to a number of other propositions that are broad but not exceptionless generalizations, and reflect on one of the major dynamical consequences of these theoretical propositions. Fourth, I discuss important issues that have arisen within key subdomains of the general theory, namely predator-prey and host-pathogen theory. Some key issues are still unresolved, even after much debate. I conclude with a discussion of how some propositions of the theory need to be qualified when considering generalist natural enemies, and with pointers for future theory development.

The domain of natural enemy-victim theory

The domain of natural enemy-victim theory is a subset of resource-consumer theory. Murdoch et al. (2003, p. 1) remark: "The consumer-resource interaction is arguably the fundamental unit of ecological communities. Virtually every species is part of a resource-consumer interaction, as a consumer of living resources, as a resource for other species, or both." Resource-consumer theory legitimately includes plants as consumers of light and nutrients, and decomposers as consumers of dead organic matter, but these are not part of natural enemy-victim theory. The R^*-rule (Tilman 1982) is a proposition of general consumer-resource theory, for instance for phytoplankters competing for nutrients. Resource-consumer theory has intimate ties with other areas of ecological theory such as foraging theory, competition, and niche theory, all of which focus on the consumer.

I suggest the domain of natural enemy-victim theory spans all resource-consumer interactions where the resource is alive and is typically harmed by consumption. The focus is as much on the resource (victim) as on the consumer (its enemy). This domain includes multiple subdomains, reflecting the historical base of much ecology in pragmatic pursuits such as harvesting, pest control, and predicting epidemics. Moreover, the classical description of ecosystems in terms of primary producers and primary and secondary consumers highlights major biological differences among species in their required resources, leading to natural suture zones in theory. Because of this, there are many *particular* subtheories tailored to the exploitation of different kinds of victims by different kinds of natural enemies. Yet many common features link these particular theories. General natural enemy-victim theory tries to identify these unifying threads and provide a conceptual structure for understanding and organizing differences among these more particular theories.

The phrases "natural enemy-victim" or "exploiter-victim" provide succinct umbrellas encompassing several distinct arenas of ecological theory (Crawley 1992), including "true" predator-prey theory (Taylor 1984; Turchin 2003), host-parasitoid theory (Hassell 1978; 2000; Murdoch et al. 2003), host-pathogen interactions and epidemiology (Anderson and May 1991; Hudson et al. 2002; Keeling and Rohani 2008), and plant-herbivore interactions (Caughley and Lawton 1981; Crawley 1983; Owen-Smith 2002). Differences among these bodies of theory reflect functional details of how the exploiter engages its victim, and impacts that victim's fitness. One important question all these subdomains grapple with is "statics": what is the average effect of each species in limiting the abundance of the other? Another broad set of questions has to do with "dynamics"—is the interaction stable? If not, do unstable populations tend to cycle or show more complex behaviors?

There is an enormous, increasingly mature, and conceptually complex body of ecological theory related to natural enemy-victim interactions. There are lively debates not just about fine points, but about major and indeed central features of these interactions. This chapter can at best point to a few general principles and unifying structural features of this sprawling theoretical edifice. Much of what I shall say attempts to synthesize and reiterate perspectives ably presented at various places in the ecological literature (e.g., Crawley 1992). How can one pretend to synthesize across this vast arena, and do so in a way that does not sound like a regurgitated abstract of an encyclopedia? In empirical ecology, there is a robust discipline of meta-analysis that assesses patterns and hypotheses by bringing together results across disparate experimental studies. There is not as yet a comparable rigorous "metatheory" for gleaning (using formal protocols) conclusions from a range of specific models (although structural equations provide one promising approach; Grace et al.

2010). In the absence of such a formalized method, there are two broad approaches to synthesizing complex realms of ecological theory, both of which I think are essential steps towards general theory.

First, one can take the stance championed by Schoener (1986) towards community ecology and identify major axes of variation distinguishing kinds of natural enemy-victim interactions. May and Watts (1992), for instance, provide a table comparing life histories and other biological attributes across major categories of natural enemies. Lafferty and Kuris (2002) suggest that four axes characterize parasites and predators, including number of victims attacked in the enemy's lifetime and enemy reduction of victim fitness to zero (e.g., true predators, parasitic castrators). These key differences should be reflected in model structures (Hall et al. 2008). Such classificatory schemes are a legitimate part of theory development, just as the periodic table was a crucial step in the development of chemical theory. Space precludes my following this direction of theory synthesis.

Second, one can attempt to identify general cross-cutting themes by comparing a range of specific models (e.g., Holt and Hochberg 1998; Murdoch et al. 2003; Borer et al. 2007), highlighting overarching constraints within which all such models must operate. This relies on the instincts of a practitioner, rather than formal comparison. It can be helpful to start out by returning to fundamentals, to clarify the assumptions and flow of logic that define the lineaments of theory. This is what I attempt here. Murdoch et al. (2003) conclude their fine volume by saying: "Consumer-resource theory [which they use broadly congruent with what I mean by 'natural enemy-victim theory'] is ... quite unified and internally coherent." I think they are right—in part, and with qualifications. Not all subdomains of natural enemy-victim theory are well-developed theoretically, and potentially significant qualitative differences among subdomains may complicate unification. Even existing theory for true predator-prey and host-parasitoid interactions (doubtless the most fully developed of the lot) has key gaps, and there are basic disagreements about some important issues. Below, I attempt to identify a few general propositions that arguably should hold for all, or at least very many, natural enemy-victim systems.

Foundational concepts

To see our way forward to identify elements of conceptual unification and lines of discord, it is useful to go back to basics. All natural enemy-victim theory necessarily rests upon the foundation of population dynamic theory (Hastings Chapter 6). There are two complementary approaches to describing population processes, reflecting an essential duality in the nature of life itself (Eldredge 1985). The first half of the duality emphasizes information; the second, physical processes. Both are needed elements in the foundation of a comprehensive, even partially unified, theory of natural enemy-victim interactions.

Demography and exploitation

Life involves the replication of organisms. Natural enemy-victim theory is an elaboration of population dynamic theory that makes explicit the reciprocal dependence of births and deaths in each species upon the abundance of the other species. An explicit consideration of demography is thus essential to providing a firm theoretical foundation for natural enemy-victim theory. As discussed in detail by Hastings (Chapter 6), the basic equation of demography is that the numbers of individuals N of a given type in a particular place always changes over a small period of time according to some variant of: $\Delta N = B - D + I - E =$ births minus deaths plus immigration minus emigration. Because giving birth and dying happen to individuals, total births and deaths are best expressed as per capita rates. In continuous time in a closed population, where all individuals have equivalent traits and there are enough of them to ignore demographic stochasticity, we have

$$dN/dt = N(b - d) = Nr, \tag{7.1}$$

where b and d are instantaneous per capita birth and death rates, and r is the net per capita growth rate. If r is constant (i.e., the environment is unchanging), we have the foundational demographic principle of exponential growth (Berryman 2003; Hastings Chapter 6): $r < 0$ implies asymptotic extinction, whereas $r > 0$ leads to unbounded growth. The exponential model is the basic model on which the rest of population biology hangs (but see the exception discussed in Holt 2009). Much of the theoretical complexity of modern demography arises from dealing with internal structural heterogeneities in populations; likewise, incorporating the details of population structure is an important objective of much current work in natural enemy-victim theory (e.g., de Roos et al. 2003; Murdoch et al. 2003). For a closed population to persist, and neither go extinct nor blow up to infinity, on average its births must match its deaths (Royama 1992)—whether its dynamics involve regulation to a tight equilibrium, or cycles, or chaos, or bounded stochastic wanderings. This observation is the starting point of analyses of many important features of natural enemy-victim interactions.

The physical basis of life

Life is a physical process. Organisms are physicochemical machines, out of equilibrium with their environments, dependent upon recurrent supplies of energy and materials for survival and reproduction. Because natural enemy-victim interactions by definition involve forced transfers of energy and materials from one organism to another, a biophysical perspective is an essential complement to purely demographic models. Both are necessary ingredients in any unified theory of natural enemy-victim interactions.

A minority tradition in population dynamics emphasizes as a foundational principle the material basis of life as a set of resource transformations, rather than only through demography (e.g., Lotka 1925; Gutierrez 1992; Getz 1993; Kooijman 1993; Ginzburg 1998; Sterner and Elser 2002; Ernest et al. 2003). Owen-Smith (2005) for instance remarks: "From a biomass perspective, population growth is not the result of a difference between births and deaths... but rather of the difference between rates of uptake and conversion of resources into biomass, and losses of biomass to metabolism and mortality (Ginzburg 1998). The biological law of regeneration... underlies exponential population growth." The second law of thermodynamics implies a continual degradation of living tissues, mandating maintenance and repair. This occurs with error, so mortality is inevitable, and of course mortality arises from many other factors—the most dramatic of which are natural enemies.

Propositions and models in natural enemy-victim theory

The physical processes of life mandate resource extraction from the environment. Resource availability varies across space and through time, so birth and death rates should also vary with environmental circumstances. Some organisms (producers) meet their needs entirely from the nonliving environment. But the presence of other organisms provides a tempting packet of materials and energy to be exploited. Because of the conservation of mass and energy, such exploitation amounts to (at best) a zero-sum game—which lies at the heart of natural enemy-victim interactions, and can be stated as the first proposition of the theory (Table 7.1).

Proposition 1. *The acquisition by consumption of a material resource for the benefit of one organism (the natural enemy, or exploiter) automatically reduces the resources available to another organism (the victim) from which the resource has been taken.* This sets up an automatic conflict, one that drives behavioral struggles, population dynamics, and coevolutionary races. Much of the splendid diversity of life revolves around the manifold ways such cross-organismal

Table 7.1 The propositions of natural enemy-victim theory, both for the general theory and a more limited host-pathogen theory.

A. Universal propositions

1. The acquisition by consumption of a material resource for the benefit of one organism (the natural enemy, or exploiter) automatically reduces the resources available to another organism (the victim) from which the resource has been taken.

2. No species can rely entirely upon self-consumption and persist.

3. There is a necessary asymmetry in the "benefit" (measured in terms of energy and materials) gained by the exploiter, and the "cost" (in the same units) inflicted on the victim; in particular, the "cost" always is greater in magnitude than the "benefit."

4. On average, the energy and nutrients gained from each act of consumption must exceed the amount of energy and nutrients used by the natural enemy in that act, if the consuming species is to persist.

5. The immediate effect of each act of consumption is to reduce the biomass of the consumed victim.

B. General propositions

6. An increase in the victim population increases the rate of consumption by each individual natural enemy.

7. The increased consumption generated by increased victim abundance in turn fuels an increase in the per capita growth rate (fitness) of the natural enemy population.

8. All else being equal, the total rate of consumption of the victim (and hence the magnitude of its loss due to consumption) increases with the abundance of the natural enemy.

9. In the absence of the focal victim, a specialist natural enemy goes extinct (in a closed environment).

10. There is context dependence in natural enemy-victim interactions.

11. Predator functional and numerical responses saturate at sufficiently high prey densities.

C. Analogous propositions of the host-pathogen subdomain

7′. The total rate of infection (new infected individuals, per unit time) increases with the density of susceptible individuals.

8′. The total rate of infection increases with the density of infected hosts.

9′. For a specialist pathogen, the infection dies out in the absence of susceptible hosts.

D. A proposition specific to host-pathogen theory

12. Susceptible hosts may be generated from infected hosts (by recovery, or possibly birth).

---

exploitation and struggles play out, and the literature of natural history and ecology is replete with tales of the details of exploitative strategies and defensive countermoves (e.g., Caro 2005). If we consider "true" predators attacking prey, the primary first-order effect is to limit prey's population growth via mortality; more generally, natural enemies can reduce their victim's population growth by affecting fecundity, as well. In the next few paragraphs, I follow a traditional demographic approach to natural enemy-victim interactions and use the physical perspective to identify general constraints that should pertain to any reasonable model, cast as propositions.

Exploitation has demographic consequences for both the natural enemy and its victim. It is useful to stand back from all the details and attempt to identify general propositions that pertain to all exploitation systems. The formulation of such propositions is clearest when cast in the formal language of mathematical models. Model 7.1 generalizes naturally to natural enemy-victim systems. Consider a single natural enemy species, attacking a single victim species. In classical predator-prey theory [going back to A. J. Lotka (1925) and Vito Volterra (1931); see, e.g., Rosenzweig and MacArthur (1963)], models are cast in continuous time, with continuously overlapping generations. For a natural enemy (e.g., predator) of density P, interacting with a victim of density N, in environment E, the simplest general continuous-time model in a closed system is:

$$\frac{dP}{dt} = Pf_P(N, P, E),$$
$$\frac{dN}{dt} = Nf_N(N, P, E). \tag{7.2}$$

Here, f_P is the per capita growth rate of the natural enemy, and f_N is the per capita growth rate of the victim. The "environment" includes all abiotic and biotic factors (including other species), expressed as fixed parameters embedded in the functional form of the equations. Model 7.2 expresses one fundamental principle of Scheiner and Willig (Table 1.3), which is that interactions depend upon the environment. I will mainly assume the natural enemy is a specialist, but towards the end will also make a few remarks about trophic generalists. (Below, for the sake of variety I sometimes use "predator" and "prey" for "natural enemy" and "victim"; the claims made are meant to apply to herbivores on plants, for example, as much as to "true" predators and prey such as wolves and moose.)

There are numerous assumptions at play in models such as 7.2, matching the assumptions of simple exponential growth models (Hastings Chapter 6).

In each player, it is assumed that one can ignore variation among individuals (e.g., stage structure, genetic variation) and spatial location (a "mean-field" assumption, so each population is well-mixed); that abundances are sufficiently large to neglect demographic stochasticity; and that reproduction and mortality are continuous processes (rather than, say, seasonally pulsed). Growth rates are assumed immediately responsive to current densities, with no lags due to development or accumulated metabolic stores. Despite this somewhat worrisome list of assumptions, analysis of the family of models given by 7.2 provides an essential springboard, just like exponential growth. The model could in principle apply to many different kinds of natural enemy-victim interactions. For instance, P could denote an herbivore and N the plant it is attacking, or P could refer to infected individuals and N to susceptible individuals within a host population carrying an infectious disease.

An important qualification

It should be stressed that starting with a pair of coupled differential equations such as 7.2 assumes that the important time lag in the system arises from delays in feedback via the interaction. Other model formulations have feedbacks that arise because of discrete time, delays in direct density dependence, and delays that emerge from demography (e.g., discrete juvenile periods). Another way one could attempt to formalize natural enemy-victim theory is in terms of the assumptions made about the nature of time lags (I thank a reviewer for this observation), and this is in effect what Murdoch et al. (2003) have done in their monograph. One of the basic points discussed below is that natural enemy-victim systems are expected on theoretical grounds to oscillate. However, the way in which this happens is influenced in system-dependent ways by how populations are structured, including developmental and regulatory delays, as well as heterogeneities such as spatial structure and genetic variation. Developmental delays in consumers mediated by resource consumption can lead to short-amplitude cycles in natural-enemy victim systems, for reasons quite different from those emphasized in classical theory (McCauley et al. 2008).

Fundamental constraint theory: the "statics" of natural enemy-victim interactions

One rich vein of complexity in modeling natural enemy-victim theory arises from the expressions one uses to express births and deaths in each species as a joint function of its and the other's abundances (i.e., their interaction) and the environment. It is useful to see how far we can go without specifying more detail than is absolutely necessary. Almost by definition of the domain of this theory, the natural enemy benefits from the victim, whereas the victim suffers reduced growth because of the natural enemy. In equations, were one to monitor each species after reciprocal removal experiments, over the short run one expects that (for simplicity, we suppress E):

$$f_P(N, P) > f_P(0, P),$$
$$f_N(N, P) < f_N(N, 0). \qquad (7.3)$$

The interspecific relationship thus has a $(+, -)$ sign structure.

If the system is in demographic equilibrium (denoted by an asterisk) with both species present, $f_P^* = f_N^* = 0$, hence $f_N(N^*, 0) > 0 > f_P(0, P^*)$. In other words, at equilibrium, when one removes the predator (leaving everything else fixed) the prey increases; conversely, were we to remove the prey, the predator on its own decreases. The demographic balance is maintained by the $(+, -)$ nature of the interaction. Though 7.3 is quite reasonable, it may not always hold for generalist natural enemies (see Holt 1983 for more details and examples). If generalist predators have a suboptimal (maladaptive) diet, removing one or more prey types may actually (at times) boost predator numbers.

A series of fundamental constraints on natural enemy-victim interactions arises from considering the joint implications of the demographic and biophysical foundations of life. Once stated, these seem obvious, but one role for theory is to lay bare propositions that everyone can agree upon, which permits us to focus on issues that provide grist for the mill of continued disputation (Table 7.1A).

Proposition 2. *No species can rely entirely upon self-consumption and persist.* This follows from the second law of thermodynamics, which states that useful energy is converted to useless energy (heat) whenever work is performed, coupled with the demographic constraints required for persistence. Reproduction, simply staying alive, and consumption itself—all are varieties of work. With self-consumption, the energy placed into new biomass must necessarily be less than the energy in biomass consumed. So, if we start with a certain number of individuals in a population, and they have only themselves to consume, after each act of consumption there will be a net loss of biomass. Over time, a population feeding only on itself will inexorably decline to extinction. So one will never expect to see a community comprised entirely of cannibals—at least not for very long! This principle is the ecological analogue of the impossibility of a perpetual motion machine.

Proposition 3. *There is a necessary asymmetry in the "benefit" (measured in terms of energy and materials) gained by the exploiter, and the "cost" (in the same units) inflicted on the victim; in particular, the "cost" always is greater in magnitude than the "benefit."* This may not imply an asymmetry as measured by Darwinian fitness, which is measured as descendents per organism, rather than per unit of material. (Hence the quote marks.) The part of this statement that refers to energy follows from thermodynamics. The assertion about materials (e.g., nitrogen content in body tissues) is surely true: from conservation of mass, the benefit gained by the exploiter cannot exceed the cost imposed on the victim. Moreover, there is always wastage in consumption, so the gain in, say, nitrogen for a consumer must always be less than the loss experienced by the consumed.

In model 7.2, if P and N are measured in equivalent physical units (e.g., energy content in each population), this proposition implies:

$$P(f_p(P, N) - f_p(P, 0)) < \left| N(f_N(P, N) - f_N(0, N)) \right|. \tag{7.4}$$

Inequality 7.4 can be viewed as a constraint on models of the form of 7.2, if these models are to be physically sensible.

Proposition 4. On average, the energy and nutrients gained from each act of consumption must exceed the amount of energy and nutrients used by the natural enemy in that act, if the consuming species is to persist. Proposition 3 is a statement about the aggregate consequences of exploitation across trophic levels. Proposition 4 is superficially similar, but subtly different; it applies to average acts of consumption, within the consumer level itself. Before any act of consumption, an individual natural enemy starts with a certain amount of "capital" energy and nutrient reserves within its body. Some of this capital is expended during consumption—after all, consumption and assimilation into body tissues are both forms of work, and so operate within the constraint of thermodynamics, which implies a fundamental rock-bottom inefficiency. If these expenditures are not matched by gains from consumption, there will be a loss in biomass with each act of consumption, and if this loss continues, act after act, extinction is inevitable. Proposition 3 is a necessary, but not sufficient, statement about consumer persistence; the net amount of energy and nutrients gained per act of consumption must not only be positive, but be sufficiently great to exceed other costs (e.g., inherent mortality, losses to disturbance, etc.). I return to Proposition 4 below.

Proposition 5. The immediate effect of each act of consumption is to reduce the biomass of the consumed victim. This proposition is almost a simple restatement of part of the definition of resource consumption, yet it has important consequences worth focusing on explicitly. With continued consumption, for victims initially at equilibrium (where births match deaths), without compensatory feedback permitting an increase in birth rates, there is an inexorable decline towards extinction. Characterizing compensatory feedbacks that affect victim species is thus a vital part of understanding the dynamics of natural enemy-victim interactions.

General (but not universal) propositions

To go beyond these schematic assertions and craft models related more closely to concrete empirical systems, more assumptions are needed (Table 7.1B). The next two propositions are generally assumed in natural enemy-victim theory, but do not necessarily hold.

Proposition 6. *An increase in the victim population increases the rate of consumption by each individual natural enemy.* In the subdomain of predator-prey theory, the functional response of an individual predator to its prey is the rate of consumption of the prey, as a function of prey abundance (Holling 1959), so proposition 6 states that the functional response is an increasing function of prey density.

Proposition 7. *The increased consumption generated by increased victim abundance fuels an increase in the per capita growth rate (fitness) of the natural enemy population.* In predator-prey theory, the numerical response of the predator describes how its own population size increases with prey density, so proposition 7 states that the numerical response is an increasing function of prey abundance. More specifically, the predator's numerical response to its prey should be a function of its functional response (the "biomass conversion" principle proposed by Ginzburg 1998). Given proposition 7,

$$\frac{\partial f_P}{\partial N} > 0. \tag{7.5}$$

Going back to the basic models of Lotka (1925) and Volterra (1926; 1931) (in continuous time), and Nicholson and Bailey (1935) (in discrete time), inequality 7.5 has been a basic assumption of natural enemy-victim theory for a very long time. Inequality 7.5 is not a universal "law" of trophic ecology, but it is a plausible generalization that applies to many natural systems. Without specific knowledge to the contrary, proposition 7 is a reasonable starting point for modeling.

But there are significant exceptions. One comes from large grazing herbivores whose energetic intake rates can decline at sufficiently high grass biomass because of foraging constraints, and a correlation between forage digestibility and biomass (Wilmshurst et al. 1999; 2000). Other exceptions arise when prey can engage in group defenses, which may be more effective at high prey densities (Caro 2005). Finally, an important class of exceptions comes from host-pathogen systems involving vectors, where an increase in host numbers, given a relatively fixed number of vectors, can dilute the likelihood of pathogen transmission to susceptible hosts (Ostfeld and Keesing 2000), leading to a nonmonotonic relationship between host population size and disease transmission. Large host numbers may even cause the pathogen to die out, because the dilution of vector attacks provides insufficient transmission between infected and susceptible hosts (Randolph et al. 2002).

Proposition 8. *All else being equal, the total rate of consumption of the victim (and hence the magnitude of its loss due to consumption) increases with the abundance of the natural enemy.* In terms of model 7.2, this implies

$$\frac{\partial f_N}{\partial P} < 0. \tag{7.6}$$

Again, this is found in nearly all models built on the Lotka-Volterra framework. And again, this inequality is a reasonable generalization, but it is not a universal law. There may be exceptions; for instance at high predator numbers there may be strong direct interference among predators, so that the total mortality imposed upon prey declines with increasing predator numbers. Inequality 7.6 may not hold with ratio-dependent predation (see below). Victim behavioral responses can sometimes imply that the victim suffers fewer losses at higher densities of the natural enemies (Abrams 1993), and that when both species have adaptive behaviors, each can at times benefit the others (Abrams 1992).

Proposition 9. *In the absence of the focal victim, a specialist natural enemy goes extinct (in a closed environment).* We noted above that at equilibrium in 7.2, were we to carry out removal experiments, we would observe that $f_N(N^*, 0) > 0 > f_P(0, P^*)$. Because we have assumed that the natural enemy is a specialist, given that reproduction requires energy and materials, there will be no births in the absence of the required victim, and because all organisms are mortal, there will still be deaths, hence the right hand inequality holds, for all P. This implies the natural enemy goes extinct.

Proposition 10. *There is context dependence in natural enemy-victim interactions.* The quantitative relationship between the aggregate rate of consumption of the victim and natural enemy density, and between victim abundance and consumption, will nearly always vary with environmental context, and with the presence and abundance of other species. This proposition differs from propositions 1–5 because it does not follow from more basic lawlike propositions in demography and physical biology. Comparable to propositions 6 and 7, it is an inductive generalization across many empirical studies (and matches a principle in Table 1.3). It differs from 6 and 7 because it is schematic, and without further fleshing out, it cannot be used to inform dynamical theory, other than to caution that "different things will happen in different circumstances." However, stating this as a formal proposition is useful, as it alerts researchers to the importance of identifying the major dimensions of potential context dependence in their empirical studies. As an example of contingency, consider the next proposition.

Proposition 11. *Predator functional and numerical responses saturate at sufficiently high prey densities.* The contingency here is contained in the word "sufficiently"; if this is true, but outside the actual variation in prey densities, then it is irrelevant to the dynamics of the system. By contrast, if saturation holds over nearly all pertinent prey densities, there are important consequences (see below).

Outcomes of natural enemy-victim theory

With the above propositions in hand, one can derive theoretical outcomes that characterize a broad swath of this theoretical domain. In certain contexts (proposition 10), predators strongly limit prey growth (proposition 5), and the prey strongly boost predator growth (proposition 7). This leads to the question of whether or not the predator will regulate that prey (i.e., keep prey numbers within bounds), and in turn the prey regulate the predator. When the natural enemy is ineffective at limiting its victim, this issue of mutual regulation is not really an issue: the regulation of the system is governed by bottom-up forces or by factors intrinsic to the prey. If, by contrast, a natural enemy effectively limits its prey to low levels, a generic outcome of simple models of these interactions is that they tend to oscillate, because of delayed density dependence. An increase, say, in predator numbers will eventually depress predator growth rates, but only after the prey are depressed in abundance, which takes some time; in like manner, an increase in prey numbers leads to an eventual increase in prey mortality, but with a lag dependent upon the timescale of the predator's numerical response. Turchin (2001) suggests that one of the general laws of ecology is that *population cycles are likely when effective consumers exploit living resources.* Murdoch et al. (2003, p. 70) likewise list a tendency towards oscillatory behavior as first in their list of "fundamental properties" of predator-prey models. I think these authors are right, and that a propensity for oscillatory dynamics should be viewed as a basic principle of natural enemy-victim interactions, emerging as an outcome of the propositions of the basic theory (which are not universal; see above).

The most important general outcome of natural enemy-victim theory is that *natural enemy-victim interactions have a propensity towards unstable dynamics if the victim is strongly limited by the interaction and the natural enemy is a specialist.*

Such oscillations can arise at many temporal scales. A prey individual may have behavioral responses to predators, and the predator may likewise respond. Such behavioral responses can set up strongly unstable spatial dynamics (Abrams 2007). The simplest models of predator-prey and host-parasitoid population interactions tend to oscillate (see below). Seger (1992) has argued that the evolutionary dynamics of asymmetric antagonistic interactions, such as between a predator and its prey, or between a host and its parasite, are inherently unstable, leading to a constant coevolutionary churning in gene frequencies and traits. This is particularly the case if fitnesses in both species are determined by one to a few major loci and are interspecifically frequency-dependent (e.g., the fitness of defense morphs in a prey species may vary with the frequency of attack morphs in a predator).

Oscillations may damp out following a perturbation or instead settle into stable cycles or more complex behavior (e.g., chaos), dependent on many system-specific details. To see why there is a generic tendency towards oscillations, we start with a system in equilibrium and examine how it responds to small perturbations in abundances (and here I follow a path that has been well-trod in the literature, e.g., Murdoch and Oaten 1975; Murdoch et al. 2003). By linearizing around the equilibrium, the local stability of the system given by model 7.2 is determined from the eigenvalues of the Jacobian matrix J (where the partial derivatives are evaluated at the equilibrium):

$$J = \begin{bmatrix} N^* \dfrac{\partial f_N}{\partial N} & N^* \dfrac{\partial f_N}{\partial P} \\ P^* \dfrac{\partial f_P}{\partial N} & P^* \dfrac{\partial f_P}{\partial P} \end{bmatrix} = \begin{bmatrix} ? & - \\ + & ? \end{bmatrix}. \tag{7.7}$$

What the left-side matrix does is compactly describe the impact of small perturbations in abundance on the growth rate. The diagonal terms describe the impact of each species upon itself (viz., direct density dependence), and the off-diagonal terms, the effect of each species on the other (cross-species density dependence, one measure of interaction strength). The properties of this matrix provide a linear approximation of the full model 7.2 (which may be highly nonlinear) near equilibrium, and determine whether small perturbations dampen away, or instead grow, and also whether or not the dynamics near equilibrium show oscillations. The matrix on the right of the equal sign describes the qualitative sign structure of interactions for this species pair. The signs of the off-diagonal terms follow from propositions 7 and 8; as we will shortly see, the signs of the diagonal terms may emerge from the natural enemy-victim interaction itself, as well as from extrinsic forces.

Following well-known recipes (e.g., Otto and Day 2007), if the trace of J,

$$T = N^* \frac{\partial f_N}{\partial N} + P^* \frac{\partial f_P}{\partial P}, \qquad (7.8)$$

is negative, and the determinant D is positive, such that

$$\frac{\partial f_N}{\partial N} \frac{\partial f_P}{\partial P} > \frac{\partial f_N}{\partial P} \frac{\partial f_P}{\partial N}, \qquad (7.9)$$

then the real part of the dominant eigenvalue of J is negative, and the system is locally stable. The quantity T measures the overall strength of direct density dependence added up over both species, weighted by their equilibrial abundances. Inequality 7.9 in a sense states that direct density dependence is stronger than the strength of the interspecific interactions. Thus, stability hinges on the nature of direct density dependence in one or both species, and for two separate reasons (corresponding to 7.8 and 7.9). If both predator and prey experience direct negative density dependence, expression 7.8 is negative. Moreover, the lefthand side of 7.9 is positive, and from propositions 7 and 8, the lefthand side is negative, so 7.9 holds. Hence, the interaction is locally stable. If neither holds, then it is unstable. If either T is negative or 7.9 is violated, the equilibrium is unstable. Stability requires both that negative density dependence exist (7.8), and that it be overall stronger than cross-species density dependence (7.9).

Whether or not oscillations occur after a perturbation from equilibrium is determined by the sign of the quantity

$$Q = (N^* \frac{\partial f_N}{\partial N} - P^* \frac{\partial f_P}{\partial P})^2 + 4 N^* P^* \frac{\partial f_N}{\partial P} \frac{\partial f_P}{\partial N}. \qquad (7.10)$$

If $Q < 0$, oscillations will arise following a perturbation (the eigenvalues of J have complex parts). If neither the natural enemy nor the victim has a direct effect upon themselves (i.e., direct density dependence is negligible), the diagonal terms are zero, implying that T (and the first term of Q) $= 0$. From propositions 7 and 8, the remaining term is negative, so $Q < 0$, and oscillations in abundance will surely occur following perturbation. In this limiting case, the oscillations have no tendency to return to the equilibrium. If at a stable equilibrium density dependence is relatively weak, Q will still be negative, and damped oscillations arise following a perturbation.

Thus, a generic property of natural enemy-victim systems is that they harbor a tendency to exhibit oscillatory behaviors following perturbation because of the $(+,-)$ structure of the interaction matrix. Kolmogorov (1936; May 1973, pp. 86–92) elegantly went beyond the linearized analysis presented above to show that given a broad set of reasonable conditions, including propositions 7 and 8 (May 1973, pp. 87–88), models of the form given by 7.2 above settle into either a stable point equilibrium or a stable limit cycle. More realistic models that include time lags in predator responses to their prey (e.g., due to developmental lags) often make such oscillations more likely (Murdoch et al. 2003). Specialist natural enemy-victim interactions can thus be strongly destabilizing. Whether or not these oscillations dampen out depends upon the strength of direct density dependence in each species, and whether or not this density dependence is sufficiently negative at equilibrium. In some circumstances (as elaborated below), such density dependence emerges from processes within the domain of natural enemy-victim theory itself; in others, the density dependence needed for stability is imposed, as it were, from outside the domain.

For bioenergetic reasons it is often the case in predator-prey interactions that $P^* < N^*$, unless prey are highly productive and predators have much longer generation times. Empirical studies of vertebrate predator and prey abundances show that typically (again, not universally) for a given body size, predators are rarer than prey (Marquet 2002). When measured in equivalent units (biomass), mammalian predators are on average two orders of magnitude rarer than prey (Carbone and Gittleman 2002), and since mammalian predators tend to be larger than their prey, the difference in abundance is even greater. By inspection of T and Q, these observations suggest that density dependence in the prey should often be much more important in determining stability of the trophic interaction than is density dependence in the predator. However, in other natural enemy-victim interactions, such as parasitoids attacking hosts, this asymmetry in abundance and hence contribution to stability may not hold.

As a cautionary note, proposition 11 does not imply that when one sees a natural enemy-victim cycling, the delayed feedbacks embodied in 7.7 are the reason. As noted above, the cycles might instead be due to intrinsic processes in the victim or natural enemy (e.g., single-generation cycles due to stage structure; Murdoch et al. 2003), or the victim's interaction with its own resources, or other factors such as maternal effects (Inchausti and Ginzburg 2009).

Steps toward specificity

This is about as much as we can squeeze out of the general model provided by
7.2. We could just stop here, but this would be a bit of a cheat, since some of
the most contentious and interesting issues in natural enemy-victim theory
arise when one adds specificity to this general model. The goal of more detailed natural enemy-victim theory is to characterize how specific mechanisms
of interactions translate into average effects of each species on the other, as well
as positive and negative density dependence within and across species. Much
of the development of specific theory in this area has been guided by the desire to develop a systematic theory for understanding how effective specialist
natural enemies manage to coexist with their victims, given the inherent instability that seems to lurk at the heart of these interactions. In addition to its importance for addressing basic ecological questions, this issue is at the heart of
devising effective and sustainable biological control programs, predicting and
mitigating epidemics, and many other themes of great practical importance.
Crawley (1992) suggests that a minimal list of desiderata for understanding
the dynamics of a specialist predator-prey interaction would include the prey's
intrinsic rate of increase; the predator's functional response and spatial foraging behavior; the nature of prey density dependence and predator density dependence; the densities at which the predator and prey isoclines intersect, and
their relative slopes at the intersection; the rate of predator immigration and
emigration; the size of the prey refuge; and the magnitude of between-patch
prey dispersal. There are few empirical systems where we have all this information. Given the importance of this topic, this is a sobering statement about the
adequacy of the knowledge base of our field

Sources of direct density dependence in natural enemy-victim systems

Before considering specific models, one can broadly divide the sources of direct density dependence (the diagonal terms of 7.7) into two classes. On the
one hand, density dependence may occur for reasons that are *extrinsic* to the
predator-prey interaction itself. For instance, prey may compete for limiting
resources, or predators may experience direct aggressive interference. Such
mechanisms of direct density dependence can stabilize the trophic interaction.
Alternatively, there may be positive density dependence in either species, due
to Allee effects (e.g., the need to find mates in sexual species), which is destabilizing. On the other hand, direct density dependence in both species may arise
for reasons *intrinsic* to the predator-prey interaction, if (for instance) at low
prey numbers predators accelerate feeding rates with increases in prey densities.

A hierarchical family of Lotka-Volterra models
for the predator-prey subdomain

To become more concrete, we return to the traditional starting point for predator-prey interactions in spatially closed systems, namely the Lotka-Volterra model. This can be viewed as the simplest member of a hierarchical set of nested families of related models. We will progress through a series of models that make increasingly elaborate and general assumptions about how each species affects the other. These models are cast in a manner comparable to the basic equation of demography, equation 7.1, separating births and deaths.

The classical Lotka-Volterra predator-prey model (no direct density dependence) is:

$$\frac{dP}{dt} = P[baN - m]$$
$$\frac{dN}{dt} = N[r - aP].$$

(7.11)

We start with the simplest, classical Lotka-Volterra model, where the bracketed terms in 7.11 correspond to f_P and f_N in 7.2. The most straightforward interpretation of 7.11 is that the two terms within brackets for the predator correspond to predator births and deaths. The quantity a is an attack rate, aN is the functional response of each predator, and baN its numerical response (i.e., birth rate). The quantity m is the predator's death rate. The prey has an intrinsic birth rate and death rate, with the difference between them being its intrinsic growth rate r, and it suffers additional mortality from predation, at rate aP. As is well-known, this model leads to neutrally stable cycles (with $T = 0$), delicately sensitive to small changes in model assumptions. However, the qualitative message it delivers—that there is a propensity to oscillate in natural enemy-victim systems—applies robustly across a wide range of models. Moreover, for the purposes of our overview of the theory of natural enemy-victim interactions, the Lotka-Volterra model provides an entrée into interesting theoretical and empirical issues, having to do with the nature of the interaction between predators and prey.

A predator-prey model with a nonlinear functional response is

$$\frac{dP}{dt} = P[bf(N, P) - m]$$
$$\frac{dN}{dt} = Nr - f(N, P)P.$$

(7.12)

The quantity $f(N,P)$ is the functional response of the predator to its prey, which in principle can depend upon both prey and predator density. Now, there can be emergent direct dependence in each species (the diagonal terms in J), and so the equilibrium can be either locally stable or unstable, depending upon the particular mathematical form for $f(N,P)$.

A fundamental determinant of stability in predator-prey interactions is thus the quantity $f(N,P)$, which is the functional response of C. S. Holling (1959), broadened to include possible dependencies of feeding rates on predator density. Model 7.12 assumes that the predator's numerical response (its birth rate) is directly proportional to the functional response. This fits many examples and is a reasonable starting point for theory (Arditi and Ginzburg 1989), though there are certainly counterexamples.

An ongoing debate about a key component of natural enemy-victim theory

A robust debate in natural enemy-victim ecology arises from different views about how one relates detailed mechanisms of interactions at the level of individuals between predators and their prey to the functional and numerical responses in 7.12. It has been over fifty years since Holling highlighted the role of the functional and numerical responses in predation, so it is striking that there is still lively disagreement about this basic issue, leading to the verbal equivalent of fisticuffs in the ecological literature.

Prey dependence

In one corner of the ring, we have proponents of the "classical" perspective, in which the predator is strictly food- or prey-limited; thus, the functional response depends solely on prey numbers, $f(N)$, and the numerical response depends in turn purely on the functional response, $g(f(N))$. For a fixed prey density, the total rate of predation upon the prey should then increase linearly with predator density.

Holling (1959) notes that a very general feature of predator feeding responses is that they asymptote with increasing prey numbers (his type 2 functional response), but sometimes accelerate over a range of low densities (the type 3 functional response). In the former case, this implies that the per capita effect of predation declines with increasing prey abundance, so $\partial f / \partial N < 0$. This is destabilizing; without other sources of density dependence, the equilibrium is locally unstable. Murdoch and Oaten (1975) show that for the type 3 response to be stabilizing (when the predator numerical response is proportional to the functional response), it is not enough for the functional response to be accelerating, but it must be so with sufficient magnitude. The precise stability condition is

$$\partial f / \partial N|_{N^*} > f^* / N^* \tag{7.13}$$

Much of the literature on functional responses has explored how behavioral and physiological mechanisms in the predator (e.g., adaptive foraging behaviors; see Sih Chapter 4) govern the detailed form of the functional response (Whelan and Schmidt 2007), and hence the stability of predator-prey interactions. Prey behavior can also influence the shape of the functional response (Brown and Kotler 2007). For instance, if prey actively hide in refuges from predation, but refuges are in limited supply, there can be direct density dependence via the functional response emerging from competition for refuges. At low prey density, most prey find refuge, and so per capita mortality from predation is low. As prey increase, more are forced out of the refuges, so mortality rates rise. This translates into an accelerating functional response for the predator, and a negative trace of J. Refuges are thus broadly stabilizing (Murdoch and Oaten 1975).

Predator dependence

In the other corner of the ring, we have ecologists who challenge the hegemony of prey dependence, for instance by asserting the primacy of a law of diminishing returns, such that the resource available per consumer declines with increasing consumer numbers (Ginzburg and Colyvan 2004). The assumption that functional responses depend solely on prey numbers sometimes does hold. But often this is not true, for instance because predators interfere with each other while searching for prey (Hassell and Varley 1969; DeAngelis et al. 1975; Free et al. 1977). Skalski and Gilliam (2001) surveyed a number of examples, and found that in the majority of cases, strong effects of predator density upon the functional responses were observed, though caution should be exercised by considering alternative hypotheses to direct predator interference, such as prey behavioral responses and spatial processes (Sarnelle 2003).

Interference among predators has both stabilizing and destabilizing effects. In terms of the formalism presented above, one element in the trace of J becomes negative because of such density dependence; this effect is stabilizing. However, if predator dependence greatly reduces per capita consumption at high predator numbers, the total impact of the predator on the prey is constrained, making it harder for the predator to keep up with a growing prey population. Basically, strong interference makes natural enemy-victim systems more dominated by "bottom-up" forces (Arditi et al. 2004).

Conversely, in some circumstances increased predator abundance *facilitates* attacks. This is expected when predators forage in groups but can also occur in less obvious situations. Miller et al. (2006) for instance report that the number of Canada goose nests attacked per bald eagle increased with eagle density. They suggest this is because goose defense strategies become much less effective at higher predator numbers. A prey individual that escapes by scuttling away from one predator may simply move into the path of another predator at high predator densities. Alternatively, grouping behavior by prey may reduce predation and help stabilize predator-prey interactions (Fryxell et al. 2007).

A variety of functional forms have been proposed that display predator dependence. One that has sparked much controversy is "ratio dependence" (Arditi and Ginzburg 1989), where the functional response depends upon the amount of prey per predator, N/P, i.e., $f(N/P)$, rather than just N, as in $f(N)$. In the Lotka-Volterra model, the functional response is aN, and the total mortality imposed by predators upon their prey is aNP. With linear ratio-dependent predation, the functional response is instead $a(N/P)$, and the total mortality inflicted upon prey by predators is $a(N/P)P = aN$, and so the per capita rate of predation experienced by the prey is a constant, a, independent of predator abundance. The truth is likely bracketed by these extremes, as represented for instance with the Beddington-DeAngelis functional response (Beddington 1975; DeAngelis et al. 1975):

$$f(N,P) = \frac{aN}{1 + ahN + iP},$$
(7.14)

where a and N are as above, h is a handling time, and i a measure of predator interference. Using 7.14, the total predation experienced by the prey at low predator and prey numbers is $f(N,P)P = aNP$, as in the Lotka-Volterra model, but at high predator numbers (for fixed prey abundance) the total amount of predation becomes aN/i, and so becomes independent of predator abundance.

There are empirical examples that fit both prey-dependent and predator-dependent scenarios, or that lie in between (e.g., Schenk et al. 2005). In systems that are spatially well-mixed and with continuous reproduction, prey-dependent models often perform well (e.g., lake plankton, Carpenter et al. 1993; bacteria and virus in a chemostat, Bohannan and Lenski 1997). But in systems with spatial heterogeneity, pulsed reproduction, and other complexities, alternative formulations may be required, and ratio dependence may at times provide a reasonable approximation. Vucetich et al. (2002) argue that on Isle Royale, the interaction between wolves and moose reflects features of both prey- and ratio-dependent models, and Jost et al. (2005) in reexamining the data suggest the patterns even tilt toward the latter. Among theoreticians, the majority opinion nonetheless appears to be that prey dependence should be the starting position in developing theory, with specific mechanisms that can lead to predator dependence incorporated as needed. This indeed tends to be my own stance, but it should be emphasized that some ecologists strongly disagree. I refer the reader to the thoughtful review of Abrams and Ginzburg (2000) that lays out some of the elements of agreement and disagreement on both sides of this issue. There is thus an important lack of consensus on a fundamental issue at the very heart of natural enemy-victim theory.

What is the natural generalization of the single-species exponential growth model to interacting natural enemy and victim species? I think most ecologists would say the Lotka-Volterra model. The trophic term in 7.11 mimics the law of mass action for bimolecular reactions in liquid media, where the rate of the reaction is proportional to the product of the concentrations of the two interacting "species" (atoms, molecules, ions). This would seem at first glance to be the simplest way one could incorporate trophic interactions into a term that could be spliced onto equations for exponential growth for each species. To play devil's advocate, I would like to consider the possibility instead that maybe the appropriate generalization is a ratio-dependent model. [I am here motivated by a discussion in Ginzburg and Colyvan (2004).] Recall from above that the essential hallmark of exponential growth is that a population either goes extinct (if $r < 0$), or grows to infinity (if $r > 0$). So maybe the comparable model to exponential growth for a trophic interaction should also have these dynamical outcomes. The Lotka-Volterra model has (neutrally stable) *bounded* growth, so does not display this behavior. But consider the following model with "pure" linear ratio-dependent predation:

$$\frac{dP}{dt} = -mP + ba\left(\frac{N}{P}\right)P = -mP + baN$$

$$\frac{dN}{dt} = rN - a\left(\frac{N}{P}\right)P = (r - a)N.$$

(7.15)

The quantity a is now the *total* amount of predation imposed upon the prey population, which (because of ratio dependence) is independent of the number of predators present.

This is an intriguing and (to me) counterintuitive prediction, violating proposition 8 above. Little is actually known empirically about how total predation pressure scales with predator abundance. In some systems, the relationship between total predation and predator abundance can differ among prey species for the same predator species (Essington and Hanson 2004). Abrams (1993) argues that for many reasons (e.g., adaptive prey responses; see Sih Chapter 4), one should not expect to see a linear relationship between predation and predator abundance, and might even at times observe a negative relationship. The relationship between total predation rate and predator abundance needs much more focused empirical study across a wide range of taxa and systems.

If $r < a$, both N and P decline deterministically to extinction. If $r > a$, both N and P grow to infinity. Deterministic extinction of both predator and prey, or unbounded growth, are found in a wide range of more general ratio-dependent models where the functional response is described by some $f(N/P)$ (Jost et al. 1999; Lev Ginzburg, pers. comm.). Thus, a ratio-dependent model such as the Arditi-Ginzburg model (7.15) has the same dynamical outcomes as does exponential growth, and so one might argue that from an abstract point of view it, rather than the Lotka-Volterra model, may actually be the natural generalization to trophic interactions of exponential growth for a single population. In the real world, of course, populations do not continue to grow to infinity, so the exponential model and the ratio-dependent model clearly must break down at high abundances. But note that species really do go extinct, and a model of exponential decline is a reasonable starting point for understanding that process. And predators really do sometimes overexploit their prey and then go extinct themselves, so the prediction of extinction by itself is not a fatal weakness of ratio dependence (Ackakaya et al. 1995; Ginzburg and Colyvan 2004). Lev Ginzburg would argue that this general line of reasoning provides a rationale for using ratio dependence, rather than Lotka-Volterra-like prey dependence, to provide the logical basis for developing predator-prey theory (Lev Ginzburg, pers. comm.).

I should stress that these reflections are philosophical in nature, and are not necessarily a compelling argument in favor of using ratio-dependent models as the routine starting basis for trophic theory (e.g., for food webs; Abrams 1994). A number of authors (e.g., Getz 1984; 1999; McKane and Drossel 2006) argue that the logical starting point of predator-prey theory is not the Lotka-Volterra model (as suggested by the hierarchical set of models of 7.11 through 7.12), but alternative formulations such as the ratio-dependent model (7.15). The jury is still out on this very important issue [e.g., see the recent exchange between Fussmann et al. (2007) and Jensen et al. (2007)], and ratio-dependent models are consistent with some empirical patterns, such as patterns in abundances of different trophic levels along productivity gradi-ents. Conversely, models that take the form of 7.2 assume spatial homogeneity, no seasonality, etc., and so circumvent some of the rationales suggested in the literature for ratio dependence. In any case, all such simple models in ecology are in my view probably what Quine (1960) would call "limit myths" about nature, rather than representations of natural law.

The problem of scaling from individuals to populations

There are subtle theoretical issues related to the debate over ratio dependence that are not yet resolved. The functional response is an attribute of individual predators. Models such as 7.2 and 7.11 are at the population level. So much of the argument revolves around the most appropriate way to scale up from individual processes to population and community dynamics, which operate at different spatial and temporal scales (Arditi and Ginzburg 1989; Abrams and Ginzburg 2000). The issue of how one constructs an ecological theory that coherently embodies scaling among different levels of ecological organi-zation is a central challenge, one that goes well beyond natural enemy-victim theory (O'Neill et al. 1986). Moment closure techniques (Keeling et al. 2000; Englund and Leonardsson 2008) are one promising approach for connecting among scales in natural enemy-victim theory, and are revealing fresh and unex-pected insights. For instance, the functional response which best describes pre-dation at a regional scale may differ between increasing and decreasing phases of predator-prey cycles, because of shifts in the covariance of local predator and prey densities (Englund and Leonardsson 2008).

Alternative modes of density dependence in predator-prey interactions

Models 7.11 and 7.12 assumed no direct density dependence, within or be-tween species, other than those mediated through the interspecific interac-tion. Adding such density dependence leads to

$$
\begin{aligned}
\frac{dP}{dt} &= P[g(f(N,P),P) - m(N,P)] \\
\frac{dN}{dt} &= Nr(N,P) - f(N,P)P.
\end{aligned}
\tag{7.16}
$$

Now, we have direct density dependence in recruitment of the prey via the term $r(N,P)$ (e.g., due to competition for resources), and in predator mor-tality in the term $m(N,P)$ (e.g., due to aggressive interference), each with causes extrinsic to the trophic interaction. In the consumer, direct density dependence can depress the numerical response below that expected from the functional response alone [e.g., as in house mice feeding on beech seeds; see Ruscoe et al. (2005)]. In general, adding extrinsic negative density dependence to continuous-time predator models such as 7.2 helps stabilize the interaction (Rosenzweig 1971; Deng et al. 2007).

In addition to such extrinsic density dependence, and density dependence via the functional response, several other modalities of density dependence may be at play in predator-prey interactions, contained in the expressions $r(N,P)$ and $m(N,P)$. For instance, prey recruitment may depend directly upon predator density, for reasons other than direct consumption, if prey reduce foraging efforts in response to the presence of predators (Brown and Kotler 2007). The functional response of the predator may then indirectly reflect resource levels for the prey, leading to potentially complex relationships between prey density and predator feeding rates. Or, in escaping predation, the prey may become exposed to alternative sources of mortality. Some of the most important directions in contemporary natural enemy-victim theory have to do with elucidating the consequences of changes in prey behavior and other traits in the face of predation (Bolker et al. 2003; Sih Chapter 4). Brown and Kotler (2007) suggest that "the ecology of fear" can resolve classical conundrums in community ecology. For instance, when predators become common, herbivores often reduce their feeding rates (e.g., to become more vigilant), and in effect trade off reduced fecundity for increased survival (Creel and Christianson 2008). This both buffers herbivores and relaxes consumption on plants. If prey vigilance increases with predator abundance, the feeding rate per predator should decline with increasing predator abundance (see Sih Chapter 4). Approximate ratio dependence thus may emerge from an intrinsically prey-dependent system because of adaptive flexibility in the prey. Brown and Kotler (2007) go on to observe that "paradoxically, efficient predators produce intrinsic instability to predator-prey dynamics, while inefficient predators produce vulnerability to extrinsic variability in prey population numbers. Fear responses by prey can break this paradox. At low predator numbers, the predator can efficiently catch unwary prey. At higher predator numbers, the predator becomes less efficient as the prey become increasingly uncatchable." Thus, prey behavioral responses may be broadly stabilizing, in the sense of bounding trajectories away from boundaries.

In the predator component of the general model given by 7.16, two further extensions to the basic Lotka-Volterra model are included. The extra P found as an argument inside g indicates that for a given rate of prey capture, predator birth rates may vary directly with their own density (e.g., due to competition for limited nest sites). Predator death rates (the $m(N,P)$) may also depend on prey and predator abundance (e.g., higher feeding rates may lead to lower predator deaths, not just higher predator births). For instance, in warm-blooded vertebrates, if the feeding rate goes below a certain level, death may rapidly occur. A number of forms relating predator numerical to functional response have been suggested in the literature (Ramos-Jiliberto 2005). It would be useful for ecologists in developing theories to disentangle distinct birth and death responses in addressing these relationships. More broadly, much more attention needs to be paid to characterizing the mathematical form of predator numerical responses to their prey.

The subdomain of host-pathogen dynamics

As noted earlier, different subdomains of natural enemy-victim theory have had somewhat different theoretical traditions. Consider models for interactions between hosts and pathogens such as bacteria and viruses. A pathogen is a natural enemy (usually a parasitic microorganism) that lives and reproduces as a population inside the victim, its host. The mathematical theory of epidemiology provides a theoretical framework for exploring the persistence and epidemic behavior of pathogens, and of the impact of pathogens upon their hosts. In a tradition going back to Kermack and McKendrick (1927), and summarized in a magisterial fashion by Anderson and May (1991) and more recently by Grassly and Fraser (2008) and Keeling and Rohani (2008), one subdivides host populations into classes such as susceptible, infected, and recovered individuals. For microparasites such as viruses and bacteria, this theoretical maneuver means that one initially ignores the dynamics of the pathogen within individual hosts, and so pathogen abundance is not tracked directly, but instead the infection status of the host.

For simplicity, and to be consistent with the natural enemy-victim model given by 7.2, I will assume that we can use a two-class model of just S (susceptible host density) and I (infected host density), implying that recovery by an infected host leads to readmission into the susceptible class. An SIS model has the following general form:

$$\frac{dS}{dt} = R - i(I,S) - mS + \gamma I$$
$$\frac{dI}{dt} = i(I,S) - m'I - \gamma I. \tag{7.17}$$

Here, R is the rate of recruitment of fresh susceptibles into the host population (which may depend upon host population size), $i(I,S)$ is the total rate of infection, m and m' are per capita mortality rates for susceptible and infected individuals, respectively, and γ is a per capita rate of recovery. It is useful to briefly compare the predator-prey model 7.12 with 7.17. The "prey" here are susceptible hosts, and "predators" are infected hosts. The total rate of infection is analogous to the total mortality imposed upon its prey by a predator, i.e., the functional response times predator density. It is also the numerical response of this "predator" to susceptible hosts. These analogies suggest several analogues of propositions 7–9 (Table 7.1C; the primes indicate the corresponding general propositions).

One proposition does not have such a ready predator-prey analogue (Table 7.1D)—Proposition 12: *Susceptible hosts may be generated from infected hosts (by recovery, or possibly birth)*. This last proposition has an important implication in terms of population regulation and stability. In human epidemiological models, it is often assumed that infection has no impact upon host population size, either through death or fecundity. If recruitment $R = bN$, where $N = I + S$, and mortality rates are $m = m' = b$, the host population stays at its initial size (Swinton et al. 2002, Box 5.1). This assumption requires that bioenergetic and fitness impacts of infection be trivial relative to other factors limiting host numbers; thus, infectious disease is irrelevant to host population regulation.

A contrasting assumption is for the pathogen not only to influence host mortality or fecundity, but actually to be the sole factor regulating the host. If we replace R with a constant per capita birth term, bS, and assume that infected hosts give birth at rate b', and recover at rate γ, 7.17 becomes:

$$\frac{dS}{dt} = bS + (b' + \gamma)I - i(I,S) - mS$$
$$\frac{dI}{dt} = i(I,S) - m'I - \gamma I. \tag{7.18}$$

Adding the above equations leads to

$$\frac{dN}{dt} = (b - m)S + (b' - m')I. \tag{7.19}$$

Setting this equation to zero shows that a *necessary* condition for regulation of a host by a pathogen, when the host is otherwise unregulated, is that the birth rate (of healthy hosts) from infected individuals must be less than their own death rates (Holt and Pickering 1985).

Whether or not regulation actually occurs depends upon the form of the transmission function $i(I,S)$, which formally matches both the total mortality term in the prey equations and the numerical response in the predator equations of 7.11. Different functional forms for transmission arise from different assumptions about the biology of transmission (Hochberg 1991; McCallum et al. 2001), analogous to how predator functional responses reflect the detailed behaviors of predators and their prey. A very active area of infectious disease ecology is focused on refining models for transmission (e.g., Keeling 2005), for instance using network models (Brooks et al. 2008). There are two idealized forms of transmission—density-dependent and frequency-dependent:

$$i(I,S) = \beta IS$$
$$i(I,S) = \frac{\beta IS}{N} . \tag{7.20}$$

The first equation in 7.20 may be reasonable for airborne pathogens, but the second is often more sensible for diseases where infected individuals come into contact with an approximately fixed number of potential hosts (e.g., sexually transmitted diseases).

Note that with density-dependent transmission, equation 7.18 closely resembles the Lotka-Volterra predator-prey model, but with extra terms ($b'I$ and γI) representing recruitment into the susceptible portion of the population from the infected portion of the population. So even if susceptible hosts go to zero, the population can recover, because some are regenerated from infected hosts. The disease can in this case regulate the host, although when this extra term is small, the system exhibits weakly damped oscillations (Figure 7.1); as this term approaches zero, the model becomes identical to the classical Lotka-Volterra predator-prey model, and so has neutrally stable cycles.

Density-dependent transmission implies a threshold host density for disease invasion and persistence, and also permits host regulation by parasitism. By contrast, if the frequency-dependent term (the second expression in 7.20) is introduced into 7.18, there is no longer a threshold host population size, and the disease on its own is not able to regulate the host to a stable equilibrium (Getz and Pickering 1983). A mixture of density-dependent and frequency-dependent transmission can imply infection dynamics where a pathogen invades, increases in prevalence, and in so doing drives both the host and itself extinct (Ryder et al. 2007).

This difference between the outcomes of parasitism for host regulation, depending on the functional form of transmission, corresponds, if not in detail,

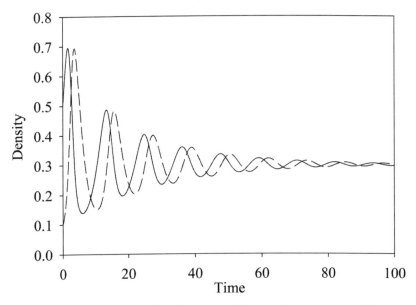

Figure 7.1 An example of damped oscillations in an SI host-pathogen interaction, following a perturbation. The model is 7.18 in the main text, and transmission is assumed to be
density-dependent (βSI). The parameters are $b = 1$, $b' = 0$, $m = m' = 0.5$, $\gamma = 0.1$, $\beta = 2$.
The infection is assumed to completely suppress host fecundity. A small amount of recovery
by infected hosts provides a weakly stabilizing effect, which slowly damps the oscillations.
(Solid line = S, healthy hosts; dashed line = I, infected hosts.)

at least in some features, to the difference between prey-dependent and ratio-
dependent functional responses in general predator-prey theory. Frequency-
dependent transmission can be written as follows (recalling that $N = I + S$):

$$\frac{\beta IS}{N} = \left(\frac{\beta S/I}{1 + S/I} \right) I. \tag{7.21}$$

The term in parentheses in 7.21 is the pathogen's "functional response,"
i.e., the rate of new infections per infected host (recall the functional response
is the rate at which prey are consumed per predator), expressed as a function
of the ratio of healthy hosts (prey) to infected hosts (predators). Frequency-
dependent disease transmission can be viewed as a form of ratio dependence.
A functional form for trophic linkages controversial in the subdomain of
predator-prey theory turns out to be a run-of-the-mill assumption in another
subdomain of natural enemy-victim theory, namely host-pathogen theory.

Note that if we look at model 7.18 with frequency-dependent transmission and no recovery in the limit when the infection is rare, the equations are approximately as follows:

$$\frac{dS}{dt} = bS - \frac{\beta IS}{N} - mS \approx (b - m)S$$
$$\frac{dI}{dt} = \frac{\beta IS}{N} - m'I \approx (\beta - m')I.$$

(7.22)

This is simply a pair of equations for two populations growing exponentially. Model 7.22 matches the "devil's advocate" position about ratio dependence and exponential growth in predator-prey theory, noted above.

Reflections on the differences between host-pathogen and predator-prey subdomains

At first glance, one sharp difference between infectious disease models 7.18 and predator-prey models (as in 7.11) is that recruitment into the susceptible class of the population can come from the infected class. But recall that an abstraction in *SI*-style models is that one simply counts the pathogen as present or absent, rather than enumerating them within each host. Hosts, in effect, are patches (with their own dynamics) colonized by pathogens. A very active area of research focuses on accounting for within-individual host population dynamics of pathogens. The appropriate analogy to be made among models may be found only by shifting scales. The predator-prey model that in some ways is more truly analogous to an *SI* model is a metapopulation model, where predators and prey are in patches coupled by dispersal, and patches transition among the states of empty, prey alone, or predator together with prey (Nee et al. 1997). If prey in patches with predators still produce propagules that can reach and colonize empty patches, the prey equation in the metapopulation model has a term matching γI in the *SI* model (Holt 1997b). This metapopulation model is stable. Likewise, in general predator-prey theory, when models such as 7.2 are extended to multiple patches connected by limited dispersal, asynchronous fluctuations among patches can contribute to indirect density dependence and overall population stability (Murdoch et al. 2003; Briggs and Hoopes 2004).

One difference between traditions in predator-prey and host-pathogen ecology is that the former has a robust subtradition of models grounded in bioenergetics (see above), whereas scant attention has been paid to resource relationships, energetics, and stoichiometry in host-pathogen dynamics. But there is increasing evidence that resources modulate many aspects of the interactions between hosts and pathogens (Smith and Holt 1996; Smith 2007; Hall et al. 2009), and I suspect that this seeming difference in emphasis will change in the near future. One deep-rooted difference between host-pathogen systems and predator-prey systems has to do with basic biology, rather than mathematical abstractions, and has not yet been addressed in depth in the literature. Namely, pathogens and parasites are symbionts, living intimately with their hosts, so they can evolve into mutualists (Hochberg et al. 2000). Some taxa can even be facultative, a parasite in one setting, but a mutualist in others. This part of the domain of natural enemy-victim interactions shades into the domain of interspecific mutualisms. Proposition 10 applied to host-pathogen systems includes the fact the qualitative nature of the symbiosis may shift from $(+,-)$ to $(+,+)$ as a function of environmental conditions.

Towards multispecies natural enemy-victim theory

We have focused on the interaction between a specialist natural enemy and its victim. Other species may be present and have an impact on the two focal species, but it is assumed these effects can be swept into model parameters or functional forms (e.g., density-independent mortality rates for a predator might reflect its use of alternative food resources other than the focal prey species); what is important in this assumption is that the abundances of these other species do not need to be explicitly tracked, and do not provide alternative modes of dynamical feedbacks. Understanding multispecies interactions is an important and vibrant area of natural enemy-victim theory (and indeed has been the theme of most of my own papers in this area), shading into other domains in ecology [e.g., food web and network theory, metacommunity theory (see Leibold Chapter 8), and succession theory (Pickett et al. Chapter 9)]. Assumptions one makes about the basic interaction between a single natural enemy species and a single victim species (e.g., prey-dependent vs. predator-dependent functional responses) can have profound consequences for the dynamics of complex food webs (e.g., Drossel et al. 2001; McKane and Drossel 2006). I here merely make a few remarks about how the propositions and models presented above need to be modified when there are multiple species present.

Consider proposition 4. There are complexities and subtleties even here. First, even for a single victim species, the statement applies *on average*. In spatially and temporally varying environments, or if the consumer can exploit multiple types of victims, it may not (and probably does not) pertain to each and every act of consumption. A generalist predator sustained by many prey species could include in its diet a few prey types that are a net loss, per act of consumption (Holt 1983). Likewise, proposition 9 surely holds for a generalist only if *all* of its victims are absent; the natural enemy may simply equilibrate at a lower (but nonzero) density in the absence of any particular victim species.

If one assumes that propositions 6 through 8 hold for multispecies natural enemy-victim models, this has implications for the domain of community theory. In Holt (1977), I showed that if one assumed that a predator feeding on two prey types has a positive numerical response to an increase in the abundance of each (i.e., propositions 6 and 7), the predator is food-limited (so has no direct density dependence), and the system settles into an equilibrium, then an alternative prey should always depress the equilibrial abundance of a focal prey species. So $(-,-)$ interactions in a prey trophic level can emerge via impacts of a higher trophic level, viz., apparent competition, which can constrain coexistence. This is an example of how propositions and theories in one domain (basic natural enemy-victim theory) can provide building blocks for constructing theory in another domain (community theory; see also Appendix to this chapter for community-level implications of the debate about predator- vs. prey-dependent functional responses). Given the basic assumptions of the Lotka-Volterra model 7.11, one can add a third species to this system (considering only trophic interactions) in three ways: (i) a second predator species, consuming a shared resource, (ii) a second prey species, sharing a generalist predator, and (iii) a species at another trophic level, i.e., a three-species food chain. The first two cases predict the exclusion of one species by another, via exploitative competition (for i), or apparent competition (for ii). The Lotka-Volterra food chain is dynamically unstable; the top level constrains the abundance of the middle level, so the bottom level grows unchecked. By working through the hierarchy of models presented above (nonlinear functional responses, direct density dependence, etc.), one can systematically examine the implications of these generalizations of the basic model for species coexistence and food chain (and ultimately food web) regulation and stability. For instance, direct density dependence in a predator can stabilize a predator-prey interaction. Given two predators competing for one shared prey species, one normally would expect competitive exclusion, but this may not occur if the superior competitor exerts strong density dependence on itself via interference, freeing resources for the inferior competitor.

In effect, this conceptual protocol provides a structured approach for identifying mechanisms of coexistence and exclusion, which are key desiderata in the domain of community ecology, using as a platform the propositions and models crafted in another domain, namely natural enemy-victim theory. In turn, some conundrums in one domain may be resolved by consideration of themes in another. As we have seen, specialist predator-prey interactions tend to be unstable, but given multiple interacting species additional forces can operate (e.g., switching by generalist predators) that help stabilize any given pairwise interaction, or at least keep fluctuations within reasonable bounds. Moreover, the instability generated by natural enemy-victim interactions creates another dimension of temporal variability in the environment, which can in turn influence the coexistence of competing species. Many of the most interesting problems in ecology involve issues that straddle domains in this fashion.

Concluding thoughts

There is much creative work ongoing in natural enemy-victim theory, and a future synthesis might look rather different from what I have presented here. It should be noted that other authors (e.g., Murdoch et al. 2003, p. 76) have provided their own renditions of the basic properties of predator-prey interactions. Some subdomains (e.g., plant-herbivore interactions) warrant much more theoretical attention. For others—continuous-time predator-prey theory, host-parasitoid theory, and infectious disease theory—recent work has highlighted the importance of key extensions for understanding issues of stability and persistence, well beyond the factors emphasized in the classical two-species theory summarized above. Using a single number (density) to characterize a population may be inadequate for most natural systems; models with additional variables describing age and the internal state (e.g., energy reserves) of both consumers and their victims might be necessary. Crucially, space, localized interactions, environmental heterogeneity, and dispersal all profoundly affect the persistence and stability of natural enemy-victim interactions (Briggs and Hoopes 2004). For instance, Keeling et al. (2000) showed for the Lotka-Volterra and Nicholson-Bailey models embedded in space that the localized covariance between predators and prey leads to equations that help unify understanding of seemingly disparate causes for stability, including localized competition among predators. There are increasing efforts to deal with the consequences of stochasticity at the level of individuals, which can scale up to entire populations. [See Chesson (1979) for a prescient overview of this issue.] Many pairwise natural enemy-victim interactions cannot be understood without careful attention to multispecies interactions in a broad community context (Holt 1997a), and indeed to ecosystem feedbacks (Loreau and Holt 2004). Future attempts toward a unified theory of natural enemy-victim interactions will need to deal with all these issues in a systematic manner.

Bioenergetic and stoichiometric perspectives potentially provide powerful tools for refining natural enemy-victim models. An influential attempt to ground predator-prey dynamics in organismal biology by Yodzis and Innes (1992) expressed attack rates and biomass loss rates as allometric relationships for production, metabolism, and maximum consumption rates, using physical principles to constrain models of trophic interactions. [Williams et al. (2007b) provide a recent example of this approach.] This work is an intellectual predecessor to recent work in the metabolic theory of ecology (Brown et al. 2004; Savage et al. 2004), which potentially provides a general conceptual framework for linking traits of individual organisms with population and community dynamic models. Across a wide range of taxa, there is a relationship between natural mortality rates and body size and temperature (McCoy and Gillooly 2008). Natural mortality includes predation; coupled with information on predator abundance, this relationship sets upper bounds on ambient attack rates. This approach potentially permits parameterization of predator-prey models in a way that does not require measuring each parameter afresh in each new setting (e.g., Vasseur and McCann 2005). A consideration of the physical dimension of life naturally leads to a focus on how feeding rates are reflected in the internal states of organisms, such as energy and nutrient reserves, which can influence decline in starving populations and offspring quality (maternal effects) (Ginzburg and Colyvan 2004). I suspect future attempts to craft a general theory of natural enemy-victim interactions will be more heavily and fundamentally rely on these physical perspectives, than just the light patina I have here provided.

One can find counter-examples or limitations to almost any generalization about natural enemy-victim theory. For instance, although I have suggested that the tendency to cycle is at a core outcome of this domain of theory, many (possibly most) predator and prey populations do not in fact oscillate (predation satiation can simply prevent predators from regulating prey at all, rather than causing cycles). Though time lags often destabilize (May and Watts 1992), certain kinds of time lags in predators can be stabilizing (Nunney 1985). Finally, spatial processes may often be stabilizing, but in some circumstances spatial heterogeneity can likewise destabilize (Hochberg and Lawton 1990; Holt 2002). Many generalizations that have been proposed for natural enemy-victim theory are not universal laws, but more akin to central tendencies in a field of theoretical possibilities, consistent with a set of foundational principles.

And finally, another crucial direction for future natural enemy-victim theory is to incorporate Darwinian evolution, which in many taxa occurs on timescales commensurate with population dynamics. Coevolutionary cycles can feed back onto population dynamics, and vice versa, so that understanding the dynamics of these systems requires paying close attention to the generation, maintenance, and reshuffling of genetic variation in each interacting player. Differences among systems in genetic variation and architecture may prove to be as consequential in determining their dynamical behaviors as the ecological factors, such as functional responses, that I have emphasized here in this attempt at a synthetic overview of natural enemy-victim theory.

Acknowledgments

I thank Sam Scheiner and Mike Willig for their invitation to participate in this endeavor, and Sam, Mike Barfield, and two anonymous reviewers for very helpful and detailed comments, Vitrell Sherif for assistance with manuscript preparation, and the University of Florida Foundation NSF, and NIH for support.

Appendix

My own interest in natural enemy-victim interactions stems from reflecting on how this theory informs our understanding of community structure. The debate between prey-dependent and predator-dependent functional responses discussed in the main text has important implications for understanding the impact of predators on species coexistence. In Holt (1977) and elsewhere I have argued that alternative prey species can experience a negative $(-,-)$ interaction mediated through their shared predator (apparent competition), due either to behavioral responses (Holt and Kotler 1987) or to the predator's numerical response. If sustaining more predators implies stronger predation pressure upon a focal prey species with a low recruitment rate, it can be excluded from a local community. Strong predator dependence in the functional response potentially weakens the indirect effect of one prey species upon another and thus reduces the potential role of apparent competition in prey community organization. Consider a community with a generalist predator at equilibrium with a resident prey (species 1, density N_1). The interaction between them is described by the Beddington-DeAngelis functional response. Because of predator interference, the resident prey typically equilibrates at higher abundance than would otherwise be the case. Assume a second prey species (species 2, density N_2) is introduced at low numbers, during community assembly, and predator attacks upon it are described by the same functional response. The rate of growth of this species when rare is

$$\frac{dN_2}{dt} = N_2 \left(r_2 - a_2 \frac{P^*}{1 + a_1 h_1 N_1^* + iP^*} \right),$$

where the asterisk designates the initial equilibrial abundances of N_1 and P. Predator interference influences the strength of predation imposed on the invading prey through three causal pathways. (1) Interference directly reduces the per predator attack rate (this is the term iP^* in the denominator). (2) Such interference weakens the ability of the predator to limit the resident prey; if this species is more abundant, this further reduces attacks upon the invader, via the functional response (the term $a_1 h_1 N_1^*$ in the denominator). (3) Interference can lower the number of predators sustained by resident prey (the P^* term in the numerator), weakening apparent competition effect. So issues in one domain, having to do with how one conceives of the interaction between a natural enemy and its victim, can have major consequences for how one conceives of phenomena and processes in another domain.

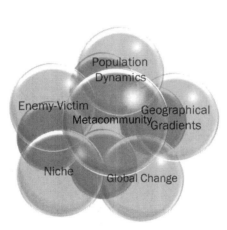

CHAPTER 8

The Metacommunity Concept and Its Theoretical Underpinnings

Mathew A. Leibold

A metacommunity is a set of local communities that are connected by the dispersal of at least one of the components (Hanski and Gilpin 1991; Leibold et al. 2004). The theory of species interactions at local scales is fairly well developed and serves as the focus of much of what we know about in community ecology. Metacommunity ecology seeks to take this understanding and use it to consider ecological dynamics at larger spatial scales by considering how dispersal among local communities affects ecological dynamics. Thus, the concept is a useful way to integrate what we know about spatial dynamics in community ecology; it expands the field of community ecology to include larger spatial scales in a couple of important ways.

First, there is the idea that dynamics that occur at one place (a single local community) are not necessarily independent of those that occur at other places. Although spatial effects have been thought about for a long time (Skellam 1951), the metacommunity concept allows us to consider a wide array of ways in which this may occur.

Second, there is an implicit idea of multiple spatial scales, and the idea that somewhat different or additional ecological principles may apply at larger scales than the local scale. Thus, the concept is one way to address how ecological processes may determine patterns of biodiversity and composition at different scales and address questions about mesoscale ecology (Holt 1993; Ricklefs and Schluter 1993). The important idea is that it is possible to derive predictions about the attributes of biotic assemblages at these mesoscales based only on ecological principles. Clearly, there are yet larger scales where historical and biogeographic effects are important, but the metacommunity concept may allow us both to understand community structure at a wide range of spatial scales that fall short of this larger scale and to explore how ecological dynamics may interact with historical or biogeographic ones at such larger scales (Cornell and Lawton 1992).

There is however, an important distinction between the "metacommunity concept" and "metacommunity theory." As a concept, the metacommunity is proving to be stimulating and useful way to organize approaches to multiscale ecology (Fig. 8.1). However, metacommunity theory is still in its infancy and consists of several different sets of ideas and model formulations. This chapter seeks to describe and evaluate a constitutive theory of metacommunities. The heterogeneity of approaches to metacommunities makes a philosophical discussion of such a theory difficult and incomplete. Furthermore, it seems likely that there are novel processes and approaches to metacommunities that will arise and not be adequately described by any current effort at describing a constitutive theory. Thus, constitutive theory is likely to change substantially in the future as thinking about metacommunities develops further.

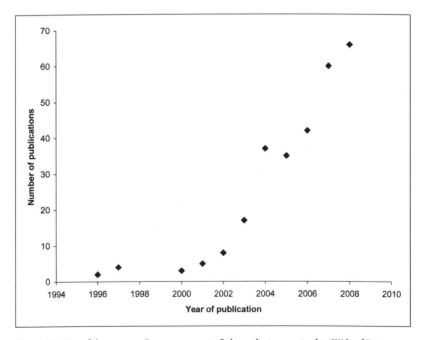

Figure 8.1 Use of the concept "metacommunity" through time as cited in Web of Science.

A well-developed constitutive theory in ecology should, to some degree, have the following features (Scheiner and Willig Chapter 1): a domain that describes the phenomena and principles involved, a set of propositions that elucidate understanding of the domain, and a set of more precise models that link this theory to data. I outline how I understand these aspects of metacommunity ecology theory and then discuss how this helps elaborate a more sophisticated view of community ecology in general.

The domain of metacommunity theory

The domain of metacommunity ecology includes explanations of how ecological communities vary in space. This domain is very similar to that of community ecology in general but focuses particularly on how communities vary from one place to another rather than on the structure of particular communities. There are numerous ecological features that form the focus of metacommunity ecology and that describe or follow from understanding how the distribution and abundance of interacting organisms is regulated by ecological processes and depend on dispersal. Some examples include:

1. Patterns of species composition (which species with which traits are found in communities) in space, in relation to environmental conditions and factors affecting dispersal.
2. Patterns of species richness (how many species are found in communities) in relation to space, environment and dispersal.
3. Consequent effects of composition, diversity, and dispersal on other attributes of communities such as food web structure, stability, and ecosystem features (e.g. productivity).

It may be useful to focus discussion of the domain by pointing out that all of these effects largely boil down to understanding how dispersal interacts with other ecological processes to affect how variation in community composition varies with respect to various landscape features. A crude way of studying this empirically is by using variation decomposition of community composition to separate components that are purely environmental, purely spatial, spatioenvironmental colinear, and unexplained residual variation. A rapidly burgeoning literature on such patterns tries to interpret metacommunity dynamics in natural and experimental systems (e.g., Cottenie 2005; Beisner et al. 2006). Although more sophisticated ways of thinking about composition are clearly needed, one might argue that the central goal of metacommunity ecology is to explain them and explore the consequences that follow from this variation.

In addition, however, metacommunity ecology also addresses similar questions at spatial scales that are larger than that of the local community. Thus, it can address questions about the composition and diversity of sets of local communities. This greatly expands the utility of the theory because it can address novel patterns. Examples include patterns related to macroecology and to paleoecology, where data often represent composite data aggregated over multiple spatial samples. In paleoecology this happens as a natural result of taphonomic processes, and in macroecology it happens as a result of data extraction from range maps. At these larger spatial scales, metacommunity ecology also begins to share a domain with evolutionary biology and historical biogeography, as these processes also explain some aspects of variation in composition and biodiversity.

The propositions of metacommunity theory

Metacommunity ecology combines classical community ecology (i.e., community ecology of closed communities) with dispersal. The central propositions of metacommunity ecology thus include all the propositions that apply to community ecology in combination with a set of additional propositions about the role of dispersal (Table 8.1).

The most important propositions that characterize metacommunity theory focus on four different ways that dispersal among local communities can affect their dynamics.

Proposition 1

Dispersal contributes to local community assembly by serving as a source of colonists for species that were previously absent. Ignoring how species originally came to exist in these communities to begin with in the absence of dispersal, local communities can change in species composition only by extinctions or in situ speciation. At such local scales, speciation would be considered sympatric, and while this can occur it is thought to be less important than allopatric speciation (Gavrilets 2003). Thus, community assembly is severely constrained if dispersal is absent. If we consider that local environmental conditions often change (e.g., succession, perturbations), it seems unlikely that species composition would be likely to track environmental conditions as well, and if these cause numerous extinctions, local diversity is also likely to be low. The only ways that biotas could respond to such changes would be via in situ evolution and while this may be important, it would not produce associations between species composition and local environmental conditions. Thus, the

Table 8.1 The propositions of metacommunity theory

1. Dispersal contributes to local community assembly by serving as a source of colonists for species that were previously absent.

2. Dispersal can generate population mass effects between communities that differ in local fitness.

3. Heterogeneity in dispersal among species can allow for novel mechanisms of coexistence in a metacommunity.

4. Metacommunities can have spatial feedbacks due to a complex suite of additional effects beyond those of recurrent dispersal between patches.

5. Within local communities organisms belonging to different species interact via a set of interactions involving both direct and indirect effects.

6. Interactions in local communities are context dependent in that they are affected by the local environmental abiotic conditions and by the existence of other species in the community.

7. These interactions affect which species can have self-sustaining populations and regulate the abundances of such populations in the absence of dispersal.

8. In the extreme when stabilizing forces are weak and fitness equalizing forces are strong, stochastic demographic dynamics may affect patterns of distribution and abundance.

9. Local extinctions are important.

observation that there are such associations indicates that this proposition is important. Evidence for dispersal playing such a role go back at least to early work on succession ecology (Watt 1947).

Proposition 2

Dispersal can generate population mass effects (also referred to as source-sink effects) between communities that differ in local fitness (Amarasekare and Nisbet 2001; Mouquet and Loreau 2002). These effects involve the maintenance of high populations in some patches due to immigration from other patches where the same species is highly productive. An important consequence is that such dispersal can allow a species to exist in a local community where it cannot have a self-sustaining population (i.e., as a sink population). Instead, that population's existence is dependent on continuous immigration from other patches where that species has high fitness (i.e., the source populations). A complication is that dispersal also involves emigration so that dispersal can also lead to extinctions of species in highly productive habitats that would otherwise be self-sustaining in the absence of dispersal. Such mass effects are often likely to be reduced by adaptive dispersal but this need not always to be so; it depends on the rules that govern such decisions (Sih Chapter 4). One important consequence of mass effects is that they may lead to reduced associations between local environmental conditions and species composition. Such dispersal instead implies that local communities that are close together in space may, to some extent, share species independently of whether they are similar in local environments.

Proposition 3

Heterogeneity in dispersal among species can allow for novel mechanisms of coexistence in a metacommunity. The most obvious case for this is the colonization-competition tradeoff hypothesis. In dispersal limited systems (where colonization limits community assembly) with a regular frequency of local extinctions, a poor competitor can coexist in the metacommunity with superior competitors that always exclude it locally if it can colonize communities that have had recent extinctions (Levins and Culver 1971; Hastings 1980). Coexistence of species with different dispersal rates is also possible under more complex scenarios, including cases where source-sink relations are present (Snyder and Chesson 2004), but their dynamics is still poorly understood.

Proposition 4

Metacommunities can have spatial feedbacks due to a complex suite of additional effects beyond those of recurrent dispersal between patches (Holt 1984). Although this proposition is still very poorly developed, the basic idea here is that dispersal may be important in providing temporally lagged feedback effects on local communities via residence in other local communities and back migration into these communities. For example, a species may exist as a sufficiently robust sink population in a community where it is maintained by immigration from other sites where it may serve as a colonist in source populations that may periodically go extinct.

Although these four propositions are distinctive to metacommunity ecology, other more conventional propositions are just as important.

Proposition 5

Within local communities organisms belonging to different species interact via a set of interactions involving both direct and indirect effects. Direct interactions include competition, exploitation , mutualisms, commensalisms, and interactions involving one-way negative effects. Indirect effects involve chains of such interactions that may also be complex by being nonlinear or interactive (Levins 1975). This corresponds to the second fundamental principle of Scheiner and Willig (Chapter 1, Table 1.3). Most of the theoretical work has focused only on competition interactions, but some emerging work incorporates consumer-resource interactions in food webs (Amarasekare 2008). Future work should explore the full scope of interactions.

Proposition 6

Interactions in local communities are context dependent in that they are affected by the local environmental abiotic conditions (the environmental template) and by the existence of other species in the community (Chase and Leibold 2003). This corresponds to the fourth fundamental principle of Scheiner and Willig (Chapter 1, Table 1.3). A large number of models concentrate on cases where local communities exist in patches that are environmentally identical, but an important synthetic point of metacommunity theory in its general form is to highlight how heterogeneity of patches can alter these results.

Proposition 7

These interactions affect which species can have self-sustaining populations and regulate the abundances of such populations in the absence of dispersal. Coexistence of populations of multiple species with such self-sustaining populations in local communities involves the interaction of stabilizing (frequency-dependent) forces as well as fitness-equalizing forces (Chesson 2000b). This expands and makes more precise Scheiner and Willig's tradeoff principle for diversity theory (Scheiner and Willig 2008). In the presence of immigration some populations will also be present despite violating this proposition at the local scale.

Proposition 8

In the extreme, when stabilizing forces are weak and fitness equalizing forces are strong, stochastic demographic dynamics may affect patterns of distribution and abundance (Hubbell 2001). This derives in part from the third, fifth, and sixth fundamental principles (Chapter 1, Table 1.3).

Proposition 9

Extinctions in local communities occur for at least three reasons. First, they may occur following environmental change or disturbance (Lande 1993); second, they may occur as community composition changes via community assembly (Law and Morton 1996); and finally, they may occur due to demographic stochasticity (Lande 1993). This derives from some complex combination of the various fundamental principles of Scheiner and Willig (Chapter 1, Table 1.3), but is also strongly supported by basic theory of populations (Hastings Chapter 6). In the absence of dispersal, local communities consequently tend to lose species and become depauperate.

Models of metacommunity theory

The previous propositions can generate a rich array of possible outcomes. In many cases, these differ substantially and sometime qualitatively from what is predicted in the absence of dispersal. Although this array of outcomes is likely to depend on different assumptions and ranges of parameter values, there is no general model of metacommunity ecology that explains this variation. That is to say that there are a number of different mathematical formulations for metacommunity dynamics that make different assumptions but no mathematical uber-model that encompasses all of them. Instead, the models that are available differ in assumptions; in a very general way these models can be classified into the four classes or archetypes described in Table 8.2 (Leibold et al. 2004).

Models of patch dynamics

Here, the focus is on colonization and extinction processes in patches. Dispersal is important because it allows recolonization following extinctions, and extinctions are stochastic and due to disturbances or demographic stochasticity. The most well-known examples include Levins and Culver (1971), Tilman and Kareiva (1997), and Hastings (1980). These models are structured in a way that completely ignores source-sink relations, thus ignoring proposition 2. Furthermore, for the most part, these models also ignore environmental differences among patches and assume that all patches are identical except for their species composition. Such models can be modified to account for

Table 8.2 A summary of the four archetypes for metacommunity theory (based on Leibold et al. 2004). (A) The propositions correspond to those listed in Table 8.1. Cases marked with a question mark indicate that the entry depends on whether patch dynamic models involve homogenous patches (the most commonly studied case) or patches that vary in environmental conditions (also possible). (B) Predictions of each model type are selected to highlight differences among models. Actual metacommunities are likely to consist of mixtures of all of these mechanisms so that some groups of species may have distributions that are more strongly related to different models.

	Patch dynamics	Mass effects	Species sorting	Neutral
A. Propositions				
1) Dispersal affects colonists	X		X	X
2) Dispersal allows source-sink relations		X		X
3) Heterogeneity in dispersal is important	X	X		
4) Dispersal allows spatial feedbacks	X	X		
5) Interactions are direct and indirect	X	X	X	
6) Interactions depend on local environments	?	X	X	
7) Coexistence requires stabilizing effects in local communities	X		X	
8) Stochastic demography is important in allowing coexistence			X	
9) Local extinctions are important	X			
B. Predictions				
Invasible local communities should be common	Y	N	N	Y
Fugitive species should exist in the metacommunity	Y	N	N	N
Composition of communities should depend on spatial effects independent of environment	Y	Y	N	Y
Composition of communities should depend on environmental effects independent of spatial effects	?	Y	Y	N
The ratio of local to regional diversity can be high	N	Y	N	Y

heterogeneity among patches (Holt 1993), and such modifications can alter our understanding of some phenomena such as the prevalence of alternate stable states in metacommunities (Chase et al. 2002). Nonetheless, much more could be done along these lines. The most important insight to come from such models is that highly dispersive but poor competitors (also known as fugitive species) can coexist in a metacommunity with superior competitors that would otherwise exclude them (Mouquet et al. 2005). Where environmental differences among patches exist, these models also predict that local community composition may not be perfectly associated with environmental differences since species may exist, albeit temporarily, in patches where they will eventually be displaced by superior competitors.

Models involving mass effects

Here, the focus is on the role of dispersal in maintaining source-sink relations among populations in different patches. The emphasis is thus on propositions 2 and 4. Such mass effects allow for more species to coexist locally than might be predicted by theory on closed (no dispersal) communities and thus produce patterns that differ from those of a community theory that ignores dispersal. As dispersal among such patches increases, however, community composition is effectively homogenized, and eventually these heterogeneities are not sufficient to allow niche partitioning by patch type. To date these models have largely ignored stochastic extinctions due to disturbances and environmental change, and they thus tend to ignore proposition 1. As in models of patch dynamics, there may be imperfect correspondence between local environmental conditions and community composition because deviations from these would result from the existence of sink populations.

Models of species sorting in metacommunities

A third set of models conforms to the most conventional perspective about local communities by examining community assembly occurs in relation to environmental variability in the absence of source-sink relations and in the absence of background extinctions. Here, dispersal is important only because it provides the stream of potential colonists that allows community composition to track environmental changes in time and space. These models thus ignore propositions 2–4 and focus on the range of parameters that prevent dispersal from being a limiting process in community assembly. The prediction is that local community composition should strongly track local environmental conditions. Although such spatial effects are not uncommon, and this may appear to make this approach unlikely, it corresponds most strongly to conventional ideas about species distributions in relations to environment. Further, this approach is the one that is most often strongly supported by observations of the relation between community composition and environment (Cottenie and De Meester 2005).

Neutral metacommunity models

A final and most recent approach to metacommunities was developed by Hubbell (2001). It focuses on the case where all species in a metacommunity have identical dispersal, extinction, and competitive abilities. It thus ignores or reinterprets most of the propositions listed above. Dispersal still acts to fuel the arrival of species into local patches (proposition 1) and migrating individuals can allow species to exist that would otherwise go extinct (somewhat like proposition 2 but without fitness differences between patches), but the important factors that regulate community composition are chance demographic events and chance dispersal events. In strong contrast to species-sorting models, there should be little correspondence between community composition and local environment because all species see environmental heterogeneities in identical ways. Again, while the common observation of environmental effects on distributions may appear to make this approach unlikely, data on relative abundance distributions show surprising fit to the predictions of this model, although other models can also show strong, and perhaps sometimes stronger, fit to the data (McGill et al. 2007).

Comparing the four archetypal models

The differences among these archetypal models thus depend on the role hypothesized for dispersal and the importance given to heterogeneity in local environmental conditions among communities. A crude conceptual landscape for these models that illustrates possible relationships among them is shown in Fig. 8.2. Here, I have identified two axes, connectivity (or inversely, dispersal) and environmental heterogeneity, as characteristics of these different models.

Connectivity and dispersal play an important role in distinguishing patch dynamic and species sorting from source-sink models when dispersal is either much less than local population turnover rates or at least comparable to this turnover rate. Only when immigrating rates are comparable with turnover rates will sink populations be regular features of local communities. Similarly, dispersal plays an important role in distinguishing patch dynamics from spe-

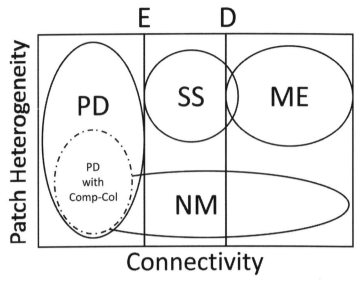

Figure 8.2 The four major archetypes of metacommunity ecology in relation to the amount of interpatch habitat heterogeneity and connectivity. Patch heterogeneity refers to the degree of interpatch differences that differentially favor different species in different patches. Connectivity can be quantified as the rate of immigration (events/time) and can be compared to two other rates, the local extinction rate (E) and the local death rate (D). PD = patch dynamic, SS = species sorting, ME = mass effects, NM = neutral model. A separate oval has been drawn for PD models that focus on competition-colonization (Comp-Col) models in the absence of patch heterogeneity to highlight the fact that most PD models are of this type. The oval for SS has purposefully been drawn so as to extend to the right of connectivity equal to D and overlap with ME to highlight that high dispersal can still produce patterns and processes similar to SS if dispersal is adaptive. See Table 8.2 for a description of the assumptions and critical processes that apply to each archetype.

cies sorting and mass effects when colonization due to dispersal is low compared to local extinction rates. This is because colonization rates that greatly exceed local extinction rates likely prevent the possibility of fugitive species and increase the rate of community assembly (i.e., succession; Pickett et al. Chapter 9) to end-state configurations (e.g., climax communities) such as those predicted under species sorting.

Environmental heterogeneity plays an important role in distinguishing neutral models and most patch dynamics models from species sorting and source-sink models. It is important to understand that in this context environmental heterogeneity is the heterogeneity that differentially affects the fitness of different species. Environmental heterogeneities may quantitatively affect species in the metacommunity by affecting total abundance of individuals or turnover (regardless of species) but would not lead to species sorting or source-sink effects if such effects are symmetrical for all the component species.

A third conceptual dimension that is not shown in Fig. 8.2 but is also important to discuss distinguishes models that are explicitly spatially structured from those in which spatial effects are implicit. To date, much of the attention given to metacommunity models treats space implicitly. They most often do this by hypothesizing that each local community contributes to a single pool of dispersers that then disperses itself back into the local communities with equal probability. A more realistic view would model dispersal as being more likely among local communities that are more highly connected. A substantial body of work in spatial ecology illustrates these effects (e.g., Gouhier et al. 2010). In the case of models without environmental heterogeneity among patches, some general results can be obtained that are more precise than those obtained by the implicit models, but this is more difficult to do with environmentally heterogeneous landscapes because the precise arrangement of sites relative to each other can matter greatly. In many cases, however, there is a general correspondence, at least in qualitative terms, between the implicit and explicit model predictions. One exception is effects that depend on spatial patterning per se that arise in explicit models, but that simply cannot be found in implicit ones. An example of this is a simple model of nontransitive competition among three species (i.e., the rock-scissors-paper game) in a spatial setting (Kerr et al. 2002), where long-term coexistence of the three species depends on patterned travelling waves of each of the species sequentially replacing each other in local communities. In an implicit model, these traveling waves do not exist and instead the overall frequencies of species in the metacommunity oscillate in heteroclinic cycles. The amplitude of these cycles is not stable so that eventually only one of the species is present in the metacommunity (Law and Leibold 2005).

Clearly there is a great need to better understand the ways these various archetypal models relate to each other and to try to synthesize them in various ways. Some progress has been made along these lines, but much more is needed. Perhaps the best synthesis has been at the interface of species sorting and source-sink models (Amarasekare and Nisbet 2001; Mouquet and Loreau 2002; Mouquet and Loreau 2003). I shall now use this interface to illustrate how theory, models, and experiments have led to improved understanding of diversity patterns in metacommunities.

Illustrating links between theory, models, and experiments

Metacommunity theory has elucidated a number of phenomena in community ecology (Holyoak et al. 2005). To illustrate how it can do so, I focus on the synthesis of the species sorting and mass effects models to explain how diversity varies with dispersal (or inversely, connectedness) developed by Mouquet and Loreau (2002; 2003). This work combined analytical and numerical methods to explore how metacommunity structure varied as a function of dispersal among local communities. A few simplifying assumptions were made including: (1) all species compete on a one-for-one basis, (2) all species have identical dispersal rates and all patches are similarly connected to the dispersal pool, (3) each species has one patch type in which it is the best competitor (there is no niche partitioning within patches), (4) initially, each species was the sole occupant of the patch in which it was dominant, and (5) there is no environmental change within patches and thus no extinctions due to processes other than competition.

This synthetic model showed that species sorting and source-sink models are closely related and can explain how different levels of diversity vary with dispersal (Fig. 8.3). In the absence of dispersal, local diversity is low (one species per patch) and regional diversity is high (as many species as there are patch types). At low dispersal, this is still true because immigrants remain rare and ephemeral in local patches. Local diversity increases when per capita dispersal rates are comparable to per capita turnover rates. This is because there are increasingly large sink populations maintained in local patches due to immigration from other patches (sources) by species that would otherwise be excluded by local dominants. However, as dispersal increases even more, sink populations become proportionally larger and affect the fitness of local dominants. This is exacerbated by the fact that local dominants also disperse more, and thus emigrate rather than contribute to sustenance of local populations. Ultimately, if dispersal is sufficiently high, the organisms involved respond to an average of environmental conditions across all patches so that individual patch effects are no longer important. Thus, the consequences of fitness differences across patches are increasingly homogenized by dispersal.

In the end, the metacommunity is essentially equivalent to a single community because individuals show no bias towards being present in local patches where they are superior. This scale transition reduces regional diversity (and consequently local diversity) to a single species, but regional diversity (sometimes called γ-diversity) declines over a large range of dispersal values. The model thus predicts that regional diversity is not affected by low levels of dispersal, but that it begins to decline at some medium to high level of disper-

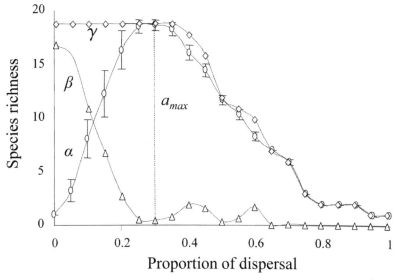

Figure 8.3 The hypothesized interaction between dispersal rate and species diversity at different spatial scales (Mouquet and Loreau 2003). Local diversity is the same as α-diversity, regional diversity is the same as γ-diversity, and compositional turnover is the same as β-diversity. The dispersal rate is conceptually identical with "connectivity" as used in Fig. 8.2. The point where β-diversity begins to decline and where local diversity begins to increase in the simulations that generated this graph correspond to a dispersal rate that is roughly equal to the per capita turnover rate of individuals in the populations.

sal. Local diversity (also called α-diversity) increases with dispersal up to the point where regional diversity begins to decline, whereupon it also declines with regional diversity. The average dissimilarity in composition among communities (also called β-diversity) starts out high and decreases in opposition to regional diversity (Fig. 8.3). In this model, dispersal cannot contribute to local community assembly (proposition 6) because local community assembly is assumed to have reached an assembly end point at the beginning of the simulations. Consequently, this model can explore the consequences only of propositions 7 and 8.

Several experiments have tested the predictions of this model on patterns of diversity (reviewed by Cadotte 2006a). The results of one of these experiments conducted with protists in microcosms are shown in Fig. 8.4 (Cadotte 2006b). Cadotte created a metacommunity consisting of small microcosms that were interconnected by tubing. Valves in the tubing allowed the manipulation of dispersal by being open for different amounts of time. The microcosms were

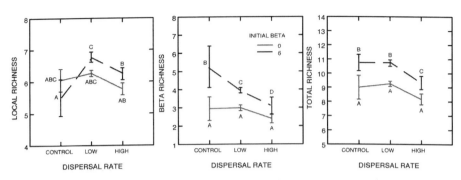

Figure 8.4 Experimental results of manipulating dispersal in protist microcosm (Cadotte 2006b). Dispersal was manipulated by connecting microcosms with tubing and was either absent (control), low (clamps on the tubing were used to control dispersal at 1 hour every two days), or high (tubing was never closed). Initial conditions either had all subpopulations started with identical communities (Initial Beta = 0) or they differed by restricting species inoculations to different microcosms with an initial Beta = 6. Multiway ANOVA indicated that means labeled with similar letters are not statistically different ($\alpha = 0.05$).

inoculated with a diverse set of protists either in a uniform fashion (all microcosms initially had all species) or variably (each microcosm differed in initial composition). Dispersal was either absent, low (presumably lower than or comparable to the population turnover rate), or high. Results show qualitative correspondence with the predictions of Mouquet and Loreau (2002) even though it is likely that organisms varied in dispersal rate and there was not likely to be strong habitat specialization.

Metacommunity theory can make predictions about numerous other features of community variability including species coexistence and similarity (e.g., Leibold 1998; Hubbell 2001), abundance distributions (e.g., Hubbell 2001; McGill et al. 2006), and species richness (e.g., Hubbell 2001; Chase and Leibold 2003). In many cases these predictions differ substantially from those predicted by models of closed communities. They thus provide a rich array of novel ways to evaluate data and experimentally study ecological processes in communities.

Spatial vs. environmental regulation of community structure in metacommunities

One of the more important ways that metacommunity theory has informed community ecology is by suggesting that species composition could variously depend on purely spatial effects as well as the more conventionally studied environmental effects (Cottenie and De Meester 2005). The four archetypal models predict that the relative magnitude of these two types of effects should vary among them. Species sorting predicts that only environmental variation among patches should determine variation in community composition, whereas the neutral models predict that there should be no such environmental regulation and that spatial effects should predominate. Mass effects predict that both environmental and spatial effects should be present. Patch dynamics models that ignore environmental variation of course predict only spatial effects but as pointed out earlier, it is possible to model patch dynamics in an environmentally heterogeneous set of patches and under conditions such that both spatial and environmental effects should be present.

A large number of studies have been conducted to evaluate the relative importance of spatial and environmental effects. These studies should be interpreted with some caution because results depend a lot on which patches are chosen for inclusion in the study, which species are chosen, which environmental factors are measured, and how spatial effects are quantified; they can nevertheless give insights into how we evaluate community regulation. Cottenie (2005) conducted a meta-analysis of 158 such studies and, inspired by metacommunity theory, there have since been quite a few more. His findings indicate that environmental effects tend to be more important than spatial ones in the majority of cases. However, he found that spatial effects were also common. The conclusion indicates that natural metacommunities vary in how they are regulated, but do not help identify just how they do so. Current work is seeking to find more refined analyses that might improve on these simple methods.

Linking metacommunity theory to other ecological theory

At a very general level, it is obvious that the links with the definition of theory in Scheiner and Willig (Chapter 1) are extremely tight. However, ecology can be viewed a bit more broadly as the interactions between organisms and their environment. Under this perspective, the links are weaker but still strong because they focus to any of the propositions that link organism-environment interactions to species distribution in space and time. At a more detailed level, there are particularly strong links with some of those described in this book

Niches (Chapter 5) and predator-prey interactions (Chapter 7)

Niche theory is concerned with describing how organisms maintain populations and how in doing so they alter the environments they live in. A critical way they do this is by interacting with their prey and their predators (Elton 1927). By and large, our current understanding of niche theory is primarily applicable at the local scale (e.g., Hutchinson 1959; Tilman 1982; Chase and Leibold 2003), and dispersal is much less often considered as part of the niche of an organism. This is not quite true since dispersal has often been considered important in niche relations in successional processes (Watt 1947; Pickett et al. Chapter 9) and has more recently been invoked in models of niche relations involving spatial storage effects (e.g., Chesson et al. 2005). However, as I have described it in this chapter, metacommunity theory consists of exploring how niche theory applied at local scales interacts with dispersal in larger spatial settings.

Population dynamics (Chapter 6)

Here the links are clearest, as connections between metapopulation theory and metacommunity theory involving patch dynamics and their domains and propositions are very similar. Indeed, such metacommunity models of patch dynamics have been extensions of metapopulation models. Metacommunity approaches may enrich metapopulation theory by focusing on the role of patch type heterogeneity. When the focus is on single species, heterogeneity in patch quality is likely to be less important than when multiple species are involved.

Succession (Chapter 9)

As was initially argued by Watt (1947), the successional process implicitly involved spatiotemporal processes that depend on dispersal among localities with different local conditions. While the propositions of successional theory are strongly linked to metacommunity theory, the domains differ somewhat in the degree of attention they give to temporal vs. spatial dynamics.

Island biogeography (Chapter 10)

A common interest in dispersal is a strong linkage with metacommunity theory, but the propositions are quite different. Island biogeography does not necessarily focus on local communities per se since it is more concerned with islands that may (and probably generally do) include multiple habitats and patches with distinct sets of locally interacting species. Thus, many of the complications and contingencies that are key propositions to at least some metacommunity models are ignored. Further, the role of dispersal in island biogeography is primarily the same as in models of patch dynamics but different than that in mass effects and species sorting models, where stochastic local extinctions are absent.

Global change (Chapter 12)

Peters et al. (Chapter 12) describe the role of dispersal and community change in global ecology. The importance is that responses of biotas (including physiological and genetic responses, as well as community responses) and their consequences for ecosystem processes are not locally constrained. Thus, while the domain is different in emphasis on anthropogenic effects at the global level and on the consequences for ecosystem processes, the propositions are remarkably similar.

Ecological gradients (Chapter 13)

Obviously, metacommunity models of species sorting and mass effects as well as patch dynamics models that include patch heterogeneity imply the existence of environmental gradients as expounded by Fox et al. (Chapter 13). The connection is particularly strong with gradient models that include species interactions. However, the domains of the two theories are a bit different. Constitutive theories of ecological gradients as described by Scheiner and Willig (2005) and Fox et al. (Chapter 13) focus exclusively on correlations between species richness (only one aspect of community composition) and environment and do not address sources of variation in communities that are due to spatial effects. Thus the domain of metacommunity theory is a bit larger than that described in Chapter 13. Further, dispersal plays no key role in the constitutive theory of gradients described in Chapter 13 although as argued earlier it is probably important in fueling community assembly and species sorting.

Biogeographic gradient theory (Chapter 14)

Although metacommunity and biogeography theory seem to share many aspects and propositions, including dispersal, environmental contingency, and spatial effects, there is a fundamental difference between the domains of the two. Metacommunity theory as formulated in this chapter focuses on purely ecological effects and primarily examines their consequences in ahistorical settings. The spatial extent that such a focus implies ignores the phylogenetic and historical biogeographic contingencies that characterize biogeographic gradient theory. An implied assumption is that all organisms have had equal likelihood of access to all sites within the metacommunity; this constrains the spatial extent of metacommunity theory to being at sufficiently small scales where this is reasonable. In contrast, biogeographic gradient theory is primarily concerned with larger spatial scales, where phylogeny and historical effects are important. An interesting possibility arises if in an area one group of organisms has a distribution that is regulated by historical processes interacts with another group that has distributions regulated by ahistorical metacommunity processes. Just such a case is suggested by recent analyses of zooplankton distributions in the northwestern United States in which calanoid copepods have distributions that seem to reveal strong historical biogeographic effects and little environmental regulation whereas coexisting daphnid cladocerans show just the opposite pattern (Leibold et al. 2010).

A prospectus for metacommunity theory

The possible synthesis of metacommunity theory suggested by Fig. 8.2 also points out that the biggest lacuna in our understanding of metacommunity theory is understanding what happens when different species differ strongly in dispersal rate. Several models and hypothesis that build on metacommunity theory show that variability in dispersal can be important in various ways.

1. Colonization-competition relations. These have been hypothesized since the verbal models of (Diamond 1975a) and the mathematical models of Hastings (1980) based on colonization-extinction dynamics. More recent work has also suggested that such a tradeoff might allow for coexistence in models based on birth-death-dispersal dynamics (Amarasekare and Nisbet 2001; Mouquet and Loreau 2002), but the mechanisms are totally different. Recent work shows that both of these types of effects can allow for coexistence in a metacommunity but that the conditions for doing so can be somewhat constrained (Yu and Wilson 2001) because they involve careful titration of the relative rates of competitive exclusions and dispersal.

2. Heterogeneous dispersal of different trophic levels or groups of interacting organisms. It has been suggested that an important stabilizing feature of food webs emerges from differences in dispersal rates of different trophic levels (McCann et al. 2005). This issue is complicated because different effects occur depending on whether dispersal is random or biased by adaptive or maladaptive behavior (Amarasekare 2008).

These findings illustrate a complex array of possibilities but it is difficult to yet see how they act in concert or how important these possibilities are in real communities. Future work on these questions will no doubt elucidate how dispersal interacts with other fundamental aspects of ecology but it is still hard to know just what this will involve.

Metacommunity theory has many elements and parallels with evolutionary dynamics in metapopulations. Evolution in metapopulations involves the role of dispersal in maintaining genetic variation on which selection can act and in homogenizing such variation in ways that roughly correspond to the roles of dispersal in metacommunities (Urban et al. 2008). A suite of novel interactive dynamics between such adaptive dynamics and metacommunity processes emerge (Holt et al. 2005; Loeuille and Leibold 2008; Urban et al. 2008). Two lines of evidence support the likelihood that there can be a strong interaction between evolutionary and ecological dynamics along these lines. The first is a growing body of work under the rubric of "community genetics" (Whitham et al. 2003) that suggests that genotypic variation within some species can alter community properties associated with that species in ecosystems. The other is another growing body of work under the rubric of "geographic mosaic evolution" (Benkman et al. 2001; Thompson 2005) that suggests that community context can strongly alter evolutionary and coevolutionary responses of species in communities over space. A few studies that explicitly explore the interface of metacommunity and evolutionary dynamics (De Meester et al. 2007) indicate that there can be important interactions between them. If this is substantiated by further work, additional propositions about the role of adaptive evolution on metacommunities will need to be explored.

Finally, a challenge for metacommunity theory emerges from its focus on intermediate spatial scales. Expanding from the local to metacommunity scale adds a number of important ideas about how purely ecological processes can explain variation in the distribution of organisms among communities. However, such purely ecological explanations will apply only up to a certain spatial scale; eventually, other processes involving biogeographical and phylogenetic dynamics will play a role and at some point overwhelm any explanations that derive from metacommunity theory. More intriguing is the likelihood that different groups of interacting organisms in metacommunities might differ in the spatial scale where this happens.

Metacommunity theory is relatively new in ecology. In this chapter I have tried to identify the elements of a constitutive theory of this topic, but there is to date no comprehensive model that integrates the main propositions. The existing work only hints at the potential for a deeper understanding of mesoscale ecology that one might expect as this theory continues to develop.

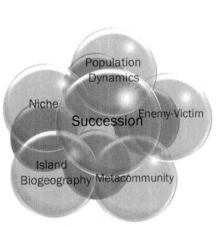

CHAPTER 9

Domain and Propositions of Succession Theory

Steward T. A. Pickett, Scott J. Meiners, and Mary L. Cadenasso

Succession is perhaps the oldest of ecological concepts, having arisen when ecology was emerging as a self-conscious discipline (McIntosh 1985). Yet it continues to address many fundamental issues in ecology, to support important applications, and to synthesize the insights and perspectives of other theories. Thus, it fulfills two functions key in assessing the utility of a contemporary ecological theory. First, it exhibits the attributes of a mature, well-developed, and intensively tested theory (Glenn-Lewin et al. 1992; Pickett and Cadenasso 2005). Second, it provides a linkage among theories and applications that have usually been considered separately (Walker et al. 2007). For example, the theory of succession or community dynamics has been applied to terrestrial and aquatic habitats (Bazzaz 1979; Stevenson et al. 1991; Biox et al. 2004), and for communities of microbes (Boucher et al. 2005), birds (Keller et al. 2003), soil invertebrates (Yi et al. 2006), and mammals (Schweiger et al. 2000).

This chapter outlines the structure of contemporary succession theory, beginning with a statement of its domain. It proceeds through an elaboration of the individual propositions of the theory. The term "proposition" is a general one that encompasses both the conceptual and the hybrid conceptual-empirical components of theory. Based on a broad view of theory (Pickett et al. 2007), propositions may be assumptions of domain, statements of concept, definitions, empirical generalizations, laws, and hypotheses or expectations derived from models. The definition of proposition used in this book is

185

Table 9.1 Contrasts between classical and contemporary theories of succession. The text parses this universe of contrast by focusing on principles of (1) domain, (2) fundamental actors, (3) environmental contrasts through succession, (4) trajectories that arise from the interactions among the actors and between actors and environment, and (5) methodology.

Attribute	Classical	Contemporary
Domain	Time > 1 yr	Any time scale
Definition	Assumptions included	Neutral definition
Actors	Community action and outcome	Individual action, community outcomes
	Plants and environmental reaction	Plants; animal vectors, consumers, and engineers; feedbacks
Causes	Facilitation predominant cause	Multiple causes: Disturbance, species availability, species performance
	Focus within community	Includes landscape and historical context
Community trajectories	Idealized; directional	Contingent; probabilistic
	Vegetation conforms to climatic ideal	Vegetation conforms to multiple, contingent factors
Theory structure	Verbal idealizations	Hierarchical causal repertoire
	Equilibrium, aspatial models	Individual based, spatial models
Methodology	Chronosequence	Long-term plots; simulation models
Theoretical context	Progressive evolution; organismal analogy with endpoint focus	Contingent evolution; process focus

narrower (Scheiner and Willig Chapter 1), so we will address domain as a background assumption. The propositions we present are simply the broad principles that embody the key aspects of knowledge about community dynamics. The propositions can be divided among those that (1) identify the basic actors in succession; (2) enumerate the higher-level environmental and contextual causes; and (3) describe the nature of successional trajectories and the outcomes of detailed causes. The propositions address issues that ecologists have struggled with and refined over decades. However, the contemporary content and assumptions embodied in the propositions differ substantially from those of earlier eras in ecology (Johnson 1979). Presenting propositions of a theory separately is an artificial device for rhetorical purposes. In fact, the propositions of a theory are closely linked to each other, and the meaning of each proposition is fully realized only when it is related to the other components of the theory (Pickett et al. 2007). To illustrate how this theory has developed, we also identify key differences between the propositions of classical succession theory as often presented in textbooks (Pickett and Cadenasso 2005; Eliot 2007) and contemporary versions of community dynamics theory (Table 9.1).

An inclusive domain

Bounding a discipline or a establishing a domain is the first job of a constituent theory. The domain statement of succession theory is a neutral definition of the concept. Neutral definitions are intended to identify the core meaning of a concept, minimizing assumptions about outcome or dominant mechanism (Jax et al. 1998). Following upon this definition of domain, it will become clear that succession is an extraordinarily broad concept, which has great synthetic power for ecology. Although most examples will come from plant communities, with which we are most familiar, it is clear that the propositions, when couched in organism-independent terms, apply to any aggregation of species (Morin 1999).

Domain

Succession is the change in structure or composition of a group of organisms of different species at a site through time. Succession or community dynamics is preeminently about the behavior of collections of species through time and can involve changes in species composition or the three-dimensional structure, that is the architecture or physiognomy, of species assemblages. Often succession has been defined only as change in species composition (Begon et al. 1996a) regardless of whether it is plants or animals that are the focus. However, the founding concern with change in dominance of plant growth form makes it clear that the architecture of a species assemblage can be of equal importance for succession (Clements 1916). Indeed, changes in plant species composition in well-watered habitats often result in dramatic architectural changes as the community shifts from herbaceous, to shrub, to tree dominance. Even in cases where one species is dominant, growth in stature, clonal growth, or canopy coalescence as individual woody plants mature are important successional changes (Muller 1952). Architectural complexity can characterize intertidal community succession after disturbance (Dayton 1975). Architectural changes such as these have implications for resource availability, niche construction for consumer species, and susceptibility to subsequent disturbances (e.g., Pickett et al. 2001a).

The neutral definition of succession or community dynamics used here focuses on the kernel of the process and does not require the assumptions that are often attached to the definition when describing particular temporal patterns. The domain of succession theory based on the neutral definition makes no judgment about progress, directionality, temporal scale, end point, or whether the species form a tightly integrated community. This approach includes both compositional and architectural changes in communities in the broadest, most generalizable way possible. This means that decisions about whether succession is progressive and directional, or takes place over temporal scales from 1 to 1000 years are the charge of specific models and not the general theory. Indeed, as we discuss later, the general mechanisms of community change are independent of temporal scale.

We suggest that the persistent controversies about directionality and the existence of an identifiable endpoint that have plagued the definition of succession be left as open issues to be dealt with by the explicit assumptions embodied in individual models. It is not succession in the broad sense that is problematic, but rather the contradictory assumptions of different models and applications that generate problematic controversy. Distinguishing between the core, neutral concepts and the specific models that translate them to real or simulated situations prevents the repeated discovery that classical succession theory does not work (McIntosh 1980) and hence should be discarded. In contrast, differentiation between the core concept, which is neutral and broadly applicable, and the more narrow specific models (Jax et al. 1998) permits appropriate application to an impressive range of real-world situations while allowing general similarities of process to be recognized (Pickett and Cadenasso 2005).

Propositions about the actors in the process

The first set of propositions identifies the nature of the fundamental units and the kinds of interactions between units through succession (Table 9.2). These propositions identify an inherently hierarchical arrangement, with the fundamental units or actors interacting on one level and the results emerging on a higher level of organization. Community dynamics theory must account for processes, interactions, and constraints that exist in linked hierarchical levels: organismal adaptation, physiology, and plasticity; population structure and density; community composition and architecture; landscape structure and fluxes; and ecosystem feedbacks and processes.

Table 9.2 The domain and propositions of the theory of succession.

Domain

> The change in structure or composition of a group of organisms of different species at a site through time.

Propositions

1. Succession is driven by the interactions of organisms with each other and the physical environment.

2. Successional patterns in communities result from the interactions of individuals.

3. Multiple trophic levels participate in the driving interactions.

4. Succession results from processes of disturbance, differential availability of species to a site, and differential performance of species within a site.

5. Successional causes can operate on any timescale.

6. The possible outcomes of interaction between individuals are tolerance (no effective interaction), inhibition, or facilitation.

7. The species composition of a site tends to equilibrate with the environment of that site.

8. The specific form of a successional trajectory is contingent on starting conditions, and the stochasticity of invasion and controls on species interactions.

9. Succession produces temporal gradients of the physical environment, biotic communities, and the interaction of the two.

Proposition 1: An individualistic process

The interaction of individuals manifests as the changes in species composition that are observed in many successions. However, architectural changes not accompanied by change in composition also reflect interaction of individuals. An example is a case of vegetation change where the stature and density of a monospecific stand changes through time. As the individuals in a collection grow, they occupy more space, until some are overtopped or otherwise disadvantaged in their interactions within the community. This is seen in the stand thinning that characterizes dense, young forests. During stand thinning, the height of the upper layer of the canopy, the depth of the canopy volume, the spread of surviving crowns, and the presence of downed debris and depth of soil organic matter may all increase. This is a structural or architectural succession with a constant composition. Bird and insect diversity, productivity, and decomposer communities are aspects of the ecosystem that may change along with the architectural changes occurring in the plant community (Odum 1969). Indeed, many of the expectations proposed by Odum (1969) can be explained by the growth of individuals, the accumulation of species with longer life spans, and the competition for limiting resources (Loreau 1998).

When succession was first conceived, much of the attention of ecologists was on the coarse scales of pattern in vegetation (McIntosh 1985). Ecology was still close to its roots in biogeography and was attempting to apply the familiar knowledge of plant physiology to explain those large global and continental patterns (Kingsland 2005). In light of this, inventing the concept of succession was a clear advance, because it exposed the dynamism that existed in vegetation at smaller scales. However, the predominance of the coarse scale focus in vegetation description led ecologists originally to articulate the causes of succession at the same coarse spatial scales upon which the changes were observed. Hence, they focused on communities. Changes in a community were presumed to be due to features of the community itself and the goal-seeking tendencies of those communities (McIntosh 1980). Hierarchy theory has since clarified the error of this approach as well as presented an alternative (Ahl and Allen 1996). Successional explanations have been vastly improved by using a hierarchical strategy of explanation (Pickett et al. 1987a), in which interactions exist at the level of individual organisms, and their direct and indirect effects emerge as a higher level of community or ecosystem patterns. There can also be downward influences from larger-scale ecosystem and landscape conditions to the community, and from community to population and individual during succession.

Although the earliest efforts to clarify the individual-based nature of vegetation dynamics (Gleason 1917) met with resistance, subsequent fine-scale, long-term, and experimental research confirmed that interactions among individual organisms as parts of populations were the core process in succession (Horn 1974; Miles 1979; Parker 2004). This is called the individualistic approach to succession. Both Clements (1916) and Gleason (1917), according to a rigorous philosophical analysis, based their explanations of successional causes on the adaptive physiology of individual organisms (Eliot 2007). Indeed, they both recognized migration and establishment and environmental sorting as key processes, as will be discussed later. According to Eliot (2007, p. 104), the difference is subtle: Gleason and Clements "differ in emphasis, in their prioritization of causal factors. While Clements emphasizes *environmental* sorting among potential individual immigrants, Gleason emphasizes environmental sorting among potential *immigrants.*" Eliot (2007) is at pains to indicate that although both Gleason and Clements recognize the interaction of organisms and environment in their successional theories, Gleason places priority on the role of immigration of organisms whereas Clements prioritizes the filtering role of the environment on the suite of organisms present. Furthermore, the two theoreticians take different scaling approaches, with Gleason emphasizing local variation, while Clements structures his arguments from adaptation to large scale variation. Eliot's (2007) paper is an important advance in recognizing the similarities and differences between the approaches of these two pioneers, which are usually represented as combative caricatures in contemporary literature. The process-based approach that is presented in the subsequent propositions combines key elements from both the Clementsian and Gleasonian traditions (Pickett and Cadenasso 2005).

In considering the individualistic basis of succession it is important to recognize that it is the interactions among individuals, not merely the existence of individuals as separate entities, that are of concern. Indeed, individuals in communities cannot be treated as independent atoms, because their density and frequency are important aspects of how they interact (e.g., Morin 1999). Likewise, the age or stage structure of the populations can be important in these interactions. The effects of one population on another can change with density and frequency, and the potential for indirect effects through third parties is also a part of the interaction environment of individuals.

Proposition 2: From individuals to communities

This proposition draws the implications of the individual focus of succession for a different realm of ecological aggregation, that of the community. The implication of the individualistic proposition 1, with its critique of the strictly community-level explanation of succession, is that community changes must result from the individualistic behavior of organisms. A point of confusion is sometimes still expressed concerning the individualistic nature of species central to the proposition. Individuality does not mean that species are necessarily acting randomly or independently, although the neutral model raises that possibility (Hubbell 2001). In fact, it is the variation in interaction among species, their direct and indirect influences on one another and on their physical and chemical environment, that embody the most general mechanism of succession (Glenn-Lewin 1980; Myster and Pickett 1988; Huston 1992). Indeed, the concept of the ecosystem was invented to acknowledge the interaction of the biotic complex and the physical complex in the context of succession (Tansley 1935). Contemporary ecology also recognizes that levels of organization larger than communities, such as metacommunities, landscapes, and patch mosaics, can also contribute to processes of community dynamics (Sousa 1984; Hansson and Angelstam 1991; Veblen 1992). This point is emphasized in later propositions.

Proposition 3: Plants, animals, and microbes

The concept of succession was invented by plant ecologists, and has been extensively used by ecosystem ecologists as well (Odum 1969). This sociological history seems to have led to an inappropriate neglect of higher trophic levels as active participants in many successions. Some early studies indicated that animal assemblages followed the compositional and structural changes wrought by the plant community, but they neglected the influences of animals in shaping succession (Smith 1928). Research in the last few decades has corrected this bias and oversight (Davidson 1993; Inouye et al. 1994; Bowers 1997; Meiners et al. 2000). The role of dispersers was perhaps the first to be recognized as important in directing succession. However, seed and seedling predators, herbivores including browsers and foliage consumers, and animals as ecosystem engineers are now widely accepted as potential agents of successional change. The potential for pathogens and mutualists to generate feedbacks affecting community dynamics is becoming more apparent (Dobson and Crawley 1994; Klironomos 2002; Bever 2003; Reynolds et al. 2003).

The incorporation of consumer trophic levels has, like the recognition of feedbacks with the physical environment, added significant dimensionality to the concept of succession. Direct and indirect effects via feedbacks with consumers, mutualists, and engineers mean that interactions are no longer conceptually restricted to the plants that so conspicuously structure many successions.

These two propositions reflect an individual-based or population approach, recognizing that multispecies assemblages of organisms of various trophic levels, including plants, animals, and microbes, generate any observed community changes (Johnson 1979). The organismal level of organization upon which succession is centered is itself a complex of interactions and feedbacks.

Propositions about the environment and context as causes

The propositions to this point set the stage for understanding how succession occurs by identifying the individuals as the core actors in a linked hierarchy of biological entities. The next two propositions indicate the nature of interactions among the actors and between actors and the abiotic environment that can develop during succession.

Proposition 4: Multiple causes

This proposition summarizes the core of community dynamics by identifying the three most general causes of the process. Disturbances are particular events that alter the physical structure of a community or cause mortality of structurally or functionally dominant organisms (White and Jentsch 2001). Such events generate opportunities for the release of suppressed residents, or the establishment of new individuals, and hence, new species. It is the behavior of specific events relative to the structure of a community, and to the requirements of species that are available to respond to it, that determines what actually disrupts a particular community and whether particular species are favored or disfavored. Disturbance as a concept is a useful generalization, but it is the characteristics and responses to particular events that contribute to the understanding of succession (Johnson and Miyanishi 2007). The general process of disturbance recognizes the particularities required through its hierarchical structure (Fig. 9.1). It has nested within it the resource base of the site, the identity of the agents of disturbance, and the intensity, size, and timing of the potentially disturbing events (Pickett et al. 1989). Differentials in disturbance influence succession by affecting contrasting site features, as well as features of the disturbances themselves. The presentation of disturbance in succession illustrates a process-based approach. It suggests a hierarchy of general causes that can be decomposed into more specific mechanisms, constraints, and enablers (Pickett et al. 1987a; Pickett and Cadenasso 2005). The general causes and their constituent mechanisms may interact.

The second process of community dynamics is differential availability of species at a site. This differential is driven by variation in species' inherent ca-

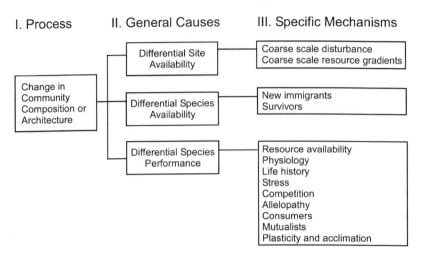

Figure 9.1 A hierarchy of successional causes, showing the most general causes on the left, and progressively more finely articulated mechanisms and constraints toward the right.

pacities for dispersal across space and, via dormancy, dispersal through time. Spatial dispersal is affected by proximity of sources of migrants and by landscape connectivity between sources and the study site (Meiners et al. 2003; Pickett and Cadenasso 2005; Hobbs and Cramer 2007). This is as true of animals as it is of plants. Landscape configuration not only determines the distance over which dispersal is required, but may also generate resistance to dispersal across boundaries (Cadenasso et al. 2003). In addition, the identity and behavior of both biotic and abiotic dispersal vectors must be recognized as a part of this process (Fig. 9.1).

The third process of community dynamics is differential species performance. This process aggregates the characteristics of individuals into those populations, and acknowledges that population features such as density, stages or ages, size classes, and frequency can influence the outcome of interactions. Differential performance encompasses all the processes of establishment, growth, survival, reproduction, and mortality that plants, animals, and microbes employ (Keever 1979; MacMahon 1980; Davidson 1993; Vitousek 2004). It includes interactions that are competitive, consumptive, and mutualistic, and the strategic, physiological, and life history foundations of these diverse interactive capacities. In addition, it includes adaptive and plastic potentials of species, and it recognizes the potential for stresses and fine-scale disturbances to interact with species performance. Variation in one or more of these processes among an assemblage can result in change in the composition or structure of the community. Any factor that alters the performance attributes of one or more species can be incorporated into a more specific model of a particular succession; we do not attempt to enumerate all the possible specific causes in this chapter. Although disturbance acts before the other two causes and sets the stage for the change within the system, either differential availability or differential performance of species is sufficient to generate successional change (Fig. 9.2).

This outline of the three causes of community dynamics is, on one hand, an

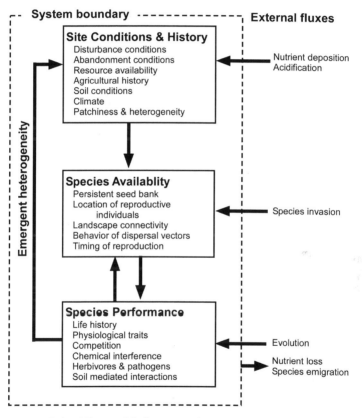

Figure 9.2 An idealized filter model of successional processes, with reference to components of the causal hierarchy in Fig. 9.1. The model focuses how individual organisms come to occupy a site, and how interactions affect community composition and structure. Although a focal system can supply species to other patches, or release nutrients, because the intention is explanation of causal interactions among organisms in a local environmental context, we do not emphasize the logically plausible imports or exports relating to the system.

empirical generalization extracted from explanations of plant, animal, and microbial successions accumulated over more than a century (Miles 1979; MacMahon 1981; Glenn-Lewin et al. 1992). On the other hand, it is a recasting of the general causes of succession that Clements (1916) and Gleason (1917) recognized within two decades of the scientific framing of the concept. The fact that the triad of successional causes satisfies both empirical and conceptual adequacy for explanation argues for its generality. We hypothesize that no specific mechanism of succession identified in the future will fall outside this causal framework, but will merely add detail to the hierarchical list of causes. We expect this to apply to macro- and microorganisms, and to aquatic and terrestrial environments.

This causal proposition also reflects the inadequacy of explanations based on only one kind of cause (Connell et al. 1987). Such one-factor explanations were encouraged by the simplistic, hypothetico-deductive philosophy of science in favor during the middle decades of succession study (Pickett et al. 2007). However, multicausal modeling and experimentation are now exposing the complexity of successional drivers (Wyckoff and Clark 2002; Clark 2007). The hierarchical framework, in which the general causes or processes of succession are disaggregated into more specific mechanisms and drivers to effectively construct working models, supports the multicausal approach (see Willig and Scheiner Chapter 15).

Proposition 5: Multiple timescales

Succession has traditionally been construed to act over timescales ranging from around 1 to 1000 years, while seasonal and paleoecological timescales have been excluded. However, we see no compelling reason for either exclusion. The causes of dynamics are common to communities on all these timescales (Brand and Parker 1995). Biogeographic patterns, postglacial vegetation shifts, and alterations of vegetation by climate change will all be explicable by differential disturbance, differential availability of species, and differential assorting of species as they respond to one another and the physical environment. For example, seasonal changes in communities may be the result of life cycle differentiation through the year of a suite of species already present, while many successions will reflect subsequent waves of immigration. Thus, although the specific mix of causes operating at these different scales can vary, the same conceptual framework can be employed for cause at any scale. This is one reason we have used a neutral definition of succession.

These two propositions identify the conceptually complete general causes of succession, and suggest that the same suite of causes operate for intraannual to paleoecological scales. One insight to emerge from the three-cause framework is that models of community or ecosystem structure may be partitioned along these three axes. Some models may emphasize disturbance, while others may emphasize the sorting processes among organisms such as competition. As mentioned earlier, Eliot's (2007) analysis of the Clements–Gleason divide shows just such a difference in emphasis. Ecology as a whole seems to employ a conceptual phase space defined by axes of disruption versus sorting. This is exemplified by the controversy (e.g., Gravel et al. 2006) over Hubbell's (2001) neutral models of succession in which temporal patterns are driven not by limiting similarity but by the vagaries of arrival at a site. Sorting, on the other hand, reflects differential species availability and differential species performance. Because succession theory incorporates both sorting and disruption, it is hospitable to models of assembly, lottery dispersal, response to disturbance intensity and frequency, and neutrality. See Chase (Chapter 5) for further discussion of the niche-based versus neutral models of community assembly.

Propositions about community trajectories

It is important to separate process- and outcome-based ideas of succession, as one necessarily generates the other. Furthermore, process-based approaches are motivated by explaining how a phenomenon occurs, whereas outcome-based approaches focus on end points. Process approaches may be especially applicable to nonequilibrium systems, whereas outcome-based approaches may be appropriate to equilibrium situations. In addition, outcome-based approaches must account for the contingencies of particular times and places. This section presents propositions about the outcomes of the causes and mechanisms of succession. These four propositions suggest the form and general content that must characterize specific models of succession to capture the richness of successional outcomes.

Proposition 6: Complex pathways

Succession was initially defined as a progressive and gradual change based on the facilitation of later species by earlier dominants. This expectation was not universally supported by succession studies, especially those encompassing the fine scale and the long term (Niering 1987; Pickett and McDonnell 1989). As a result, the roster of successional turnovers was extended to become logically exhaustive by including tolerance and inhibition as potential outcomes (Connell and Slatyer 1977). Although inhibitive interactions had been recognized for a long time, they were initially neglected in the classical successional theory (Egler 1954). Contemporary animal ecology incorporates such dynamics via the concept of priority effects (Drake et al. 1996). Likewise, the role of physiological tolerance or life cycle characteristics was recognized in some studies, but these neutral factors were not included as part of the conceptual apparatus of successional theory (Keever 1950). Incorporating the multiple causes described by Connell and Slatyer (1977) into successional theory was a crucial advance. Accepting neutrality was delayed by the difficulty of experimentally distinguishing it from niche-based interaction (see also Chase Chapter 5).

This advance required an additional insight to emerge fully as a tenet of successional theory. Facilitation, inhibition, or neutrality of species turnover are in fact net effects of particular interactions and mechanisms (Pickett et al. 1987a; 1987b). A given net effect can reflect a mixture of specific mechanisms, so it is important to discern the causes that operate in particular successional turnovers. Moreover, a given successional sequence will probably include all three net effects of facilitation, inhibition, and neutrality and tolerance (Hils and Vankat 1982; Armesto and Pickett 1986). It is naive to expect the successional dynamics of a system to operate via only one kind of net effect, although one may dominate. Different net effects may occur at different times, or at different locations within a dynamic community. However, they may well occur simultaneously. For example, while some pioneering annual plants may facilitate the invasion and survival of other species, biennial dominance in the second year of succession in abandoned fields is simply a reflection of their life cycles (Keever 1983).

Proposition 7: Organism-environment interaction

One of the fundamental principles of ecology is the tendency of biotic composition and structure to reflect adaptive matching with the environment (Chapin et al. 2002). This idea is exemplified by the characteristic faunal and floral composition of biomes as a reflection of climate (Clements and Shelford 1939), or of Holdridge's (1947) life zones as a response to temperature, rainfall, and elevation, or of C. Hart Merriam's (Merriam and Steineger 1890) elevationally delimited zones on mountains. This generalization emerged from biogeography and was well accepted by the time ecology took root as a distinct discipline.

The idea was translated into ecology as the idealized concept of climatic climax and its local variants. The justification for this concept was seen as problematical, and after long debate, was rejected by ecologists (Botkin and

Sobel 1975; Johnson 1979). However, the coarse scale tendency of species lists and vegetation structures to sort at continental and regional scales is clear (Clark and McLachlan 2003). As an ideal, it is thus still possible to posit that vegetation tends to equilibrate with the prevailing environment. At its most demanding, this proposition takes the basic assumption about plant-environment relationships in plant ecology, and applies it to a temporal context. At its most liberal, the proposition suggests that the contingent pathways observed in real, messy successions (Johnson and Miyanishi 2008) will still reflect the environmental limits that prevail over the course of time in a given location. It can be considered one of the boundary laws of community dynamics: no community will exist outside the environmental limits of its component species and ecological interactions. Thus, proposition 7 on one level reflects a fundamental assumption in the paradigm of organism-environment relations, while on another level it motivates experimental or model tests to evaluate the impact of factors and events that can deflect the composition or architecture of a community from the ideal expected in a given environment. Relic communities are a case in point that seems to violate the proposition, but they do so only because of the lag in compositional change resulting from their dominance by long-lived organisms established under a different prevailing environment. The next proposition examines the possibility of narrowing the ideal still further, excluding disturbance, heterogeneity, or other factors that can in reality determine vegetation composition and structure. We move to a proposition that focuses on the sorting component of vegetation dynamics.

Proposition 8: Successional contingency

The contingencies that can operate in succession are many. Some of these are set by the conditions that exist at the start of the succession, say a wet versus a dry year, or the identity of residual individuals that happen to be spared by a specific disturbance event. Others are the result of events and conditions that may appear stochastically at various times through succession, such as fine-scale physical disturbance events or outbreaks of herbivores. Contingencies may exist in (1) the extent of environmental contrasts or resource gradients that can exist at a site through time, (2) the adaptive repertoire, including plasticity, contained within the species seeds, eggs, or clonal fragments present, (3) the type, size, and timing of disturbances, (4) climatic shifts in the resource base, (5) the order in which species colonize, and (6) the landscape context.

This proposition accepts that succession is contingent on a large suite of interacting drivers, generating the inherent variability of successional systems. Any tendency for the composition or structure of an assemblage to equilibrate with an environment at a particular scale (proposition 7) is constrained by the strength of disequilibrating forces occurring at that scale (Huston 1994). Thus, the utility of models of an environmentally determined, equilibrated species assemblage is a particular research choice and not a universal feature of all successions. Whether the ideal environmental equilibrial community can be used as an end point of sorting depends on whether the disrupting forces are sufficiently weak relative to the prevailing environmental filters over the examined time period. Environments in which the conditions change during the time environmental sorting takes place will probably not generate a stable attractor or endpoint. If the disturbance regime, landscape context, species pool, or biophysical template change, successional trajectories may also differ from one period to the next. Such variety can, of course, lead to the potential for multiple stable states (Holling 1992).

Some models recognize that plant assemblages may respond more to the patterns of disturbance and species availability than to uninterrupted species interaction. In such nonequilibrium situations, succession will be dependent on landscape context, and the variation between sites that may reflect different successional states or histories. This means that succession cannot be understood by focusing only on a single site or set of local conditions (Pickett 1976; DeGraaf and Miller 1997; Ejrnaes et al. 2003). The dynamics and structure of adjacent sites may influence a given succession by serving as a source of potential colonists or new species with low resource tolerances.

One contingency that may determine successional trajectories is the length of the resource gradient that can exist over time. Different habitats can show long or short temporal gradients of compositional change based on the resource levels and stresses that dominate the region. The length of the resource gradient acquires further biological significance based on the maximum level of resources that can be accumulated. Species assembly in low-resource sites may exhibit a continued species accumulation with time as resources accumulate, while in high-resource sites, species richness may peak at intermediate periods of succession (Auclair and Goff 1971).

The order of species colonization is important to the details of successional trajectories because it may determine the order of facilitative, inhibitory, or neutral interactions. Priority effects are especially obvious in successions on sites having low resources (Lockwood 1997). In such sites, growth of all immigrants is constrained. Hence, maturation and senescence will be slow, providing fewer opportunities for the replacement of dominants in a given unit of time. Furthermore, with slow growth, the rate of competitive exclusion will be slow.

Proposition 9: Successional gradients

Resources and regulators change with succession. Succession is a feedback between a species pool and a complex biotic-abiotic environmental gradient (Pastor and Post 1986). Although the physical environment sets the initial resource base, physiological stress, and regulator conditions, these are subsequently modified by the species and their interactions in the succession. Therefore, succession generates a complex environmental gradient through time.

A long-term ecosystem successional gradient is illustrated by the trajectory of nutrient pools through time (Vitousek 2004). For nutrients, such as P, that are borne in the substrate in which soil development occurs, the pattern over the very long term is one of leaching and loss (Fig. 9.3). For nutrients, such as C and N, whose concentrations in an ecosystem depend on biological metabolism, the amount will at first increase and then decrease in the ecosystem. This means that over the very long term, nutrient limitation in ecosystems will shift from N to P. When subsets of these time frames are examined following discrete disturbances, shorter successional patterns can be discerned. The nu-

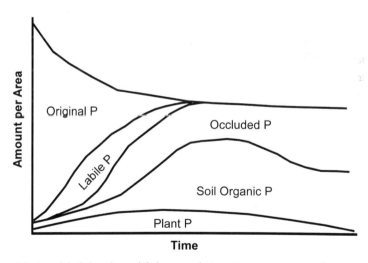

Figure 9.3 A model of phosphorus (P) dynamics during primary succession. After Vitousek (2004).

trient dynamics of these shorter successions will be determined by the relative degree of N and P limitation that has been generated by the longer succession and that are in effect for that time span. For example, long-term leaching of nutrients following glacier retreat at a regional scale may set fertility levels, while succession in individual abandoned agricultural fields in the area may generate local dynamics.

This proposition suggests that for many ecological purposes, considering resource gradients to be independent and fixed is appropriate, but for succession studies, these gradients should be considered a dynamic part of the system. The outcomes of species interaction can create environmental legacies affecting subsequent dynamics. The origin of the ecosystem concept was stimulated by the insight that succession involves the reciprocal interaction of the physical and the biological complexes at a site, following the logic of this proposition, rather than some ill-defined organismal tendencies (Tansley 1935).

Structure and output of succession theory

To this point, we have examined the individual components, or propositions, of succession theory. However, a theory is in fact an integrated conceptual and empirical structure (Pickett et al. 2007). In this section, we relate the various propositions and the roles of the theory components they represent.

We have already used one organizational component as a framework for successional theory—the general causal hierarchy (Fig. 9.1). This consisted of a statement of the domain, an identification of the general causes of succession, and subsets of more specific causes, that is, the constraints and mechanisms that drive succession in particular models or situations. The specific models, which in the minds of many are unfortunately the entirety of theory, are tools to put those causes into action. Models select the mechanisms or interactions that will be related in particular studies, set the spatial and temporal boundaries of interest, specify the inputs and results that will be addressed, and limit the kinds of outcomes. Laws are statements that specify the limits on interactions, describe the nature of trajectories, or indicate boundary conditions that cannot be violated. Models and laws are closely related, as they are both intended to describe aspects of the behavior of the system of interest (Pickett et al. 2007). However, models are often used to indicate how general laws apply to particular empirical or hypothetical situations. Furthermore, models tend to be more complex than laws, as they address many dimensions or features of a system, while laws often focus on a few simplified or general factors. We present a successional law and models as organizing and operationalizing tools of succession theory.

A law of succession

When present, an important organizational tool for general theory is its suite of laws. The law of succession parallels the laws of evolution in structure and use (Pickett and Cadenasso 2005; Pickett et al. 2007). The law of succession (Pickett and McDonnell 1989) has a universal conditional form, which indicates what the possible causes of succession are. The law of succession states that (1) if sites are differentially available, and (2a) species are differentially available at those sites, or (2b) species perform differentially at those sites, then the aggregate community structure or composition will change through time. A "zero force" form is made possible by casting the statements in terms of the absence of the phenomenon each identifies as resulting in no net change in the community. Differential availability must include the preexisting conditions at various sites (Fig. 9.2), which will interact with the various intensities, sizes, and frequencies of disturbance, to condition sites for successional change. It is a complex of conditions that constitute site availability, rather than a simple concept of a vacant site (Walker 1999). Differential species availability can include persistence through the initiating disturbance or migration from elsewhere. Hence, a patch dynamic or metacommunity perspective (Hastings Chapter 6; Leibold Chapter 8) is implied by the core processes of succession (Pickett 1976; Pickett and Rogers 1997). Finally, it is important to recognize that differential performance can be achieved based on evolved species traits, interactions at the producer trophic level, or by interaction with consumers in the form of diseases, predators, and herbivores. Differential performance therefore incorporates the potential for frequency-dependent, density-dependent, and stochastic constraints on limiting similarity, as well as limiting similarity itself.

Predictions, expectations, and hypotheses

The previously outlined propositions are primarily in the realm of principles that lay out the high-level, general structure of community change theory (Jax 2006). They focus on process and the modes by which successional outcomes occur. They provide the raw materials for the particular models from which expectations about succession emerge. Particular models will usually address some narrower domain than the entire theory of succession, or will focus on a subset of the kinds of interaction that are possible (Fig. 9.1). In other words, the specific models of succession are nested within the causal repertoire and address a certain subset of successional relationships. They operationalize the causal repertoire. A general template for such models is illustrated in Fig. 9.4.

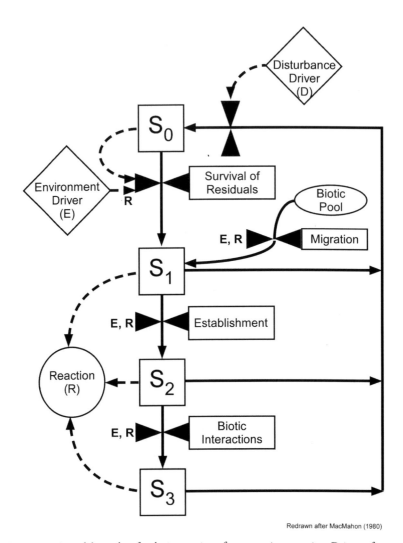

Redrawn after MacMahon (1980)

Figure 9.4. A model template for the interaction of processes in succession. Drivers of disturbance (D) and environment (E) are shown in diamonds, or shown next to control points (bowties) in the process, such as survival or residuals, migration, establishment of new individuals, and biotic interactions. S_n refers to states of the system through time. The effect of the state of the system on the environment is represented in the circle as reaction (R). Reaction becomes a component of the environment for subsequent states, shown by the proximity of R to control gates. Redrawn from MacMahon (1981).

The fact that specific models of succession will usually address subsets of the causal and contextual universe of succession means that classification of cases is required. We have already pointed to one kind of classification—emphasis on disruption of species interactions as the dominant driver of community composition versus emphasis on sorting among species as the dominant driver. This classification isolates the process of differential site availability from the remaining two differentials of successional process—species availability and species performance. One could further remove differential species availability by constructing models assuming equal access by all species or conducting experiments in which species were made uniformly available. Experimental studies have recently begun to examine the role of interactions between vertebrate consumers and dispersers with landscape structure as controls on differential species availability (e.g., Ostfeld et al. 1999; Cadenasso and Pickett 2000; Meiners and LoGiudice 2003). From such studies it is becoming clear that consumers are important and widespread causal agents in succession.

One of the most familiar classifications of succession is the dichotomy between primary and secondary successions (del Moral 1993; Walker 1999). The main difference between these two classes is in the resource availability of the sites representing each type. Primary successions are defined as those beginning on a new substrate that is often low in biologically mediated resources such as C and N, but perhaps relatively high in pools of substrate-based nutrients, such as P (Walker 1993). Primary successions can differ in the forms and identities of the nutrients available in the substrate, based on the origin of the rocks or sediment. For example, certain volcanic parent materials can yield fertile soils. Primary successions also are expected to largely lack resident propagules or survivors compared to secondary successions, in which clonal fragments, dormant seeds, or animal resting stages commonly survive the initiating disturbance (Myster 2008). This classification identifies resource pools, biological resource legacies, and biological propagule legacies as three key initial conditions for succession models. To adequately characterize different successions, it is mechanistically safer to identify the specific resource levels and species availabilities than to rely on the dichotomy of primary and secondary (Fig. 9.5). Some successions expected to behave as primary in fact possessed some attributes, such as survival of propagules or resource legacies, through the disturbance (del Moral 1993; Walker and del Moral 2003).

Succession models and models that are relevant to succession take on a vast array of forms. Lottery models in essence emphasize the role of site disruption and differential availability in contrast to emphasis on interactive species sorting (Chesson 1991). Most successional modeling focuses on the process of sorting, and therefore emphasizes differential species performance. For

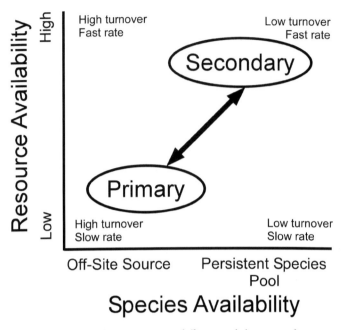

Figure 9.5 Primary and secondary succession as differentiated along axes of site resource availability and source of propagules. High resource levels permit rapid competitive exclusion and hence fast rates of succession, whereas low resource levels constrain the rate of competitive exclusion and thus the rate of succession. Dependence of succession on propagules from off site generates high turnover of species composition as new species become available, whereas persistence of propagule pools means that early and late successional species tend to be present throughout the process, yielding low turnover. Redrawn from Pickett and Cadenasso (2005).

example, Markov matrices assume stationarity in the transition probabilities from earlier to later communities (Usher 1979; 1992). Gap-based models are driven by the differential shade tolerance of species that participate in the succession (Horn et al. 1989). Individual-based models have become the predominant strategy for modeling succession, and some now incorporate neighborhood effects of dispersal or neighborhood effects on soil and aboveground resources (Canham et al. 2003).

Modeling succession hinges on whether the process or the end point is emphasized. Compositional or biogeochemical equilibria after long periods of sorting or resource partitioning between production, maintenance, and storage can be useful points of reference (Odum 1969). These states are potential end points of succession. In contrast are successional models that focus on situations that can be interrupted by disturbance before idealized sorting can be completed (Peet 1992), or on situations in which not all species are present equally (Hastings 1980; Petraitis et al. 1989; Whittaker and Bush 1993), or on situations in which consumers or stress factors intervene strongly to re-sort species (Olff et al. 1999).

There are a number of idealized expectations that apply to succession, *ceteris paribus*. One is the intermediate disturbance hypothesis, in which high intensities or frequencies of disturbance are assumed to permit the persistence of only a limited pool of species that are well adapted to the effects of disturbance (Hastings 1980; Petraitis et al. 1989). As disturbance frequency or intensity is relaxed, the number of species that can coexist in the site increases due to the release of the disturbance constraint. Ultimately, however, competition and other biotic interactions begin to impose a limit based on the exclusion of species by superior competitors. This relationship assumes a tradeoff between competitive ability and tolerance of disturbance or stress. The theoretical scheme of Grime (1979; Grime and Hodgson 1987) makes the role of stress explicit and suggests that community composition reflects the species' tolerance of disturbance on one dimension, tolerance of stress on another, and differentiation between colonization ability and competitive ability on the other. The Grimean scheme adds more complexity and reality to the expectations of the intermediate disturbance hypothesis. The intermediate disturbance hypothesis has been difficult to discriminate in the field due to the many factors that can function simultaneously with physical and biological tolerances.

Another *ceteris paribus* expectation of succession is the intermediate richness hypothesis. The expectation here is that during intermediate phases of succession in productive sites, species richness will be highest (Loucks 1970). Like the intermediate disturbance hypothesis, the intermediate richness hypothesis assumes that relatively fewer species are adapted to the stresses of early successional habitats than are adapted to the more moderate physical conditions that are generated by the ecological engineering as species invade, grow, and coalesce their biotic effects. However, under the physically more moderate conditions that often begin to appear as succession proceeds, more biotically competitive species dominate, and exclude the pioneers and earlier dominants. This process will result in lowered diversity in later succession.

The intermediate diversity or richness hypothesis led the prominent plant population biologist John Harper (1967) to ask why species do not evolutionarily climb their own successional trees? That is, why do species fail to adapt to the changing conditions during succession? There are two reasons that this provocative expectation is not borne out. First, all species are to some extent fugitive (Hanski 1995). That is, they migrate and abandon sites where they are less well adapted and establish in sites where they are better adapted. Plants probe their potential environments via dispersal, and mobile animals actively probe sites. Second, the contrasts between adaptation to high-resource levels versus adaptation to low-resource environments involves both physiological and architectural tradeoffs that are difficult to shift evolutionarily over the short term (Bazzaz 1979; 1983). Particular genera may have species that are arrayed along an adaptive successional gradient, but rarely does any single species contain within it the genetic scope—including plastic capacity—to persist through such a broad array of conditions (Tilman 1991). Putting these two generalizations together explains why species exploit shifting successional mosaics in landscapes rather than genetically adjust to the sere at any one point in space (Pickett 1976).

A modification of the intermediate richness hypothesis recognizes the constraints of xeric systems. Rather than an ultimate decrease in diversity over the span of several centuries as seen in mesic systems, xeric plant communities should show continuing increase in diversity (Auclair and Goff 1971). Such secondary successional systems may in fact be responding more like primary succession because of the stress imposed by low levels of available resources. Over very long time periods, however, such systems should show the diversity decline, just as do the very oldest primary successions (Vitousek 2004). Ecosystem parameters are also expected to change during succession. Like the ideal diversity expectations, these are based on the evolutionary tradeoffs that species exhibit. The relevant tradeoff here is between the high metabolism associated with using freely available and uncontested resources, and the slow metabolism and high storage capacity associated with exploiting contested and low levels of available resources (Odum 1969). Metabolic rate is also inversely associated with length of life. Putting these assumptions together suggests that early successional ecosystems will exhibit high productivity, low storage in soil pools, and a relatively low proportion of its total metabolism in respiratory activity. Hence, ecosystem metabolism is expected to have greater production than respiration early in succession, and to eventually come to a point after which production and respiration balance (Bormann and Likens 1979). Food webs should show concomitant shifts from grazer to decomposer dominance over this span (Quetier et al. 2007). An additional expectation is that system sensitivity to external disturbance should increase as species invest more in a complex structure that has low productive potential per unit time (Gunderson 2000).

The controversy and excitement concerning neutral theory (Hubbell 2001; Gravel et al. 2006) relate to succession theory. The neutral model posits that species have identical properties, and that local coexistence can be predicted by localized, as opposed to global, dispersal dynamics (Bell 2001). Mechanisms of coexistence can be divided among those that (1) are based on niche partitioning and competitive tradeoffs, (2) frequency dependence driven by prey-specific consumers, (3) limited recruitment to local habitats because of restricted dispersal, or (4) a dynamic equilibrium of speciation and extinction, i.e., neutral drift (Chave 2004). The two broad kinds of models based on neutrality or differentiation cannot be distinguished based on pattern data from the field (Chase 2005; Gravel et al. 2006). Indeed, Pueyo et al. (2007) showed that a large number of plausible models exist between neutrality and idiosyncratic species behavior that can generate the same diversity outcomes. An emerging view is that local community organization is open to landscape or metacommunity influences, and that niche partitioning and neutrality can be usefully treated as ends of a continuum (Gravel et al. 2006). The kind of mechanism—competitive exclusion based on niche difference vs. stochastic exclusion—that operates in a particular situation depends on whether immigration can act to prevent limiting similarity (Gravel et al. 2006). This logic parallels the division of successional causes between those that disrupt com-

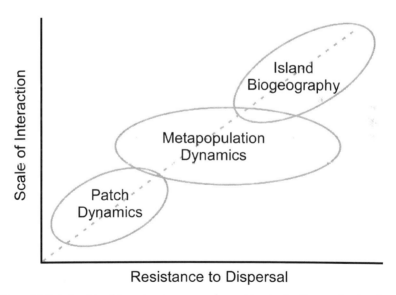

Figure 9.6 Relationship of three theories commonly considered to be discrete, showing their relationship within a conceptual space defined by resistance to dispersal versus spatial scale of interaction. Based on Pickett et al. (2007).

munity composition and open up sites versus those that operate through sorting species along the complex gradient of succession. The theoretical option of community control by migration was opened in the 1950s (Skellam 1951) and has matured through the intellectually parallel streams of island biogeography (Sax and Gaines Chapter 10), metapopulation theory (Hastings Chapter 6), patch dynamics, and metacommunity theory (Pickett and Rogers 1997; Bell 2001; Leibold et al. 2004; Leibold Chapter 8). Contemporary succession offers a framework to explicitly link these seemingly disparate views (Fig. 9.6).

Structure of succession theory: a summary and linkages

Succession or community dynamics theory is rich and well connected with other biological and ecological theories. Hereafter, we relate community dynamics theory to the fundamental principles of ecological theory (Scheiner and Willig Chapter 1).

1. Heterogeneous distributions of organisms

Succession is a case of heterogeneous distributions of organisms and their aggregations in time and space. Through time, succession illustrates either architectural or compositional changes in communities. The temporal distribution of dominance is heterogeneous, so that individual species abundances show overlapping or idiosyncratic patterns at the fine scale. Often long lags precede or follow peak abundances, having apparently stochastic minor fluctuations. Succession itself becomes a higher-order source of heterogeneity, as successions start at different times or occupy different sites across landscapes and regions. The existence across landscapes of communities, patches, and ecosystems that have different successional origins, ages, and characteristics is a dimension of environmental heterogeneity that affects other ecological patterns and processes.

2. Environmental interactions of organisms

The heterogeneous distributions of organisms during succession are the result, in part, of the interactions between the organisms and the environment that prevails at different times. Succession can be conceived as a temporal gradient of environmental change, driven by the different organisms that predominate through time. The feedback between organisms and environment is a key aspect of successional pattern and process.

3. Heterogeneity of entities

In the overview of fundamental principles, this one was considered a result of the other seven (Scheiner and Willig Chapter 1). So it is with succession. Successions themselves can contribute to environmental heterogeneity in time and in space. During succession, heterogeneity often accumulates, at least in certain ranges of the process, as clonal organisms, neighborhood effects, and architectural complexity of dominants establish and increase. The various contingencies experienced by specific successions can lead to great heterogeneity over space in the communities, ecosystems, and habitats that are available for organisms and in which biogeochemical processes differ.

4. Contingency: history and stochasticity

This fundamental principle of ecology is compellingly applied to succession. The initial and boundary conditions of a particular community determine much about its subsequent temporal trajectory. Stochasticity appears as some organisms happen to survive a catastrophic disturbance, or as the organisms that happen to be within close dispersal range dominate the disturbed site, or when stresses, resources, and disease agents infiltrate unpredictably from nearby landscape elements. Contingency is well illustrated by the increasing role of accidentally or intentionally introduced plant, animal, and microbial species into the trajectories of various successions. Likewise, historical contingencies appear in those current communities where composition and changes are conditioned by the past management by humans. These include such legacies as the fire or hunting practices of indigenous populations no longer dominant, or the nature of more recent agricultural practices on succession. The order of establishment is an important contingency in particular communities. In addition, contingencies also take the form of stochastic interactions with climate cycles and disturbance events.

5. Heterogeneity of environmental conditions

Succession responds to differences in environmental conditions from place to place. The classical distinction between primary and secondary successions is an expression of heterogeneous resource levels and stocks of resting stages, seeds, or clonal fragments in disturbed sites. Even within secondary successions, particular trajectories differ along spatial environmental gradients as a result of differences in soil nutrients, moisture, and interacting consumer populations.

6. Finite and heterogeneous resources

The limitation of resources is key to understanding succession. Limiting environmental resources change in abundance through succession and ratios of aboveground to belowground limits change. The internal limitation of resources within each organism is also an important explanatory principle in succession. Internal limits require that resources allocated to one structure or process are not available for other structures and processes. This principle of allocation is the reason that genotypes—whether within or between species—are usually not capable of exploiting widely contrasting successional environments. Differences in capacities for dispersal, competition, mutualism, defense, and food or resource acquisition all depend on this fundamental limitation of available and assimilated resources by organisms.

7. Birth and death

Succession occurs because organisms are mortal. Regardless of their inherent patterns of senescence, they may be killed by predation, disease, or physical disturbances. When large numbers of organisms occupying a site are killed, the mortal event is almost always followed by colonization, interaction, and sorting of other individuals or species. Even when a single canopy tree dies in a forest, brought low either by disturbance or by senescence, succession can be affected. In mesic forests, more shade-tolerant trees can ascend to the canopy after mortality of prior canopy dominants that had reached the canopy under less light-limiting conditions. If there is no clear heir apparent in the understory, new establishment or thinning among of a number of saplings can occur. Vegetation succession has been summarized by some as a plant-by-plant replacement process. Turnover in animal communities can also involve birth and death, but migration is a common alternative for mobile organisms. Mortality agents important to succession include both physical events and the depredations of herbivores, diseases, and predators.

8. Evolution

Evolution is the biological mechanism underlying the contrasts among species that are worked out in succession. Allocation strategies, mentioned earlier, are evolutionary products. Contrasts among the crown architectures of plants adapted to shady versus sunny conditions are shaped by evolution. Chemical and mechanical defenses reflect evolutionary histories with their metabolic costs and reproductive benefits. Indeed, the spatial and temporal patchwork of habitats characterized by contrasting successional composition and environment must be one of the important stages for the enactment of the evolutionary play and the spatial or interactive assortment of its products.

The relevance of succession to the fundamental principles of ecology is one indication of the synthetic power of the concept. Additional aspects of its synthetic significance can be understood based on its structure.

First, ecological concepts have three dimensions—meaning, model, and metaphor (Pickett and Cadenasso 2002). We believe that neutral meanings or definitions are most generalizable, but that they require models to translate them to different cases, or to expose the interactions of the causes of the process (Jax 2006; 2007). Metaphor, the third dimension of an ecological concept, has two uses in ecological theories. Metaphors can serve as an initial stimulus for a theory, or as an image to translate the theory or its components to nonspecialists or to the public. In the case of succession, the term itself is clearly a metaphor originally derived from society. When the theory was introduced, one community was envisaged to be replaced by another that is in the aggregate better adjusted to the environment created by the previous community. However, as we have seen from the causal richness and the complex reality embodied in the propositions of contemporary succession theory, this simple image is incomplete (Table 9.1). Unfortunately, there is no single image that has emerged to take the place of the royal analogy of orderly transitions. Cooper (1926) suggested that community dynamics was better imagined as a braided stream than a swift, straight channel, which implies multiple pathways (e.g., Pickett 1989) and continuous change, but at faster or slower rates depending on the nature of the particular channel. Backwaters and eddies are of course part of Cooper's image.

The contemporary theory of succession is, as noted above, causally rich. Even though Clements originally proposed a remarkably broad conception of the causes of plant community change, he left out important mechanisms such as herbivory, and seemed to set disturbance outside the process of sorting in which he was most interested (Pickett et al. 2008). Vegetation dynamics, and indeed the dynamics of all sorts of communities and at all spatial and temporal scales, is driven by the differentials between available sites, the differential availability of species to a site, and their differential performance within a site. Succession is seen to take place in a dynamic spatial context, which can be identified as a landscape approach, or patch dynamics, or metacommunity dynamics. Succession therefore has a rich, spatially explicit repertoire of causes.

Several model templates exist in succession. The one most closely related to the causal hierarchy shows how the subsets of those general processes might act to filter the species interacting through time at a site (Fig. 9.4). Other model templates exist, depending upon whether the aggregate assemblage is the model target, as it is in Markov models (Horn 1975), or gap replacement is the focus (Shugart 1984), or individual woody plants in their spatial neighborhoods are the target, as in the Sortie model (Pacala et al. 1993).

The particular models represent various specific successional causes and mechanisms from the causal hierarchy (Fig. 9.1). The models combine the specific mechanisms to precisely determine how individual organisms interact with one another and with the physical environment, thereby generating community change through time. The large computing power now available permits spatially explicit, individual-based models to address complex mixes of dispersal, resource levels, competition, and survival, for example (e.g., Clark 2007). No longer is it necessary to decide which single factor drives succession based on mutually exclusive alternative hypotheses. Rather, discovering the mix of factors and interactions that come into play has great power for understanding community trajectories, and for examining how changing land use and climate (Bazzaz 1986; Bazzaz and Sipe 1987) can modify trajectories. The plethora of models is a response to rejecting the unrealistic assumption of stationary contexts and conditions for most successions.

Succession has remarkable powers to synthesize different ecological perspectives and applications (Fig. 9.7). Oddly, this seems only rarely to be recognized, perhaps because the terminology used by the otherwise relevant disciplines and community dynamics theory overlap so little. For example, the disparate concerns of invasion ecology, focusing on conditions in potentially invasible sites, or characteristics of the potential invading species, seems locked in arguments of alternative hypotheses when in fact some complex mixture of mechanisms can operate, as suggested by the successional framework (Davis et al. 2005). Succession has the power to unify aspects of many other ecological perspectives as well. For example, the spatially explicit concerns of landscape ecology are central to the workings of disturbance and the distribution of propagules between disturbed and successional sites. Assembly rules are expressions of the interactions within communities that lead to successional sorting. Indeed, community assembly is preeminently a temporal process (Keddy 1992). Assembly emerges from the subtleties of autecology and physiological ecology, which expose adaptations that clearly contribute to the local coexistence of organisms (Bazzaz 1986). Organismal physiology, body plans, architecture and morphology, and the timings of life cycle events that are summed up in the unfortunately neglected term "autecology" are the stuff of sorting along successional gradients. Indeed, gradient theory is one of ecology's powerful and pervasive ideas that is expressed in a successional context through time as much as in a spatial context (Austin 2005; Fox et al.

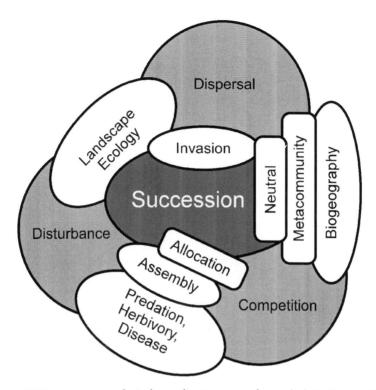

Figure 9.7 Succession as a synthetic theory, showing suggested general relationships to other ecological theories. The theories of disturbance, dispersal, and competition are shaded to represent their strong and longstanding relationship to the three successional processes of differential site availability, differential species availability, and differential species performance, respectively. Although classical succession theory emphasized those three founding theories, contemporary succession theory draws on and informs the broader range of theories illustrated.

Chapter 13). The gradients to which species respond during succession are not independent variables. Instead, they are emergent properties resulting from the interaction of organisms with one another and with the physical environment (Vitousek 2004). The ecosystem concept grew out of this recognition, and it is helpful to think of successions as complex gradients of ecosystem change. There is little of contemporary ecology that does not contribute to successional explanation, or benefit from understanding the spatial and temporal dynamics of ecological systems. That spatial and temporal dynamic is the essence of succession. Thus, one of ecology's oldest theories turns out to continue to have broad and adaptive relevance as a synthetic tool.

Acknowledgments

We are grateful to the editors and two anonymous reviewers for their helpful comments on an earlier draft of the paper. Many of the ideas in this chapter arose from interactions among the chapter authors who were participants in a Workshop on the Theory of Ecology supported by the University of Connecticut through the Center for Environmental Sciences & Engineering and the Office of the Vice Provost for Research & Graduate Education. We also acknowledge NSF Long-Term Research in Environmental Biology grant DEB-0424605.

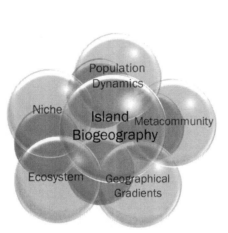

CHAPTER 10

The Equilibrium Theory of Island Biogeography

Dov F. Sax and Steven D. Gaines

The equilibrium theory of island biogeography (ETIB) is arguably the single most influential theory in the study of geographic patterns of diversity of life on Earth. Its influence is marked not just by the research it has motivated, but also by the theories and applications it has spawned. Many present-day strategies in reserve design, landscape ecology, and metapopulation theory draw significant portions of their intellectual capital from ETIB. Indeed, it is hard to imagine what the modern face of ecology, biogeography, or conservation biology would be like without ETIB.

The equilibrium theory of island biogeography was first articulated by Robert H. MacArthur and Edward O. Wilson in a short journal article (MacArthur and Wilson 1963) and later expanded to a monograph (MacArthur and Wilson 1967). It was this second publication that garnered the attention necessary for the theory to develop into the mainstream research juggernaut that we know today. By the early 1970s many of the best and, perhaps prophetically, most influential researchers in the fields of ecology and conservation biology were actively engaged in testing or expanding ETIB (e.g., Brown 1971; Terborgh 1973a; Simberloff 1974; Diamond 1975b). Through the 1980s and '90s theories influenced by ETIB became predominant in the literature. Today, the topic of ETIB is still actively investigated (e.g., Heaney 2007; Whittaker et al. 2008) and its continuing legacy is pervasive in recent literature. Indeed, MacArthur and Wilson's work (1967) was the focus of a recent symposium at Harvard University and a subsequent edited volume, which celebrates and examines the influence of ETIB over four decades (Losos and Ricklefs 2009).

MacArthur and Wilson (1967) aimed to shift biogeography from a discipline that was primarily preoccupied with natural history to one that was also focused on theory and experiments. Specifically they wanted to construct a "theory of biogeography at the species level" (p. 5)—one that deemphasized the traditional focus of biogeography on the composition of biotas, the distribution of higher taxa, and the role of geological change. They believed that biogeography could "be reformulated in terms of the first principles of population ecology and genetics" (p. 183). This perception suited their conception of how theories should be developed, as they suggested that "a theory attempts to identify the factors that determine a class of phenomena and to state the permissible relationships among the factors as a set of verifiable propositions" (p. 5).

In this chapter we outline the conceptual basis, predictions, impact, and future of the study of ETIB. ETIB is the best-known and most investigated portion of the total body of theory described by MacArthur and Wilson in their monograph (1967). Many other topics were also discussed at length in their book (e.g., the strategy of colonization, invasibility, and the variable niche). We restrict our focus in this chapter, however, to ETIB itself. Further, we do not attempt to provide a review of the work that preceded ETIB and laid its conceptual groundwork (e.g., Dammermann 1948; Darlington 1957; Wilson 1961), as thorough reviews on this topic are available elsewhere (Whittaker and Fernandez-Palacios 2007; Lomolino and Brown 2009; Lomolino et al. 2009). Here, we begin by describing the domain and propositions of ETIB. We then explore how these propositions are built from deductive reasoning and fundamental principles in ecology and the physical sciences. Next we examine the degree to which evidence has supported or refuted the theory. We also consider how ETIB relates to other models of species richness and how ETIB has influenced the development of other theories in ecology. We consider the case study of anthropogenic invasions of species to islands with respect to ETIB. Finally, we consider how ETIB can be improved and what areas of research ETIB can productively lead us towards in the future.

Domain

ETIB provides an explanation for variation in patterns of species richness across space and over time for islands and insular habitats. MacArthur and Wilson (1967) believed that insularity is "a universal feature of biogeography" (p. 3) and should be applicable in a broad array of habitats. They suggested that the principles one could derive from the study of islands and island archipelagos should apply to at least some degree to all natural habitats. Because insularity is not a discrete trait of habitats, the degree to which a given habitat should fall within the domain of ETIB should be related to its degree of isolation from other suitable habitats (i.e., the degree to which the matrix that fills the space between focal habitats impedes dispersal by those species that occupy the focal habitats). Further, whether or not islands or insular habitats are natural or anthropogenic should not influence whether they fall within the domain of ETIB. All areas that are discretely partitioned from a surrounding matrix should have characteristic patterns of species richness and species turnover that are coincident with the size and isolation of those partitioned areas. MacArthur and Wilson (1967) suggested therefore that these same principles should apply to habitats fragmented by human activities. Consequently, the domain of ETIB can be envisioned to include much of the natural and human-modified habitats that occur globally.

The spatial and temporal extent of the domain of ETIB, however, is not universally agreed to be so extensive. Many authors (e.g., Whittaker 1998; 2000; Whittaker and Fernandez-Palacios 2007; Lomolino et al. 2009) have suggested that the spatial extent of ETIB should be restricted to a much narrower set of conditions than originally envisioned. For instance, Lomolino et al. (2009) argue that only islands at relatively intermediate levels of isolation and at intermediate sizes should be well-characterized by dynamic equilibrial processes described in ETIB; more specifically, Lomolino et al. (2009) posit that on very small, near islands species richness should be effected primarily by stochastic processes and little influenced by equilibrial ones. We discuss this perceived contraction of the applicable domain of ETIB in the section "Evaluating ETIB." Additionally, disagreement arises over the temporal domain of ETIB. In its most basic form ETIB applies to patterns of species richness and turnover with respect to relatively short ecological-timescales. Many authors consider ETIB solely in this temporal context (Whittaker and Fernandez-Palacios 2007). However, one can also consider the longer-term processes that MacArthur and Wilson (1967) postulated would impact equilibria over geological and evolutionary timescales.

Fundamental principles underlying the propositions of ETIB

ETIB, as originally envisioned by MacArthur and Wilson (1963; 1967), can be partitioned into a basic model that operates over ecological time and a more inclusive model that operates over evolutionary time. It is the basic model, over ecological time, that has been influential and remains well known.

Principles underlying the ecological model of ETIB

There are seven propositions of the basic ecological model of ETIB (Table 10.1). Proposition 1 is derived from deductive reasoning. The rate of immigration (i.e., arrival of new species) to an island must decrease as more and more of the species that could potentially arrive have done so. Further, this rate must reach zero once the entire pool of species that could colonize are present. If species are considered as equivalent units, i.e., with equal probabilities of arrival, then this rate should decrease linearly with increasing richness

Table 10.1 Propositions of the equilibrium theory of island biogeography.

Ecological propositions

1. The rate of immigration of species (i.e. the arrival of new species) to an island decreases as the number of species that have arrived on an island increases; the rate reaches zero when all species that could colonize from an available pool of species have done so.

2. The rate of immigration of species to an island decreases with increasing isolation from a pool of potential colonists.

3. The rate of immigration of species increases with increasing island size.

4. The rate of extinction of species established on an island (of a given size) increases with increasing numbers of species.

5. The rate of extinction of species established on an island decreases with increasing size of an island.

6. The number of species on an island will be determined by an equilibrium between rates of immigration and extinction.

7. The rate of species turnover (i.e., change in species composition) will be determined by an equilibrium between rates of immigration and extinction.

Evolutionary propositions

8. In addition to immigration, the number of species on an island can be increased by speciation.

9. Speciation will only be important to an equilibrium in species number in a "radiation zone" found at the outward limits of species capacity for natural dispersal.

10. The distance to a "radiation zone" is taxon specific.

11. An equilibrium in species number reached ecologically is a "quasi equilibrium" that can be increased over evolutionary time.

(Fig. 10.1A). This proposition is refined, however, by MacArthur and Wilson, who invoke differences among species, wherein some species are more likely to arrive to an island per unit of time than others. This results in the classic downward bow in the immigration curve depicted in ETIB (Fig. 10.1B). Note, however, that as long as the immigration curve decreases monotonically the precise form of this curve will not change the qualitative predictions of ETIB.

Propositions 2 and 3 are in accord with the physical process of diffusion. With respect to proposition 2, for example, objects diffusing from a source will reach a near object more frequently then they reach a far object. This means that rates of immigration should be higher on near islands than far islands, as species by chance should more frequently encounter near islands (Fig. 10.1B). Nevertheless, both far and near islands could be colonized by all species in the pool of potential colonists given sufficient time, such that the rate of immigration for near and far islands both reach zero once all species from the pool are present (Fig. 10.1B). Proposition 3 is based on the same process of

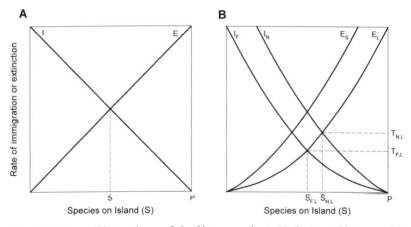

Figure 10.1 The equilibrium theory of island biogeography. A. The basic equilibrium model. An equilibrium number of species (S) is set by two opposing processes, immigration (I) and extinction (E). The rate of immigration decreases and the rate of extinction increases linearly with increasing richness; the rate of immigration reaches zero when the entire pool (P) of potentially immigrating species have arrived. B. The full equilibrium model. Rates of immigration and extinction are bowed downward to reflect various suppositions of natural history and population ecology (see text). Immigration rates on islands far (I_F) from a source are expected to be lower than those on near (I_N) islands. Extinction rates are expected to be higher on small (E_S) islands than on large (E_L) islands. In addition to different equilibrium numbers of species, differences in rates of species turnover (T) are expected to vary with the combination of immigration and extinction rates that characterize any given island.

diffusion, but with regard to target size—such that arrival of colonists will be more frequent to a large island than to a small one. This proposition was described by MacArthur and Wilson (1963); in the monograph (MacArthur and Wilson 1967), it was excluded from their introductory description of ETIB, but included in their description of stepping stone islands, and thus retained within the larger suite of factors postulated to influence rates of immigration.

Propositions 4 and 5 are derived from deductive reasoning and knowledge of natural history (particularly the study of population ecology). If all species have an equal probability of going extinct per unit of time, and this process is independent of the total number of species, then the rate of extinction will increase linearly with an increasing number of species (Fig. 10.1A). More realistically, however, species may vary in their individual probability of extinction, which may impact the rate of extinction for any given number of species. Further, as the number of species that are using a given set of resources increases, the average number of individuals per species must decrease, so long as resources are being fully (or nearly fully) used; because smaller populations are more likely to go extinct than are larger populations, the rate of extinction may not be independent of the total number of species. Additionally, increasing the number of species may increase the probability that species will interfere with each other. These factors may result in an extinction rate with a downward bow from a straight line (i.e., one that increases nonlinearly with an increasing number of species, as in Fig. 10.1B). Note, however, that MacArthur and Wilson (1963; 1967) do not incorporate concepts of positive species interactions, such as facilitation and mutualism, into their rationale for extinction rates (but see Wilson 1969); doing this could alter the relative position of the extinction curve either upwards or downwards, but should not change its basic shape (Wilson 1969). As with the immigration rate, however, so long as the extinction curve increases monotonically, the precise form of this curve will not change the qualitative predictions of ETIB. Finally, for any given number of species, a larger island should have a lower rate of extinction, since average population sizes would be larger and hence these species would be less likely to go extinct (Fig. 10.1B).

Propositions 6 and 7 are in accord with equilibrium processes observed in chemistry, physics, and population biology. For example, the total number of individuals in a closed population will be determined by an equilibrium between opposing rates of birth and death. By altering either of these rates, the total number of individuals in the population can be altered. MacArthur and Wilson (1967) similarly reasoned that the number of species supported on an island will depend on an equilibrium between the rates of immigration and extinction (Fig. 10.1A). By varying the rates of immigration and extinction, the resulting equilibrium number of species can be shifted (Fig. 10.1B); these shifts in equilibrium points provide an explanation for species-area and species-isolation relationships. Rates of species turnover are also derived from equilibrium processes, such that islands with the highest rates of immigration and extinction will be expected to have the highest rates of change in species composition even though species number remains constant (Fig. 10.1B). Although MacArthur and Wilson (1967) predicted that turnover rates should vary with island characteristics, the graphical depiction (as shown in Fig. 10.1B) appeared in later publications (e.g., Simberloff 1974).

Principles underlying the evolutionary model of ETIB

As suggested previously, not all authors agree that evolutionary considerations should be included as part of ETIB. Here, however, we discuss those propositions described by MacArthur and Wilson (1967) that are expected to directly influence the predictions of ETIB over evolutionary timescales (Table 10.1). The principles underlying evolutionary components of ETIB are based on deductive reasoning (proposition 8), observed patterns in the geographical distribution of endemic species (propositions 9 and 10), and principles of evolutionary biology (proposition 11). Deductively, it is clear that species could be added to an island by either a process of immigration or by in situ speciation events (proposition 8). Geographic patterns in the distribution of endemic species, particular of ones believed to have evolved unique species status on the island or archipelago in question, show that the importance of speciation increases with isolation (Mayr 1965). MacArthur and Wilson (1967) reasoned that speciation will be most important at the outward margin of the zone that species are capable of dispersing to, presumably because subsequent gene flow from mainland sources would be quite low; this should result in a radiation zone, where speciation is more likely to occur (proposition 9). Further, because taxonomic groups differ in their ability to disperse, the location of radiation zones should be taxon specific (proposition 10). Finally, MacArthur and Wilson (1967) believed that "an equilibrium can be defined in either ecological or evolutionary time" (p. 176). Consequently, an ecologically observed equilibrium is really a "quasi-equilibrium" (p. 176) that can be increased over evolutionary time (proposition 11). They expected this to occur because of evolutionary adaptation among species and to the environment over time (MacArthur and Wilson 1967; Wilson 1969).

Evaluating ETIB

Despite its iconic nature, ETIB generally has been difficult to test. Because the equilibrium model was designed to provide an explanation for increases in richness with increasing island area and decreases with increasing isolation, confirmation of these patterns is not strong evidence in support of ETIB, particularly as other models (see section below) share these predictions. Nevertheless, it is worth noting that these patterns (even with 40 years more data) often do bear it out as expected. For example, a recent study of the vascular floras of 488 islands worldwide found that area alone predicted 66 percent of the variation in species richness, and that the next most important predictors were isolation, temperature, and precipitation (Kreft et al. 2008). A thorough evaluation of ETIB, however, and the hundreds of papers that consider its predictions is beyond the scope of this chapter (for an overall review see Whittaker and Fernandez-Palacios 2007). Here, we provide a brief sketch of how the predictions of ETIB have fared in the literature, in particular with respect to species richness and turnover.

There is mixed support for several of the specific predictions made by ETIB for species richness. First, ETIB predicts that the slope of species-area relationships will increase with increasing isolation. This has proven true in some comparisons, particularly between classes of islands (e.g., vascular plants on landbridge versus oceanic islands; Sax and Gaines 2006), but has not proven true in many other cases (Whittaker and Fernandez-Palacios 2007). Second, ETIB predicts that an integration of immigration and extinction curves will produce a colonization curve, representing the total accumulation of species on an island, that should approach an asymptote over time. Evidence that is arguably consistent with this has been observed on small, experimental islands (e.g., Simberloff and Wilson 1970), but few studies are available for larger islands. Some of the best evidence on larger islands comes from Krakatau—but even in this case evidence of an asymptote in species numbers is difficult to evaluate, because the timescale of an expected asymptote is not well defined by ETIB. Consequently, it is unclear how to evaluate the shallowly increasing slope of bird richness or the more sharply increasing slope of plant and butterfly richness over time on Krakatau (Whittaker 1998). Third, ETIB predicts that declines in area should drive species losses in habitats whose spatial extent has been reduced. This process is referred to as relaxation in species numbers. Evidence consistent with this expectation does exist (e.g., for the loss of species in Singapore; Brook et al. 2003). In many cases, however, this expectation is not well supported. For example, many oceanic islands have lost much of their habitable area to agricultural and urban areas, yet few plant species have gone extinct (Sax et al. 2002). Here too, however, this evidence is difficult to interpret, because the timescales necessary for relaxation to occur are poorly defined by ETIB; it is conceivable, for example, that extinctions for certain taxa may only manifest after many hundreds of years (Sax and Gaines 2008).

The primary focus of testing and falsifying the ecological model of ETIB has been in evaluating its predictions of species turnover. Here too, the evidence in support of ETIB has been mixed at best, with a few studies in apparent (but debated) support, but many other studies that do not or only partially support the theory's predictions. Two of the best known studies posited to support ETIB are the study of insect immigration following the experimental defaunation of small mangrove islets (e.g., Simberloff and Wilson 1969) and the turnover of land birds on the Channel Islands of California (Diamond 1969). Both showed turnover of species with relatively stable equilibria in species richness. Evidence in support of ETIB from Simberloff's work has been questioned because of the difficulty in distinguishing transient species from actual immigrants and for being done on small, experimental systems that may not scale up to larger islands (Whittaker 1998). Diamond's work has been critiqued because estimates of turnover may have been inflated by incomplete census data and by anthropogenic effects (Lynch and Johnson 1974). Mixed support for ETIB can be seen in a wide variety of other studies. For instance, Cody (2006) in a 25-year study of nearly 200 continental islands in the Barkley Sound, British Columbia, found that rates of immigration and extinction on islands were highly correlated, such that stable equilibria in species numbers were maintained (in support of ETIB). Yet rates of immigration and extinction were highest on large islands, regardless of isolation. As a result, turnover was highest on large islands—in contrast to predictions of ETIB; although in this case because the absolute difference in isolation among these islands is relatively small it is possible to conclude that these findings are not in conflict with ETIB. Finally, there are cases where the predictions of ETIB do not hold well at all, as for orb-weaving spiders in the Bahamas (Toft and Schoener 1983).

Overall, several inherent difficulties exist for testing turnover rates. First, some of the best evidence comes from small experimental systems, which present challenges in generalizing to larger scales. Second, studies performed at larger spatial scales are generally performed over relatively short periods of time (but see Cody 2006), making it difficult to generalize to longer time periods. Third, complete and comprehensive census data are rarely available, making it difficult to evaluate actual turnover rates (Whittaker and Fernandez-Palacios 2007). Fourth, the definitions of immigration, colonization, and extinction are poorly articulated, in that it is unclear what determines whether a species is established on an island, which in turn influences whether a species is classified as a failed immigrant or instead as an established species that went extinct. Further complicating these evaluations are differences in how ETIB is defined. Some authors consider only the ecological model of ETIB, while other authors consider the evolutionary model and other refinements discussed in the 1967 monograph. Consequently, the same evidence posited against the equilibrium model by one individual might be held in support of the broader, more inclusive view of ETIB by another (Whittaker 2000). Given these many complications, Whittaker (1998) has suggested that it is difficult to find any convincing, unequivocal evidence of turnover that matches the predictions of ETIB.

Given the many difficulties with testing ETIB, one might ask why it has been so influential and why it remains such an active focus for research. The answer has much to do with the compelling simplicity of a model that offers to explain complex patterns in species richness and turnover. As Heaney (2000), Lomolino (2000a), and others have suggested, it is ETIB's heuristic features that make it so powerful. It provides a frame of reference for observations about the natural world, particularly ones that differ from the expectations of ETIB. Undoubtedly, its influence is further heightened by two factors. First, no alternative model has gained primacy in the literature. Second, ETIB has influenced the development of many other theories in ecology and conservation biology.

Contrasting ETIB with other models of species richness

Since a central goal of ecology and biogeography is to explain the distribution and diversity of life, it is not surprising that many alternative theories exist for patterns of species richness. Some of the most influential include: neutral theory (Hubbell 2001), stochastic niche theory (Tilman 2004), and species-energy theory (Wright 1983). Other unconsolidated theories also exist, such as the idea that richness patterns on islands are largely driven by nonequilibrial aspects of dispersal limitation. We do not attempt to provide an exhaustive review of this topic, which would include discussion of longstanding models by Lack (1947), as well as more recent work by Heaney (2000), Lomolino (2000b), O'Brien et al. (2000), and others. Here, we highlight the importance of dispersal limitation in influencing richness patterns, along with just two of the more influential, codified theories so as to provide a context for better understanding ETIB.

Dispersal limitation and a nonequilibrial view of island richness

The ideas comprising a nonequilibrial view of island richness have not been codified into a named hypothesis or theory. Nevertheless, this view is prominent in the literature. It involves a similar and overlapping set of ideas often referred to with slightly different language: undersaturated (Lawlor 1986), nonequilibrial (Whittaker 1998), not saturated (Sax and Gaines 2008), and so forth. Put simply, this view holds that richness is not a consequence of an equilibrium between opposing forces. Like ETIB, this view posits that richness could be increased if dispersal limitations were removed or if rates of speciation were increased, but unlike ETIB, it does not posit that these factors are necessarily opposed, at least in part, by increases in extinction. In principle, distinguishing this view from ETIB is difficult in many cases because the timescale of the key processes are defined poorly. So, for example, if immigration rates are rapidly increased by anthropogenic means to an island, how long will it take to see increases in extinction rates, as predicted by ETIB? Will some of the increases in richness be transient and offset by future extinctions, or will the increases in richness be persistent? See the following section on anthropogenic invasions for a more complete discussion of this issue. In other cases, however, it appears certain that a dispersal-limitation, nonequilibrium model is an appropriate way to understand patterns of richness. For example, natural rates of immigration for mammals on remote oceanic islands appear to be close to zero (e.g., Lawlor 1986). Indeed, a nonequilibrial explanation is consistent with observed patterns for many dispersal-limited groups: nonvolant mammals, freshwater fishes, amphibians, and others.

Species-energy theory

Species-energy theory as proposed by Wright (1983) is a direct derivation from ETIB, although the precursors to this theory include much older thoughts on the importance of energy in determining numbers of species (e.g., Forster 1778). In Wright's (1983) model, the principal difference from ETIB is the substitution of energy for size of an island. In relation to ETIB, this means that extinction rates would be set by available energy instead of island size. Similarly, Wright substituted species-energy relationships for species-area relationships. He described energy with actual evapotranspiration (AET). This metric of energy has the advantage of being readily measurable for most places. With these substitutions, Wright found that he increased the amount of variation typically explained by species-area relationships (SARs). He also suggested that this would allow the framework of ETIB and SARs to be applied across disparate regions of the world (e.g., in temperate and tropical regions), while using a single explanatory framework. Wright's theory has been adopted and investigated by many other researchers (e.g., Turner et al. 1988). The most influential papers on species-energy theory, however, appear to be motivated independently from Wright's work. These papers by Currie and Paquin (1987) and Currie (1991) examine the relationship of species richness of plant and animal groups on continental areas. However, Wylie and Currie (1993) apply these same ideas to mammals on land bridge islands, and like Wright (1983), who examined plants, find that available energy is an excellent predictor of species richness. Similarly, Kalmar and Currie (2006), who explicitly set out to test ETIB and species-energy theory, find that area, climate, and isolation are all important predictors of bird richness on islands worldwide, but in contrast to predictions of ETIB, isolation does not influence the slope of the species-area relationship (SAR); Whittaker (2006) suggests, however, that the failure to find an effect of isolation on the slope of SARs may be due to mixing islands from different climatic zones together into a single analysis. Much research is continuing along these lines currently (Fox et al. Chapter 13), and this work offers the potential for major advances in our ability to predict patterns of species richness.

Stochastic niche theory

Stochastic niche theory (Tilman 2004) is a modern reformulation of resource use among species that provides an explanation for patterns of species invasion or colonization and observed levels of species richness. It is similar to ETIB in that its core model effectively treats species as equal entities, while it also provides a larger construct in which species differences could allow for refinements of the model (Stachowitcz and Tilman 2005). The model differs from ETIB in having priority effects in which species that arrive first are most likely to retain their place in a community (or on an island). So instead of having a dynamic equilibrium set by opposing rates of immigration and extinction (in which the addition of a species should result in the loss of an existing one), stochastic niche theory posits that species that colonize should continue to be added so long as there are sufficient available resources. Once those resources start to become limiting, then the probability of adding additional species should decline in proportion to species richness; at some point of resource use the total number of species should be very difficult to increase—this level of richness may appear as a saturation point, even though it would be generated by nonequilibrium processes. This model has been relatively influential to date, but more time will be needed to judge its eventual impact on the field.

The relation of ETIB to other theories in ecology

First and foremost, ETIB is an extension of basic population biology theory. Population biology theory often treats all individuals within a population as equals, with respect to growth rates, carrying capacity, and so forth. This is a simplifying assumption, as clearly not all individuals are truly equal—some are larger, some more fecund, and so forth. Nevertheless, this simplifying assumption allows for models explaining general patterns and trends among individuals within a species to be understood and predicted (Hastings Chapter 6). ETIB borrows concepts from population biology theory that are applied to individuals and instead applies them to species. For instance, ETIB treats species as largely equivalent units. Further, instead of a carrying capacity for individuals set by opposing rates of birth and death, ETIB posits a saturation point in species number set by opposing rates of immigration and extinction. Like population biology theory, wherein the carrying capacity for individuals can be increased, so too can the saturation point for species be increased in ETIB, whenever changes in equilibrium processes are changed. Certainly, ETIB goes beyond population biology theory in many ways, but its similarities are readily apparent.

ETIB is also intricately related to many other theories in ecology. Indeed, the ultimate legacy and influence of ETIB is perhaps not found in the model's predictive power, but instead in the great diversity of theories that ETIB has influenced or helped to generate. The influence of ETIB on all of these models illustrates the power of clearly articulating a simple heuristic theory, as this provides the foundation for further advances. In many cases, these models and theories have perhaps superseded the influence of ETIB itself. In an earlier section we described the relationship between ETIB and species-energy theory, stochastic niche theory, and nonequilibrium views, but several other prominent theories have been influenced by ETIB. We highlight four of these below.

ETIB has influenced the development of metapopulation models. Although the origin or metapopulation models predate ETIB (see discussion in Hanski 1991) the modern conception of these ideas was developed by Richard Levins (1969; 1970) after the development of ETIB. Levins codified the definition of a metapopulation as "a population of populations which go extinct locally and recolonize." Consequently, at its core, a metapopulation model borrows basic ideas from ETIB of immigration and extinction dynamics affecting a network (or archipelago) of local patches; a good example of this is Hanski's incidence function model (Hanski 1994). Metapopulation models have had broad and wide-ranging influence on the fields of ecology and conservation biology, influencing subfields like population ecology (Freckleton and Watkinson 2002) and metacommunity theory (Leibold Chapter 8).

ETIB has also influenced the scientific underpinnings for reserve design. This began with a series of publications that followed the publication of Mac-Arthur and Wilson's work, including Terborgh (1974) and Wilson and Willis (1975), but was strongly influenced by the publication of Jared Diamond's (1975b) seminal work on reserve selection. Diamond's work built on Mac-Arthur and Wilson's supposition that ETIB should apply to habitat fragments; he extended this idea to consider the pros and cons of alternative reserve designs. This resulted in the famous SLOSS (single large or several small) reserve design debate (Gottelli 2004), but also formed the basis for consideration of issues in reserve design and planning that continue to this day (e.g., Pressey et al. 2007). Some authors believe, however, that while ETIB has been important in the development of this field, current work transcends the boundaries and explanations originally proposed by the theory (e.g., Laurance 2008).

ETIB strongly influenced the development and codification of the field of macroecology, a term coined by Brown and Maurer (1989). Macroecology examines the statistical properties between the dynamics and interactions of species populations (Brown 1995). It posits that advances in our understanding of the natural world can be advanced by observations of general trends, and a search for emergent properties, particularly when comparing species within given taxonomic groups (such as birds or plants). Consequently, macroecology as an approach has proven to be a powerful tool, one that extends well beyond the domain of ETIB, to consider patterns of body size, range size, allometric scaling, speciation, and other critical concepts in ecology and evolutionary biology (e.g., Brown 1995; Enquist et al. 1998; Blackburn and Gaston 2003; Allen et al. 2006).

ETIB has influenced the development of "neutral theory," as outlined by Stephen Hubbell (2001). Like ETIB, this theory is inherently a dispersal-assembly model (built on concepts of immigration, speciation, and extinction). Perhaps even more than ETIB itself, neutral theory extends the concept of symmetry (i.e., equivalency) between species as a simplifying assumption that allows important patterns of species distribution and abundance to be examined against a null hypothesis. Like ETIB, this model has generated great debate and in so doing it appears to have pushed the field forward. It is too early to judge what its ultimate influence will be, but it has the potential to have a strong, multidecadal trajectory, not unlike that of ETIB.

A case study of ETIB: anthropogenic species invasions of islands

Human-facilitated colonization events (i.e., species invasions) have provided researchers with a novel source of insight into ecological, evolutionary, and biogeographical questions since at least the 19th century (Sax et al. 2007). Darwin (1859), Grinnell (1919), and Baker and Stebbins (1965), as well as many recent investigators, have viewed invasions as a set of unplanned experiments. For instance, Huey et al. (2000) used invasions of fruit flies in North and South America to examine the speed and predictability of evolution of clines in wing size across geographic gradients. Similarly, Bruno et al. (2005) used invasions as a source of evidence to examine the relative roles of competition, predation, and facilitation in newly assembled species associations. It is perhaps not surprising that researchers have also used invasions to better understand patterns of species richness and turnover on islands (e.g., Sax et al. 2002; Sax and Gaines 2008).

MacArthur and Wilson (1967) were not silent on the topic of species invasions, but instead made a set of observations and predictions regarding them. First, they conjectured that species addition (whether through natural or human-facilitated immigration) must lead to a reduction in the population size of existing species, unless newly added species were able to utilize resources that had been previously unused. Second, they reasoned that "by increasing the immigration rate, an impoverished biota can be changed into a richer one, yet without altering its equilibrial condition in the end. It will merely shift from one saturated state to another" (p. 176). This suggests that increased introduction rates of species (whether caused by humans or not) could lead to increases in the stable equilibrium number of species that inhabit islands. Third, they predicted that such increases would not be permanent if immigration rates were subsequently reduced. Consequently, they suggest that "if the new inpouring of immigrants were to be held constant, the number of Hawaiian bird species—native plus introduced—would move to a new, much higher equilibrium level. . . . If, on the other hand, all further importations were strictly forbidden so that the immigration rate returned to the old, natural level, the number of species might gradually decline to a third equilibrium not radically different from the pre-European level" (p. 177).

Given MacArthur and Wilson's (1967) predictions, as well as the general usefulness of invasions for testing theory, it seems worthwhile to consider the implications of species invasions for the evaluation of ETIB and reciprocally for understanding the ultimate impact that invasions may have on native taxa. ETIB, as described by MacArthur and Wilson (1967), needs to be modified only slightly to consider human-mediated invasions. One change to classical

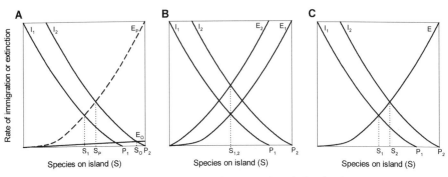

Figure 10.2 The equilibrium theory of island biogeography in light of anthropogenic species invasions. Prehistoric immigration rates (I_1) are presumed to be much lower than the human increased rates of immigration (I_2) now observed on islands, in part because the total pool size has been increased from those species able to reach islands before human influence (P_1) versus the greatly expanded number of species currently in the pool (P_2) of potential immigrants. A. Extinction rates observed (E_O) for vascular plants and freshwater fishes have been extremely low to date following increases in immigration rates, such that the observed (and inferred) equilibrium number of species (S_O) is much higher than the historic equilibrium (S_1). It is conceivable that the potential extinction rate (E_p) is actually much higher, but that long time-lags exist before extinctions are manifest, making it difficult to know the actual shape of the extinction curve or how many species will potentially exist at equilibrium (S_p) in the future. B. The number of land bird species currently occurring on oceanic islands (S_2) is similar to historic numbers (S_1), in spite of a large turnover in species composition. ETIB can account for such a pattern if the increased immigration rate (I_2) has been matched by an increased extinction rate (E_2). C. If land birds and nonvolant mammals are considered as single combined group then the net increase in richness (S_2) observed on islands can be interpreted in light of an increased immigration rate (I_2).

ETIB is needed because human activities can increase not just rates of immigration, but also the total potential number of colonists, such that the pool of potential colonists is increased (Fig. 10.2). This occurs because humans are transporting species not just from the source pool that originally stocked a given island, but also from various other parts of the world, where species were previously unable (or at least extremely unlikely) to be drawn from by natural processes of dispersal. Some authors, such as Whittaker and Fernandez-Palacios (2007), have suggested that human alterations of the environment might also have increased the total carrying capacity for species on islands (e.g., by removing forest cover and creating room for smaller herbaceous species to invade). This is certainly conceivable, in which case this could be modeled, with respect to ETIB, by sharply lowering the extinction curve—as the effective size of an island would be larger. Here we review the evidence of species invasions and turnover, with respect to ETIB, on oceanic islands in four taxonomic groups: freshwater fishes, vascular plants, nonvolant mammals, and land birds.

Freshwater fishes and vascular plants

Freshwater fishes and vascular plants show qualitatively similar patterns of invasion and species turnover on islands. Both groups have seen dramatic increases in total number of species established on oceanic islands, with many successful invasions and few extinctions (Sax et al. 2002; Sax and Gaines 2003). This has occurred because many nonnative species have become established, while few native species have gone extinct. In freshwater fishes, the most extreme example is in Hawaii, where all 5 native species have persisted in spite of the successful invasion of 40 nonnative species—such that total species number has increased by 800% (Eldredge and Miller 1995). Vascular plants have also experienced a large increase in net richness of species, with an extremely consistent near-doubling in species number across oceanic islands of the world (Sax and Gaines 2008). In New Zealand, more than 2000 nonnative plants have become established, largely matching the slightly more than 2000 native plant species extant, with only a handful of native extinctions. Further, in New Zealand there is a sufficient fossil record to ascertain that a large fraction of the flora has not recently gone extinct, that is, the few recorded extinctions are not an artifact of incomplete data (Sax et al. 2002).

With respect to ETIB, for both fishes and plants, these patterns are best fit by an increased immigration rate (and pool size), as well as by an empirical extinction rate that is nearly flat, but increasing very slightly with increased richness (Fig. 10.2A). The relatively flat extinction rate could be due to an overall increase in the carrying capacity of islands caused by anthropogenic increases in habitat heterogeneity. Alternatively, the few extinctions observed to date could be due to long time lags. Consequently, if there are significant time lags associated with relaxation (i.e., change from a supersaturated back to a saturated state), then it is possible that the ultimate extinction rate could be much higher (Fig. 10.2A). If this were the case, the current increases in species number would be transient and species richness should decline as a result of processes that have already been set in place. This relaxation would not be expected to occur all the way back to predisturbance levels, unless the rate of immigration were also reduced. However, a reduction in immigration rate in the near term seems extremely unlikely, at least for plants, because a large pool of potential colonists have already been transported to many islands. For example, in New Zealand there are more than 20,000 nonnative plant species, which have not yet become established, that currently reside in gardens around the country (Sax and Gaines 2008).

Although the patterns observed with fishes and plants can be viewed within the context of ETIB, they can also be considered within the context of a non-equilibrium state in which species richness on islands has been undersaturated (sensu Lawlor 1986) because of dispersal limitation. The extreme dispersal limitation experienced by freshwater fishes in crossing saltwater barriers is consistent with this alternative explanation for observed patterns of species richness. Similarly, the observation that the number of naturalized plant species has been increasing linearly over the past few hundred years, with no signs of an asymptote on more than a half-dozen islands that have adequate historical records to make such determinations, further conflicts with expectations of ETIB (Sax and Gaines 2008). Additional work will be necessary to fit the best theoretical framework for interpreting anthropogenic patterns of invasion in fishes and plants, as well as future extinctions in native species.

Nonvolant mammals and land birds

Mammals and birds have shown very different patterns of invasion on islands. Nonvolant mammals have historically been absent from most oceanic islands because of dispersal limitation; their richness has increased greatly with human introductions (Lever 1985). Consequently, mammals appear to fit a nonequilibrium model on oceanic islands. In contrast, land birds, which are effective dispersers, are species rich on many oceanic islands. Their pattern of change in richness following human introductions and habitat disturbance has been more complex. Richness of land birds initially decreased following colonization of islands by humans and their mammalian commensals (mice, rats, and cats), with large numbers of bird extinctions. Human introduction of birds, along with natural colonization following habitat transformation (e.g., conversion of forest to pasture) has increased bird richness back to levels that existed prior to human colonization, i.e., prior to colonization by either aboriginal or European peoples (Sax et al. 2002). This has occurred in a highly nonrandom manner such that the number of birds driven extinct on oceanic islands is closely correlated with the number that have subsequently become established ($r = 0.88, p < 0.001$). This pattern of change in birds can be envisioned in a ETIB context if an increased immigration rate has been matched by an increased extinction rate (Fig. 10.2B). An increased extinction rate for birds is conceivable if introduced mammals, which can compete for resources with birds or prey upon them, have reduced the carrying capacity for birds on islands (i.e., if the size of islands for birds has affectively been reduced).

Another way that these data could be viewed to be consistent with ETIB is if we consider birds and mammals together, as a single group. Given the many mammal-like roles that birds have filled on oceanic islands that were historically lacking mammals, such an amalgamation may be reasonable, although certainly unconventional. In this case, when viewed collectively as a single group, the net change in richness and turnover (of birds and mammals) fits well with an ETIB model (Fig. 10.2C), as net richness and turnover have increased. Additional work will be needed to ascertain how best to encapsulate future changes in richness on islands for birds and mammals.

Mapping a route forward to improve ETIB

The ETIB, as originally conceived, is not sufficient to provide a universal model for understanding patterns of species richness or turnover on islands (Whittaker 1998; Lomolino 2000a; Heaney 2007). Whittaker (1998; 2000), Lomolino et al. (2009), and others have suggested that ETIB, as a dynamic equilibrium model, may be sufficient only within a subset of the larger range of conditions that exist in nature, from equilibrial to disequilibrial and from static to dynamic systems. Nevertheless, as a heuristic device ETIB has been and continues to be extremely important in organizing thought on patterns of species richness and in providing a framework against which to consider other models. The question remains whether modifications and improvements on the basic ETIB model can help to provide a more robust framework for evaluating and predicting patterns of species richness or whether sharper departures from ETIB are needed. Here we outline three areas of research that may help to determine the most productive roads forward.

Integrating ecological and evolutionary processes

Ecological and evolutionary process are both considered within MacArthur and Wilson's work (1963; 1967) and by Wilson (1969). This was done in two ways. First, MacArthur and Wilson (1963; 1967) suggested that species addition to islands could occur from immigration and speciation. As discussed above, they believed that the latter should be important only in radiation zones far from mainland sources of species. Second, they suggested that over long periods of time, an increase in species number could be achieved through adaptations of species to each other and to the local environment. Consequently, they suggest that an ecologically based, immigration-driven equilibrium can be increased through evolutionary processes over time. This second view was further developed by Wilson (1969), who argued that the relative influence of this adaptation over evolutionary time could account for the more species-rich assemblage of native ants versus nonnative ants on islands in Polynesia. Neither MacArthur and Wilson (1963; 1967) nor Wilson (1969) provide quantitative estimates of how long these evolutionary processes may take to occur. Further, through a process of domain contraction, the evolutionary components of ETIB were largely set aside in the 1970s and 1980s (Lomolino et al. 2009).

A new paradigm appears to be emerging (e.g., Lomolino 2000a; Heaney 2007; Whittaker and Fernandez-Palacios 2007; Lomolino et al. 2009) that an improved theory of island biogeography must more explicitly consider the simultaneous influence of speciation, immigration, and extinction, as well as ecological interactions and differences among species in all these features. Such a paradigm would more explicitly recognize the important role that adaptive radiation plays in increasing species richness on islands (Heaney 2007). Over the last 40 years there have been important advances in our understanding of speciation on islands (e.g., Funk and Wagner 1995; Gillespie 2004), but these improvements have remained divorced from integrated ecological and evolutionary models of species richness. Consequently, there is much need for work on this topic, which offers the potential to more robustly describe the distribution of life on Earth.

Incorporating island ontogeny

ETIB considers few characteristics of islands beyond their size and isolation. This simplifying step has provided the basis for model development, but is perhaps not sufficient for the integration of ecological and evolutionary processes described earlier. Recently, Whittaker et al. (2008) have suggested that it may be important to consider island ontogeny (i.e., the life history of an island itself). Consider, for example, that a volcanic island's size, topographic complexity, ability to support life, and potentially its likelihood of promoting speciation may all change over geological time as the island forms out of the ocean, grows larger and higher, then begins to shrink, lose elevation, and eventually become eroded below the surface of the sea. Such a model of island ontogeny explicitly considers the possibility of within-island allopatry, i.e., allopatry that occurs within a topographically complex island. This is distinct from MacArthur and Wilson's (1967) description of allopatry, in which only among-island allopatry is considered. Models that incorporate island ontogeny would also fit well within the framework suggested by Heaney (2000) in which species richness was regularly within a dynamic state of disequilibrium, always a few steps behind constantly changing geological or geographical circumstances. This promising avenue of research offers one important inroad to improving ETIB.

Borrowing aspects of ETIB to study mainlands

A variety of recent work on mainlands has shown that species richness can often remain remarkably stable over time or across space despite large changes in species turnover. In contrast to ETIB, these mainland environments occur in habitats that are artificially delimited (i.e., they are not marked by discrete physical boundaries like islands or insular habitats). For example, Brown et al. (2001) have shown within delimited areas that birds in Michigan, mammals in Arizona, and trees in Europe have retained relatively stable numbers of species through time despite large changes in species turnover. Similarly, across space, Sax et al. (2005) have shown that analogous communities on mainlands that are dominated by nonnative species often support very similar numbers of species as native communities, despite having very different species compositions. Brown et al. (2001) have suggested that equilibrial numbers of species may be maintained, despite turnover in species composition, by interaction between a carrying capacity for species set by local ecological conditions and by opposing rates of immigration and extinction. This is intriguing, in part because Rosenzweig (1995) and others have suggested that immigration pressure will generally not be limiting in mainland environments. Ironically, if the model proposed by Brown et al. (2001) bears out then it may suggest that cases of dynamic turnover in species identity with relatively stable numbers of species may be most common in places where immigration pressures are not limiting (i.e., on mainlands and not on islands). Further work is needed to better explore this possibility.

Implications of further exploration of ETIB for ecology, evolution, and conservation

ETIB has served as a powerful heuristic tool for advancing our understanding of patterns of species richness in time and space. Recent attempts to test and expand the theory have raised two issues that are particularly pressing for conservation biology, but also for the continued advance of ecological and evolutionary theory. First, we need to better characterize the rates at which species extinctions are likely to occur once processes have been set in motion that may commit species to eventual demise. The time lags and extinction debt involved in such extinction processes are still poorly explored and in need of much attention (Tilman et al. 1994). Indeed, Whittaker et al. (2005) have said that "it is disappointing that we still know so little about the power and timescale of 'species relaxation.'" We agree. Our failure to answer this question makes predictions of species loss, as a consequence of habitat destruction and species invasions, difficult to determine (Sax and Gaines 2008). Second, conflicting opinions about the future rate and speed of speciation as a consequence of habitat fragmentation and species invasions abound (e.g., Rosenzweig 2001; Vellend et al. 2007). Determining how likely we are to maintain the process of speciation, and how it may be impaired or facilitated by human actions, is of pressing concern, particularly in light of the many extinctions that we anticipate will occur as a consequence of human actions. Ultimately, understanding both sides of the same coin (i.e., speciation and extinction) are fundamental to understanding ecological and evolutionary processes, but are also fundamental to effectively conserving and promoting biological diversity. We believe that the continued study of ETIB, along with its embellishments, improvements, and derivative theories, is key to integrating ecological and evolutionary perspectives needed to best manage biological resources in the future.

Acknowledgments

This chapter benefited from detailed comments by Mark Lomolino, Rob Colwell, an anonymous reviewer, and the editors. Consideration of ETIB in light of anthropogenic species invasions benefited from conversations with Jim Brown.

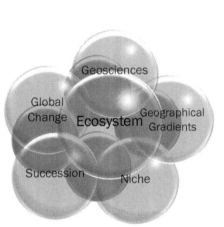

CHAPTER 11

Theory of Ecosystem Ecology

Ingrid C. Burke and William K. Lauenroth

Where does the "ecosystem ecology" section of book about ecology be-long—at the beginning, or the end? Ecosystem ecology is frequently viewed as the "largest level" organizational scale of ecology that fits logically after individuals, populations, and communities. For many, ecosystem ecology is the least appealing part of ecology because it includes more chemistry, phys-ics, and math than many of the other subfields of ecology, few if any familiar or charismatic individual organisms, and an entire vocabulary that refers to the arcane details of nutrient cycles. We think this perception can likely be credited to the way ecosystem ecology is taught rather than to its content. Our interests and enthusiasm for ecosystem ecology are directly traceable to the fact that it intellectually embraces all of the subdisciplines in ecology—including evolutionary ecology—and that it addresses the energy and matter that are exchanged among all organisms and their environment: it is the fabric that ties together all of ecology. Furthermore, and particularly relevant to the 21st century, ecosystem ecology is critical for understanding our relationships with the environment. The ecological importance of many current environ-mental problems, including disturbance, ecosystem restoration, and global warming, can best be understood within the context of ecosystem ecology.

Reiners (1986) suggested that there are at least two models or theories of ecosystem ecology. One addresses ecosystem energetics and is largely based on the second law of thermodynamics. The second deals with ecosystem stoichio-metry, or how the fundamental ratios of elements in organisms control the distribution of elements in the environment. Reiners further speculated that a third theory might address the response of ecosystems to disturbances.

In this chapter, we seek to integrate ideas that relate to ecosystem energetics and stoichiometry, and to provide a framework for predicting the effects of disturbance on ecosystems. Our domain includes energy and matter cycling at all spatial and temporal scales, and our approach is grounded in systems theory.

Ecosystem ecology is firmly rooted in physics, chemistry, and biology. All ecological processes are constrained by the laws of thermodynamics. But because ecosystems are open with respect to energy; for example, through the process of photosynthesis, they store energy in carbon–carbon bonds against the forces of entropy. Nearly all of the energy available for organisms and their interactions is provided in these carbon–carbon bonds. Both autotrophs and heterotrophs rely on these compounds as inputs to energy-releasing metabolic cycles, so the spatial and temporal pattern in the net production of reduced carbon compounds is the most fundamental attribute of ecosystems and of the biosphere, driving trophic dynamics, the cycling of all biologically active elements, and the distribution of key chemical and physical conditions including oxidation, temperature, and acidity. Thus, carbon balance represents the currency of both energy and matter in ecosystems.

We propose that the single most important process for understanding ecosystem structure and function is net ecosystem carbon balance (NECB). NECB is the net result of carbon inputs and losses in a system over a particular time period; it may be conceived as the net change in carbon (or organic matter) storage in any particular system (as defined by the user, from a specific ecosystem to the globe) due to all biological and physical processes and human management practices (Fig. 11.1) (Chapin et al. 2006). For the biological components of NECB (often called net ecosystem exchange), the carbon balance is determined by inputs through autotrophic carbon fixation (generally photosynthesis, but also some chemoautotrophy), and outputs via respiration by all organisms (autotrophic and heterotrophic). Physical and human processes that influence NECB include lateral transport (erosion, organic matter redistribution including via harvest), losses of volatile CH_4 or CO, or leaching losses of soluble carbon. There have been many iterations of terminology associated with carbon balance, but we use a recent review that defines NECB (Chapin et al. 2006). Thus,

$$NECB = (\text{Photosynthesis} - \text{Respiration}) \pm F_{CO}$$
$$\pm F_{CH_4} \pm F_{VOC} \pm F_{DIC} \pm F_{DOC} \pm F_{PC} \quad , \quad (11.1)$$

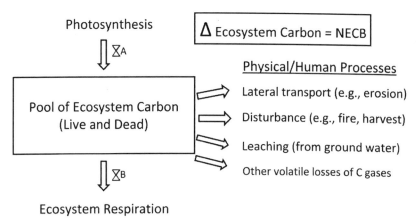

Figure 11.1 Net ecosystem carbon balance (NECB) represents the net storage of energy as reduced carbon in ecosystems and integrates understanding of matter and energy in ecosystems. Relationships such as these above have been synthesized in many ecosystem models, from the site-specific scale to regional and global scales (Canham et al. 2003), representing the application of ecosystem theory to predictions of future conditions. In general, the environmental conditions controlling photosynthesis (A in figure) and those controlling ecosystem respiration (B) differ in their importance (see text).

where, F_{CO} is the flux of carbon monoxide, F_{CH_4} is the flux of methane, F_{VOC} is the flux of volatile organic carbon, F_{DOC} is the flux of dissolved organic carbon, and F_{PC} is the flux of particulate carbon. All fluxes may be negative or positive. NECB is closely related to disturbance and successional cycles (Gorham et al. 1979), such that recently disturbed ecosystems (e.g., those impacted by a fire or hurricane) are losing carbon because photosynthesis is low and decomposition high; these systems have negative NECB and represent net carbon sources to the atmosphere. Ecosystems that are undergoing later stages of succession, in which photosynthesis is more rapid than respiration (or decomposition), have positive NECB and represent carbon sinks relative to the atmosphere. Ideas associated with NECB integrate key components of ecosystem ecology, including stoichiometry, energetics, and predicting and understanding the consequences of disturbance. Some ecosystems are defined by the extent to which they receive allothonous sources of organic carbon (wetlands, estuaries, some streams); and for these, the "F" terms above are particularly important in determining NECB (Chapin et al. 2006).

Systems theory and ecosystem modeling are critical approaches and tools for ecosystem science (Lauenroth et al. 1998; Lauenroth et al. 2003). In this chapter, we elucidate ideas and their logical structure that have been embedded in ecosystem models since early in the history of the discipline (Fig. 11.1; Canham et al. 2003). Today, ecosystem models represent the major way in which ecosystem theory is synthesized and applied to predictions of future conditions.

Below, we introduce two theories and their constituent propositions that focus on energy and matter cycling, using NECB as the integrating function. Each theory is linked to Scheiner and Willig's general theory of ecology presented in Chapter 1. For each proposition, we present a brief synopsis of the historical importance of the idea and a case study for the application of the idea to current ecological science.

Theory 1: Net ecosystem carbon balance, energetics, stoichiometry, environment, and evolution

The rates of net ecosystem carbon balance and its distribution in time and space both respond to and control patterns and levels of oxidation, temperature, ultraviolet radiation, and the distribution of elements; NECB has ultimately constrained the evolution of metabolic pathways (Table 11.1). In turn, patterns of NECB respond to and drive element distribution and environment.

Proposition 1

The heterogeneous distribution of elements and chemical features of Earth and ecosystems (oxidation, temperature, acidity, and nutrient retention) causes a separation in time and space between organisms that are dominantly autotrophic vs. those that are heterotrophic. Patterns of NECB are complex, taking place on temporal scales of seasons to millennia, and occurring on a scale that ranges from individuals to the entire biosphere. The key components of NECB, photosynthesis and respiration, are each controlled to a different degree by the physical environment (water, temperature, light, oxygen, other nutrients, gravity, etc.). For instance, total rates of heterotrophic respiration are most sensitive to oxygen availability (and other electron acceptors) and temperature. Photosynthesis, or net primary production (gross photosynthesis minus plant respiration), is most sensitive to light, water, and nutrients (Fig. 11.1). Because physical conditions vary in time and space, the balance of heterotrophic respiration and autotrophic production do as well, resulting in variable accumulation of organic matter, as well as the consumption and production of oxygen and carbon dioxide.

Table 11.1 Propositions of the theory of net ecosystem carbon balance

1. The heterogeneous distribution of elements and chemical features of Earth and ecosystems (oxidation, temperature, acidity, and nutrient retention) causes a separation in time and space between organisms that are dominantly autotrophic vs. those that are heterotrophic.[a]

2. The separation in time and space of autotrophic and heterotrophic processes drives the distribution of biologically active elements and many key chemical features of Earth and ecosystems.[a]

3. Ecosystem function, through NECB, has constrained evolution. Rates of NECB and its distribution control rates of oxidation, global temperature, and the levels of ultraviolet radiation reaching Earth's surface, and have ultimately constrained the evolution of metabolic pathways.[b]

4. Because organisms have relatively constant stoichiometry (ratios of carbon to other elements), the pattern of carbon storage by ecosystems (NECB) provides the fundamental control over spatial and temporal patterns of element retention in ecosystems, element release from ecosystems, and the distribution of elements among ecosystems.[c]

5. Each element has unique chemistry that contributes to its biological function, stoichiometric relationships, and distribution in the environment.[d]

[a] Related to Scheiner and Willig, Chapter 1, Principles 1, 3, 4, 5, and 6.
[b] Related to Scheiner and Willig, Chapter 1, Principle 8.
[c] Related to Scheiner and Willig, Chapter 1, Principle 2.
[d] Related to the Laws of Thermodynamics and Conservation of Matter.

Proposition 2

The separation in time and space of autotrophic and heterotrophic processes drives the distribution of biologically active elements and many key chemical features of Earth and ecosystems. Perhaps the best example of the effects of the separation in time and space of autotrophic and heterotrophic processes is during Earth's early history, when organic matter accumulated on the surface, an oxygenated atmosphere formed, a stratospheric ozone layer was produced, incident ultraviolet radiation was reduced, greenhouse gas carbon dioxide was removed from the atmosphere, and global temperature dropped; in these changes are the roots of the "co-evolution of climate and life" (Lovelock and Margulis 1974; Margulis and Lovelock 1974; Schopf 1983; Schneinder and Londer 1984). The separation of autotrophic and heterotrophic processes continues to both respond to and drive element distribution and the key chemical features of Earth and ecosystems (oxidation, temperature, acidity, and nutrient retention). Areas with low water availability (deserts), low light (at depth in lakes), or low nutrient availability (open oceans) are limited in autotrophic potential and all biological activity. Of all of the times and places where autotrophic and heterotrophic processes are differentially favored, perhaps most important are those in which autotrophic activity is high, but limitations in oxygen or alternate electron acceptors result in low rates of respiration; this leads to accumulations of organic matter (positive NECB), which occurred during the Paleozoic Era.

Application and advancement of these ideas at this time are most evident in studies of land-atmosphere interactions related to ecosystem influences over global and regional climate change and variability. Net ecosystem carbon balance is strongly influenced by climate, by land use change (including changing patterns of disturbance, such as fire), and by nitrogen deposition; as the theory states, these changes also have strong feedbacks to atmospheric dynamics and climate. Over the past two centuries, land use change and resulting negative terrestrial NECB on a global scale have been responsible for an estimated 25–49% of the increase in atmospheric CO_2 concentration, with a decrease in relative impact today, which is closer to 12% (Brovkin et al. 2004). Human use of fixed carbon compounds for food, fuel, and building materials represents ~24% of potential net primary production (Vitousek et al. 1986; Haberl et al. 2007). Future warming of the global climate is predicted to decrease terrestrial ecosystem NECB, but much is unknown (Field et al. 2007). The interactions of climate change and ecosystem dynamics will partly depend on vegetation responses to what may be novel climates for species assemblages (Williams and Jackson 2007). Further, biophysical feedbacks strongly influence terrestrial ecosystem structure and functioning on latent and sensible heat exchange, the hydrologic cycle, and resulting regional-scale weather and climate (Bonan 2008). The answers to some of these key questions of land-atmosphere dynamics will advance not only ecosystem science, but also influence human responses to compelling global environmental issues.

Proposition 3

Ecosystem function, through NECB, has constrained evolution. Rates of NECB and its distribution control rates of oxidation, global temperature, and the levels of ultraviolet radiation reaching Earth's surface, and have ultimately constrained the evolution of metabolic pathways.

Many authors have written about the coevolution of climate and life (Lovelock and Margulis 1974; Margulis and Lovelock 1974; Schopf 1983;

Schneinder and Londer 1984), but few represent this as a consequence of ecosystem processes. Clearly, the metabolic pathways of the first organisms evolved under the constraints of an environment characterized by high ultraviolet radiation, few energy sources, and low availability of electron acceptors (e.g., oxygen). The key feedback to altering the environment and changing the constraints over evolution of new metabolic pathways was positive global NECB.

Since the beginning of life, many important alternate pathways for autotrophy and heterotrophy have evolved. There are numerous pathways for heterotrophic respiration using electron acceptors other than oxygen, and these are a good deal older than aerobic respiration (Lurquin 2003). Early autotrophic pathways (chemoautotrophs) and heterotrophic pathways (anaerobic) have both evolved under low oxygen environments. But the dominance of aerobic environments on Earth occurred because of the positive NECB since photosynthesis developed. The associated buildup of oxygen, according to current theory, generated an ozone layer that reduced incident ultraviolet radiation and mutation rates and led to the current biochemical dominance of aerobic respiration (Cockell and Blaustein 2001). Further, the positive NECB since Earth formed has altered global temperatures by lowering greenhouse gas concentrations.

There are numerous current applications and advances in ecosystem science related to the evolutionary significance of biochemical pathways. Many of the capabilities of microorganisms that developed during earlier environmental conditions may be critically important for understanding and managing greenhouse gas concentrations from an ecosystem perspective. For instance, methane, an important greenhouse gas, is oxidized by microorganisms that likely evolved very early in the Earth's history, when the atmosphere was predominantly reducing (Hallam et al. 2004). Understanding the molecular genetics associated with this process could be important not only for understanding evolutionary ecology, but for understanding ecosystem functioning and long-term greenhouse gas concentrations (Hallam et al. 2004). Understanding the genetic structure that gives rise to many other microbial processes may have great importance for managing ecosystem greenhouse gas fluxes (Zak et al. 2006).

Proposition 4

Because organisms have relatively constant stoichiometry (ratios of carbon to other elements), the pattern of carbon storage by ecosystems (NECB) provides the fundamental control over spatial and temporal patterns of element retention in ecosystems, element release from ecosystems, and the distribution of elements among ecosystems. Small differences in stoichiometry among organisms feed back to influence ecosystem function.

An important component to ecosystem theory since the 1950s has been that organisms have relatively constant element stoichiometry, and that through their activities they control the element distribution within their environments (Redfield 1958; Reiners 1986; Elser et al. 2000).

We assert that the critical mechanism of this stoichiometric feedback between organisms and their environments occurs through NECB. The distribution of all biologically active elements can be explained by the fact that they are bonded to carbon skeletons and travel through the biota in concert with carbon. Thus, the rate at which carbon skeletons are formed and broken (NECB) controls element cycling. Reiners (1986) described two major categories of organic compounds, protoplasmic and mechanical, which have distinct element ratios. For instance, protoplasmic compounds (enzymes, nucleic acids, etc.) have a high concentration of nitrogen, while mechanical or structural components (wood, bone, shell) have higher concentrations of materials that are resistant to decay (e.g., lignin), which tend to accumulate in soils and sediments. Thus, the type of organism and its tissue allocation influences element distribution. However, the key notion is that all elements, when biologically active, are bonded to carbon skeletons formed through autotrophy and broken through heterotrophy. Thus, when NECB is positive, biologically active elements are stored, and when it is negative, they are released to inorganic forms, leaving them vulnerable to movement through gaseous or soluble form or to being bound in mineral form. In other words, elements cannot accumulate in ecosystems unless NECB is positive. As Aldo Leopold wrote in his essay "Odyssey,"

> An atom at large in the biota is too free to know freedom; an atom back in the sea has forgotten it. For every atom lost to the sea, the prairie pulls another out of the decaying rocks. The only certain truth is that its creatures must suck hard, live fast, and die often, lest its losses exceed its gains. (Leopold 1949)

Over the past several decades, there have been many important advances related to the idea that disturbance to the spatial and temporal distribution of NECB alters fundamental environmental characteristics such as oxygen concentration, temperature, and element distribution and retention. Alternatively, disturbance to those environmental characteristics alters the spatial and temporal distribution of NECB. We offer three examples that span local, regional, and global scales to elaborate these ideas.

In the mid-1970s, forest ecologists and managers proposed that the most important predictor of nitrates in stream water was watershed NECB (Vitousek and Reiners 1975; Gorham et al. 1979). This hypothesis has been tested and extended to other nutrient elements (e.g., phosphate) in forests as well as stream, desert, and urban ecosystems (Gorham 1961; Lewis and Grimm 2007). In general, reductions in NECB allow inorganic forms of nutrients to be exported from ecosystems. By contrast, disturbances that increase NECB, such as water addition in dry areas or nitrogen deposition in areas with adequate water, promote storage of nutrient elements in organic form.

Ideas regarding stoichiometry have recently been important for advancing the understanding of terrestrial-aquatic linkages with respect to a key environmental problem: large-scale eutrophication of fresh waters and resulting hypoxia of coastal waters (Schindler 2006; Diaz and Rosenberg 2008; Conley et al. 2009). Eutrophication is well known to occur because increases in phosphorus or nitrogen disproportionately increase autotrophic activity, subsequently leading to net heterotrophic activity that results in oxygen deficits (Howarth 2008). Although reducing anthropogenic inputs of either element would seem to be a high priority for reducing the possibility of environmental collapse (Conley et al. 2009), recent results suggest that the balance between N inputs and P inputs is critical for predicting appropriate management strategies. Schindler et al. (2008) used long-term whole-ecosystem lake experiments to show that reducing N inputs alone does not reduce eutrophication; cyanobacterial N fixation increases in response to reduced N, maintaining N:P ratios and the eutrophic condition. Carpenter (2008) and Schindler and Hecky (2009) suggest that studies of N:P stoichiometry over large temporal and spatial scales, particularly in estuarine systems, are at the scientific forefront for understanding how to better mitigate hypoxia.

Proposition 5

Each element has unique chemistry that contributes to its biological function, stoichiometric relationships, and distribution in the environment. Three aspects of elemental chemistry (below we refer to them as elemental observations) may be used to predict the differences among elements in their ecosystem cycling, their accumulation in ecosystems, and their impact on net ecosystem carbon balance.

There are interesting and important distinctions among the biologically active elements; fascination with these differences has resulted in the plethora of terminology surrounding element cycles that our students and occasionally our colleagues in different subdisciplines find to be dense and difficult (e.g., nitrification, denitrification, ammonification, dissimilatory sulfate reduction). However, if one focuses on three simple observations, the element cycles become highly tractable and closely connected to a theory of ecosystem ecology. The key that connects these observations to ecosystem theory is that the most important differences among biologically active element cycles occur when elements are in their inorganic phase. This proposition explicitly connects the ideas of NECB and stoichiometry (Theory 1) with disturbance (Theory 2), and provides an explanation for the connections of nutrient cycles with carbon.

Elemental observation 1: Elements that are highly soluble or tightly captured by soils or that have a gaseous phase may be lost from actively cycling in ecosystems. These are the elements whose availability most frequently constrains the C cycle.

Thus, nitrogen, when not bound to carbon in biomass, is not conserved in ecosystems (Gorham et al. 1979; Vitousek and Melillo 1979; Vitousek et al. 1997a), and as a consequence nitrogen is the most commonly cited limiting element in terrestrial ecosystems, for instance (Attwill and Adams 1993). In its form as nitrate (a monovalent anion), nitrogen is easily leached from soil, and in its gaseous form as nitric or nitrous oxide or ammonia it may be volatilized and lost as well. Alternatively, phosphorus, because of its lack of a gaseous form and its capability as phosphate to bind with metal cations at low pH or calcium at high pH, is conserved in soils, but it is also limited from actively cycling in ecosystems, thereby sometimes constraining NECB.

Elemental observation 2: Elements (other than oxygen) that can serve as electron acceptors influence NECB by allowing heterotrophy to occur in aerobic places, frequently resulting in the production of greenhouse gases.

Elemental observation 3: Elements, in their reduced form, can serve as energy sources where reduced organic carbon is not available.

These distinctions among elements based upon their electron affinity are a critical component to understanding propositions 3, 4, and 5, and tie together chemistry, evolutionary biology, ecosystem functioning, and global change. Because nitrogen, sulfur, and carbon, in their oxidized forms (nitrate, sulfate, and carbon dioxide), as well as some metals, can serve as electron acceptors when oxygen is in short supply, substantial heterotrophy can occur. When the electron acceptor is nitrate or carbon dioxide, products may be nitrous oxide or methane, which are both active as greenhouse gases with significantly higher per molecule radiative warming potential than carbon dioxide. Sulfur can be particularly important in both roles: sulfate is a key electron acceptor in anaerobic marsh environments (Howarth 1984), allowing a large proportion of stored energy to be biologically accessible. Further, sulfide (elemental observation 3) is utilized as an energy source in amphi-aerobic environments by chemolithoautotrophic bacteria, potentially contributing a significant amount of carbon balance in some environments (Howarth 1984).

Theory 2: Disturbance and ecosystems

Among the most important areas for ecosystem ecology theory to contribute to problems today is in understanding and predicting the effects of disturbance on ecosystems (see Pickett et al. Chapter 9 for a further discussion of impacts of disturbance and Peters et al. Chapter 12 for human impacts on ecosystems). White and Pickett (1985) define disturbance as "any relatively discrete event in time that disrupts ecosystem, community, or population structure and changes resources, substrate, or the physical environment." When focusing on energy and matter dynamics, disturbance can be defined as any alteration in NECB, including intentional land or water management and unintentional impacts or changes in the natural disturbance cycle. The effects of disturbance on matter and energy dynamics in ecosystems may be predicted by three fundamental characteristics: the turnover time of matter and energy storage compartments, the above- and belowground distribution of biomass, and the evolutionary history of the dominant organisms (Table 11.2).

Ecosystem ecologists conceive of storage and cycling of matter and energy using systems approaches (e.g., Odum 1983). We conceptualize storage compartments, or pools, as somewhat arbitrary collections that are connected by flows of material or energy (fluxes). We say arbitrary because often our conceptualization of components or fluxes is not strictly measureable or even

Table 11.2 Propositions of the theory of disturbance and ecosystems.

1. The resistance or resilience of a pool is determined in part by the turnover time [a]

2. The distribution of biomass (dead or alive) above- and belowground is an important determinant of the resistance or resilience of an ecosystem to any particular disturbance.[b]

3. Ecosystems subjected to disturbances that fall within the evolutionary history of the dominant organisms are characterized by resistance or resilience to change; those subjected to disturbances outside the evolutionary history of the dominant organisms respond with either altered steady states or instability.[c]

[a] Related to the laws of thermodynamics and conservation of matter.
[b] Related to Scheiner and Willig, Chapter 1, principle 2.
[c] Related to Scheiner and Willig, Chapter 1, principle 8.

biologically accurate. For instance, the separation of live biomass from dead is not practically possible in soil systems, which are complex assemblages of root material, exudates, live and dead microbial biomass, micro and macro invertebrates, and detritus. However, the construction of simple and complex conceptual and simulation models of ecosystems has led to important theoretical advances and practical understanding.

We identify four general responses to ecosystem perturbation: resilience, resistance, altered steady states, and instability (Fig. 11.2; Holling 1973; Pimm 1984; Aber and Melillo 2001; Carpenter et al. 2001). In this paper, we define resilient ecosystems as those that are easily removed from their initial state but return to it quickly (most akin to Pimm 1984). Systems characterized by resistance are difficult to remove from their original state, but once removed, recover very slowly (again, Pimm 1984). Altered steady states occur in ecosystems when irreversible changes occur, but changes lead to reasonably "stable" new situations. Instability of ecosystems or system components may occur when changes are wrought that lead to the collapse of food webs (Post et al. 2002) or the reductions in capability of systems to maintain key struc-

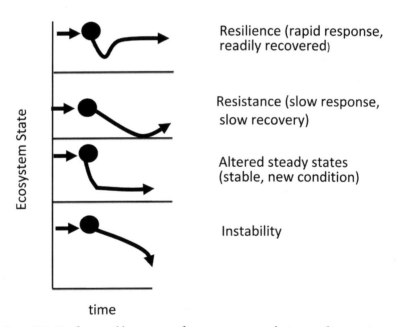

Figure 11.2. The four possible responses of ecosystems to perturbations: resilience, resistance, altered steady states, and instability. The responses may be predicted by the turnover time of the affected pools and the evolutionary history of the organisms.

tural components (e.g., catastrophic soil loss). Avoiding irreversible changes requires understanding what types of process may lead to each of these four responses.

Among the most useful concepts for quantifying and describing ecosystem function is turnover time (also called residence time), which is an estimate of the rate at which units of energy or matter pass through pools. Turnover time can be estimated as the size of the pool divided by either its rate of input, or its rate of output (often the pool is assumed to be at steady state, so that the calculation is the same). Temporal dynamics of systems or their components are, as we will see below, closely linked to the turnover time of the compartments.

Proposition 1

The resistance or resilience of a pool is determined in part by the turnover time. Pools with short turnover times tend to be resilient, that is, easily removed from their initial state but quick to recover; those with long turnover times tend to be resistant, or more difficult to perturb, but recover slowly from disturbance (Odum 1983; Aber and Melillo 2001).

Disturbances may decrease ecosystem storage by reducing primary productivity through killing live plant parts or whole plants (harvest, logging, fire), through export of live material (fuel or fodder), or through increasing ecosystem respiration (increase in temperature, draining to stimulate aerobic decomposition, etc.). All of these disturbances can result in negative NECB depending on the relative magnitudes of effects on autotrophy and heterotrophy. Ecosystems with a relatively long carbon turnover time (e.g., alpine tundra, deserts, semi-deserts, boreal forest) will recover slowly, perhaps over hundreds of years. Alternatively, ecosystems with rapid turnover times, such as tropical forests or annual grasslands, may recover more on the order of years to decades.

The theory is useful for predicting many other biogeochemical phenomena as well. Atmospheric pollutants with turnover times on the order of days (SO_2, 1.5 days) may be removed at a rate proportional to turnover time. It follows that if inputs ceased, acid deposition components would take only weeks or less to disappear, showing a high level of resilience. By contrast, tropospheric dimolecular oxygen (O_2, turnover time 4,500 years), characterized by a very large pool and very long turnover time, is highly resistant to change and is of essentially no concern for being substantially influenced by human activities over several centuries.

Interestingly, the same processes that have minimal impact on oxygen concentrations—fossil fuel combustion and land clearing—have major proportional impacts on the concentration of carbon dioxide (CO_2, turnover time ~4–25 years) in the atmosphere (Wigley and Schimel 2000). Pools with intermediate turnover times are of most concern, because they are responsive enough to be influenced by humans, yet resistant enough that they recover slowly. Most of the important greenhouse gases (N_2O, CH_4, CO_2, O_3) fall into this category of intermediate turnover, with recovery times substantially longer than the duration of any particular political administration.

This traditional view of disturbance and maintenance of ecosystem function has become particularly relevant as ecologists have begun to focus on ecosystem services, or the importance of ecosystem processes such as carbon storage (NECB) to human welfare (Carpenter et al. 2001; Millennium Ecosystem Assessment 2005). With increasingly intensive extraction activities for fossil fuels, forest products, water, and other natural resources, ecosystem restoration is critical to maintaining such ecosystem services (Hall and O'Connell 2007). Incorporating ecosystem theory into restoration and reclamation remains a substantial challenge for ecosystem ecology (Day et al. 2009).

Proposition 2

The distribution of biomass (dead or alive) above- and belowground is an important determinant of the resistance or resilience of an ecosystem to any particular disturbance. Disturbances may be characterized by their degree of effect on aboveground vs. belowground biomass; for instance, fire, logging, and hurricanes primarily impact aboveground plant and detrital components. Ecosystems with a large proportion of stored biomass belowground, or with short turnover times, are resilient to the impacts of aboveground disturbances (Burke et al. 2008). Fire or harvesting in perennial grasslands, resulting in removal of only aboveground biomass, generally have minimal impacts on carbon storage or nutrient cycling. In contrast, those same disturbances in forested systems have major, long-lasting impacts. Similarly, perturbations that include plowing, mining, urbanization, or focus on belowground or whole ecosystem components have long-term and practically irreversible impacts on carbon storage for all ecosystems, even those with a large proportion of biomass stored belowground.

Proposition 3

Ecosystems subjected to disturbances that fall within the evolutionary history of the dominant organisms are characterized by resistance or resilience to change; those subjected to disturbances outside the evolutionary history

of the dominant organisms respond with either altered steady states or instability. Ecosystem managers have realized that land use management strategies
that are tuned to the evolutionary history of organisms may be sustainable relative to those that are not. A single management practice, for instance grazing
by large herbivores, has different influences in ecosystems with long evolutionary histories of grazing (Serengeti, North American Great Plains; Milchunas
et al. 1988), compared to ecosystems with plants not adapted to heavy grazing
(e.g., North American Great Basin) where livestock grazing may result in irreversible changes in species composition. Similarly, ecosystems whose fire history includes long fire intervals (e.g., *Pinus contorta* in the western U.S.) may
be well adapted for fire suppression within timeframes of several centuries.
However, lower montane forested ecosystems of the Rocky Mountain Region
of the U.S., with a historical range of variability of fire frequency of 8–12 years
(Schoennagel et al. 2004), may be dramatically altered by long-term fire suppression, the buildup of fuels, and resulting catastrophic fires.

This proposition is closely linked to recent work that evaluates the relationship between ecosystem function and biodiversity and asks the question,
what are the impacts of changes in species composition on ecosystem function (Tilman and Downing 1994; Tilman et al. 1997; Tilman et al. 1998)?
Introductions of exotic species or losses of keystone species are examples that
represent changes outside the evolutionary history of ecosystems. Following
introductions by invasive species, food web dynamics, dominant plant life
form, fire dynamics and disturbance patterns, and the entire structure of ecosystems can be changed to alternative steady states (D'Antonio and Vitousek
1992; Sala et al. 1996). Clearly, inclusion of evolutionary history and natural
disturbance in land management provides predictions: ecosystems subjected
to disturbances that fall within the evolutionary history of the dominant organisms will be characterized by resistance or resilience, while those subjected
to disturbances outside the evolutionary history of the dominant organisms
will respond with either altered steady states or instability.

Linking the theory of ecosystem ecology to other constituent theories

As this chapter shows, ecosystem ecology is dependent upon other constituent theories regarding species distribution, foraging (Sih Chapter 4), niche
formation (Chase Chapter 5), metacommunity ecology (Leibold Chapter 8),
succession (Pickett et al. Chapter 9), and biogeographical gradients (Colwell
Chapter 14). As Chapter 9 and this chapter explain in detail, the parameters
of the physical environment (most importantly NECB) set the initial resource
base, physiological stress, and regulator conditions of an ecosystem, and these

parameters subsequently impact and are modified by species and their inter-actions within ecosystems. Ecosystem ecology thus provides an integrative framework that assists with understanding other aspects of ecology, as it presents the basis of energy flows and nutrient cycles, which necessarily impact all other dynamics within ecosystems.

Summary

Ecosystem theory, which embraces and integrates biology, evolutionary ecology, chemistry, physics, and global ecology, has had a strong historical influence on natural resource management. Net ecosystem carbon balance (NECB)—or the net result of carbon inputs and losses into a system over a particular time period—is the most important process for understanding ecosystem structure and function. This chapter shows: (1) how the rates of net ecosystem carbon balance and its distribution in time and space respond to and control patterns and levels of oxidation temperature, ultraviolet radiation, and the distribution of elements and constrain the evolution of metabolic pathways; and (2) how the effects of disturbance on ecosystem matter and energy dynamics may be predicted by three fundamental characteristics: the turnover time of matter and energy storage compartments, the above- and belowground distribution of biomass, and the evolutionary history of the dominant organisms. At this critical juncture in the Earth's history, when human societies are placing large demands on ecosystem services at local to global scales, ecosystem theory has much to offer for understanding ecosystem sustainability. Today, this theory is often applied through integrative tools such as conceptual and simulation models that are used to assess vulnerabilities to future changes.

Acknowledgments

Ideas in this chapter are the result of decades of interactions with valued colleagues. The authors would like to particularly recognize Bill Reiners for his early inspiration regarding theory of ecosystem ecology. The University of Connecticut supported travel to workshops to interact with authors about other areas of ecology, through the Center for Environmental Sciences & Engineering and the Office of the Vice Provost for Research & Graduate Education. Anne Jakle provided valued assistance in manuscript preparation.

CHAPTER 12

Perspectives on Global Change Theory

Debra P. C. Peters, Brandon T. Bestelmeyer, and Alan K. Knapp

Human influences on ecological drivers are increasingly recognized as dominant processes across a range of spatial and temporal scales. Regional to global-scale changes in drivers and important resources, such as atmospheric carbon dioxide concentrations, climate, and nitrogen deposition, are known to alter biotic structure, ecosystem function, and biogeochemical processes with feedbacks to human activities and the atmosphere (Petit et al. 1999; Grimm et al. 2000; Fenn et al. 2003; IPCC 2007). Human activities also directly affect ecosystems at finer scales through urbanization, species movement and extinction, and changes in land use that, in aggregate, have global impacts as human populations continue to increase and migrate (Alig et al. 2004; Theobold 2005; Grimm et al. 2008b). Although large volumes of data on global change drivers and ecological responses to them have been synthesized (e.g., Heinz Center 2002; Millennium Ecosystem Assessment 2005; Canadell et al. 2007; IPCC 2007), a coherent body of ecological theory focused on global change is lacking (Peters et al. 2008).

The U.S. Global Change Research Act of 1990 defined global change as: "Changes in the global environment—including alterations in climate, land productivity, oceans or other water resources, atmospheric chemistry, and ecological systems—that may alter the capacity of the Earth to sustain life." It is now clear that direct and indirect human actions are responsible for most of the more dramatic change occurring today and forecast for the future (Vitousek et al. 1997b). Global change is an aggregate of different forces that operate across all scales; many of these forces will be discussed in this chapter. In contrast, drivers historically studied by ecologists were assumed to occur locally and were ecosystem-specific, such as fire in forests and floods in streams. For contemporary ecological systems, a framework is needed to integrate across scales and to better understand the effects of multiple interacting drivers on ecosystem dynamics. For example, elevated concentrations of carbon dioxide (CO_2) and other greenhouse gases in the atmosphere are increasing temperatures globally whereas precipitation regimes are changing locally; ecosystem responses to these multiple, interacting drivers are often unknown (IPCC 2007). Existing ecological theories can be brought to bear on global change issues, but these theories need to be adapted and modified such that the key and sometimes unique aspects of global change drivers and responses to them are explicitly considered.

Ecosystem responses to global change drivers are often measured experimentally across a range of scales. The term "scale" generally implies a certain level of perceived detail (Miller 1978) that can be quantified using two key components: grain (the finest level of spatial and temporal resolution of a pattern) and extent (the spatial and temporal span of a phenomenon or study; Turner et al. 1989). Here we focus on a hierarchy of characteristic scales defined as the spatial and temporal scale on which ecological phenomena principally operate and can be most appropriately studied (Wu 2007). We discuss characteristic spatial scales defined primarily by spatial extent and recognize the correspondence between spatial and temporal scales (Urban et al. 1987). For example, fine scales of individuals or portions thereof (e.g., leaves) operate at small spatial (centimeters to meters) and short temporal (seconds to days) scales compared with plots or patches that contain groups of individuals, populations, or communities operating at intermediate spatial (tens of meters to ha) and temporal scales (days to years) (e.g., Shaw et al. 2002; Magill et al. 2004; Morgan et al. 2007; Siemann et al. 2007). Broad spatial scales include ecosystems at landscape scales, and biomes or geographic distributions of species at regional, continental, and global scales where pattern dynamics occur over long timescales (decades to centuries).

Fine-scale patterns of individual dynamics can be extrapolated to broader spatial extents using spatial patterns of responses combined with flows of materials (nutrients, water, propagules) and phenological responses (Walther et al. 2002; Parmesan and Yohe 2003; Root et al. 2003). Changes in regional-to global-scale patterns in vegetation structure and productivity can also be estimated directly using simulation models or remotely sensed images (Defries et al. 2000; Sitch et al. 2008). In most cases, these studies have been conducted at one scale or, in some cases, at multiple independent scales (e.g., Peterson et al. 2003). Less attention has been devoted to how patterns and processes interact across scales to generate emergent behavior (e.g., Peters et al. 2004; Allen 2007). The propagation of fine-scale dynamics to larger scales of pattern often can not be predicted using linear extrapolation as a method for upscaling. Similarly, the overwhelming effect of broad-scale drivers on fine-scale dynamics can not be understood by simply downscaling effects of these drivers (Peters et al. 2007). Alternative approaches are needed that account for cross-scale interactions (Peters et al. 2009). Because global changes in drivers and responses are inherently cross-scale and are connected spatially (Peters et al. 2008), theories of global change must account for these interactions.

There are many well-known examples of how fine-scale processes can propagate to influence large spatial extents, and how broad-scale drivers can overwhelm fine-scale variation in pattern. For example, land use practices in central Asia, including overgrazing by livestock and cultivation of marginal lands, are interacting with effects of drought to result in high plant mortality and increased soil erosion at the scale of individually managed fields (e.g., 1–10 ha). Because most farmers and ranchers are following the same practices, these field-scale dynamics often aggregate nonlinearly with thresholds to the landscape and regional scales to generate large dust storms, the frequency of which has increased from 1 in 31 years to 1 per year starting in 1990 (Liu and Diamond 2005). As these dust storms continue to expand in spatial extent, they can travel intercontinentally to influence air quality in North America (http://svs.gsfc.nasa.gov/goto?2957), and can overwhelm natural determinants of local air quality (Jaffe et al. 2003).

In this chapter, we outline the characteristics of a theory of global change that draws upon a range of existing theories, including those from foraging (Sih chapter 4) and niche theory (Chase chapter 5), population biology (Hastings Chapter 6), succession theory (Pickett et al. Chapter 9), ecosystem ecology (Burke and Lauenroth Chapter 11), and environmental (Fox et al. Chapter 13) and biogeographic (Colwell Chapter 14) theory, as well as landscape ecology (Turner et al. 2001) and other disciplines such as Earth system sciences. We develop the basis for this theory and provide supporting evidence for its utility. We also provide an example where misleading results are likely to be obtained if the underlying concepts for such a theory are not accounted for, and we discuss new research directions based on this theory.

Domain, propositions, and mechanisms

The domain of a theory of global change is the causes and consequences of ecological properties of systems when the natural and human-induced drivers interact across a spatial and temporal hierarchy of characteristic scales. As a theory of global change develops, there are four key propositions that need to be considered—two occur under current conditions with natural variation in drivers, and two are unique to systems experiencing global change (Table 12.1). These four propositions provide the context from which we derive our perspective on global change theory.

Our propositions follow from the eight fundamental principles of a theory of ecology, as proposed by Scheiner and Willig (2008; Chapter 1). In addition, spatial and temporal interactions of patterns and processes, originally articulated by (Watt 1947), are a fundamental concept in our developing theory. Our perspective integrates a number of theories operating at specific scales with theories that link scales. We also incorporate ideas from other disciplines (physical sciences, human systems) to address connectivity by water, wind, and humans. We believe that any global change theory needs to be organized by spatial scales that correspond to scales through time.

Propositions under current conditions

In our developing theory, pattern-process relationships interact across a hierarchy of scales (proposition 1) to result in spatial heterogeneity and connectivity among spatial units to be the critical system properties governing dynamics (proposition 2, Table 12.1). Our first proposition is derived from hierarchy theory (Allen and Starr 1982) and a framework for interactions across scales (Peters et al. 2004; 2007). Our second proposition combines a framework developed for heterogeneous arid landscapes (Peters et al. 2006) with an emerging connectivity framework designed to relate fine-scale dynamics with broad-scale drivers (Peters et al. 2008).

Proposition 1. Interactions across scales. Our first proposition is that there is natural variation in environmental drivers and system responses that form a hierarchy of interacting spatial and temporal scales (Fig.12.1a). This proposition expands on general principles from hierarchy theory, where a small number of structuring processes control ecosystem dynamics; each process operates at its own temporal and spatial scale (Allen and Starr 1982; O'Neill et al. 1986). Finer scales provide the mechanistic understanding for behavior at a particular scale, and broader scales provide the constraints or boundaries on that behavior. Other chapters in this book describe these fine-scale mechanisms more completely (Holt Chapter 7; Burke and Lauenroth Chapter 11) or describe the relationships among scales, such as the use of patches to explain landscape-scale dynamics in metapopulation theory (Hastings Chapter 6) and succession theory (Pickett et al. Chapter 9), and develop scaling relationships (Leibold Chapter 8)

Table 12.1 The domain and four propositions associated with a developing theory of global change. Propositions 1 and 2 occur under current conditions with natural variation in drivers. Propositions 3 and 4 are unique to systems experiencing global change.

Domain

> Causes and consequences of ecological properties of systems when the natural and human-induced drivers interact across a spatial and temporal hierarchy of scales.

Propositions

1. Pattern-process relationships interact across a hierarchy of scales.

2. Intermediate-scale properties associated with connectivity and spatial heterogeneity determine how pattern-process relationships interact from fine to broad scales. More specifically:

    a. Global-scale patterns emerge from a hierarchy of interacting processes that propagate responses from fine to broad scales (i.e., plants to landscapes and regions). Fine-scale patterns often cannot be understood without knowledge of broader-scale processes.

    b. Dynamics at any location on the globe are affected to varying degrees by transfer processes that connect adjacent as well as distant locations.

    c. Transfer processes (wind, water, biota) connect locations via the movement of organisms, materials, disturbance, and information. The loss of historic transfer processes can result in disconnected locations. Conversely, an increase in magnitude and frequency of transfer processes can increase connectivity among previously isolated locations.

    d. Spatial heterogeneity determines how drivers and transfer processes interact and feedback on one another across scales.

    e. The relative importance of fine- or broad-scale pattern-process relationships can vary through time, and alternate as the dominant factors controlling system dynamics.

3. Human activities associated with disturbance and modifications of resources are ultimately the dominant drivers of global change.

4. Global change drivers are of unprecedented magnitude.

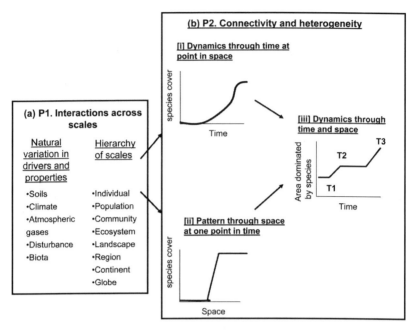

Figure 12.1. Two propositions upon which a global change theory is based are occurring under current conditions: natural variation in system properties (drivers and responses) and connectivity among spatial units that link dynamics through time and across space. Connectivity is represented by the slope of each line segment between each pair of thresholds (T1-T2, T2-T3). Species cover is used as an example; other response variables are possible.

The concept of pattern-process interactions provides a general mechanism for dynamics within scales that lead to shifts in "scale domains" (sensu Wiens 1989). Functional relationships between pattern and process are consistent within each domain of scale such that linear extrapolation is possible within a domain (Wiens 1989). Thresholds occur when pattern-process relationships change rapidly with a small or large change in a pattern or environmental driver (Bestelmeyer 2006; Groffman et al. 2006); both external stochastic events and internal dynamics can drive systems across thresholds (Scheffer et al. 2001).

Interactions among local and broader-scale processes can be important to patterns of distribution, abundance, and diversity (Ricklefs 1987; Levin 1992; Carpenter and Turner 2000). These cross-scale interactions generate emergent behavior that cannot be predicted based on observations at single or multiple independent scales. For example, human activities at local scales can drive land change dynamics at regional scales (Luck et al. 2001; Dietz et al. 2007). Cross-scale interactions can also be important to metapopulation dynamics

(Hastings Chapter 6) in that demographic and dispersal processes interacting with habitat heterogeneity can drive these dynamics across scales (Schooley and Branch 2007). A number of theories have used hierarchy theory as a basis for describing cross-scale interactions, including theories of complex systems (Milne 1998), self-organization (Rietkerk et al. 2004), panarchy (Gunderson and Holling 2002), and resilience (Holling 1992). Recently, a framework was developed to explain how patterns and processes at different scales interact to create nonlinear dynamics with thresholds (Peters et al. 2004; 2007). This framework focuses on the importance of connectivity and spatial heterogeneity in determining how pattern-process relationships interact across scales and forms the basis for our proposition 2.

Proposition 2. Connectivity and spatial heterogeneity. Our second proposition is that intermediate-scale properties associated with connectivity and spatial heterogeneity determine how pattern-process relationships interact from fine to broad scales (Fig. 12.1b). Within a domain of scale (i.e., fine, intermediate, or broad), patterns and processes reinforce one another and are relatively stable. However, changes in drivers or disturbance can modify these pattern-process relationships in two ways. (1) Fine-scale patterns can result in positive feedbacks where new processes and feedbacks become important as the spatial extent increases. This change in dominant process is manifested as nonlinear threshold changes in pattern and process rates. The nonlinear propagation of fire through time is one example where connectivity in fuel load shifts from individual tree morphology to within-patch distribution of overstory and understory plants to among-patch variability in topography and species distributions as the spatial extent of a fire expands (Allen 2007). (2) Broad-scale drivers can overwhelm fine-scale processes, such as regional drought that produces widespread erosion and minimizes the importance of local process such as competition to ecosystem dynamics. At the scale of landscapes, dispersal of invasive species can overwhelm local environmental variation in vegetation, soils, and grazing pressure to drive invasion dynamics (Peters et al. 2006).

Spatial heterogeneity and connectivity are interrelated: it is the combination of trends through time and patterns across space that lead to measures of connectivity. For example, expansion of an invasive species across a landscape can follow a nonlinear pattern where cover of the invasive increases through time for any given point within a spatial extent (Fig. 12.1b[i]). At any given point in time, invasive species cover is heterogeneously distributed across the spatial extent in a number of different ways, from uniformly high or low or with a gradient or ecotonal spatial structure (Fig. 12.1b[ii]). Combining trends through time with patterns in space leads to nonlinear changes in area dominated by the invasive species as a percentage of the spatial extent through time (Fig. 12.1b[iii]). There are three thresholds in the system that are points in time where the slope of the line changes discontinuously as the dominant process changes (T1, T2, T3). A species that initially invades a landscape will first spread within a patch such that local recruitment of seeds and competition among plants are the dominant processes. As more seeds are produced and more plants succeed within a patch, a threshold is crossed where dispersal to other patches becomes increasingly important to landscape-scale pattern. The slope of each line segment (e.g., % invasive cover/ha/y) between each pair of thresholds (T1–T2; T2–T3) is a measure of the connectivity of plants of the invasive species across the landscape. The importance of spatial context is illustrated by the adjacency of each point to other points such that points closer to the area dominated by the invasive species are more likely to be invaded than points at greater distances.

Our connectivity proposition is itself based on the following related propositions (Table 12.1) developed from Peters et al. (2007). (1) Global-scale patterns emerge from a hierarchy of interacting processes that propagate responses from fine to broad scales (i.e., plants to landscapes and regions). Fine-scale patterns often cannot be understood without knowledge of broader-scale processes. (2) Dynamics at any location on the globe are affected to varying degrees by transfer processes that connect adjacent as well as distant locations. (3) Transfer processes (wind, water, biota) connect locations via the movement of organisms, materials, disturbance, and information. The loss of historic transfer processes can result in disconnected locations. Conversely, an increase in magnitude and frequency of transfer processes can increase connectivity among previously isolated locations. (4) Spatial heterogeneity determines how drivers and transfer processes interact and feed back on one another across scales. (5) The relative importance of fine- or broad-scale pattern-process relationships can vary through time, and alternate as the dominant factors controlling system dynamics.

We illustrate changes in cover through time and across space as estimates of connectivity and a description of the mechanisms producing these changes using a landscape change scenario from the northern Chihuahuan Desert in southern New Mexico, USA (Peters et al. 2004). A combination of field survey–based maps (1915, 1928/29), black-and-white aerial photos (1948), and pan-sharpened QuickBird satellite images (2003) were georegistered for a 942 ha pasture at the USDA ARS LTER site north of Las Cruces, NM (32°30′N, 106°48′W) (Fig. 12.2a). Three classes of vegetation were digitized manually, and ARCGIS was used to obtain the area occupied by each of three classes through time: shrubs, grasses, and the ecotone between them.

In this landscape-scale example, most points on the landscape convert from

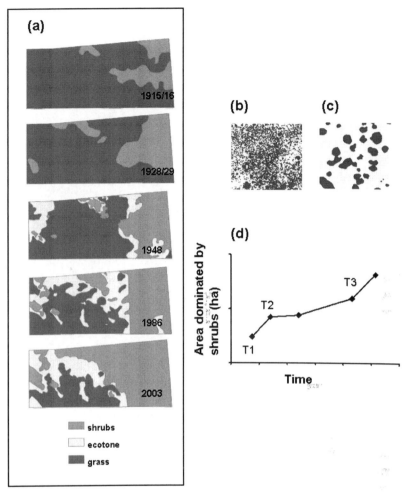

Figure 12.2 An example of a measure of connectivity in shrubs in the northern Chihuahuan Desert. Area covered by one of three vegetation types: shrubs, grasses, and the ecotone between them, was calculated for four dates using vegetation types and aerial photos on the left, and displayed through time in the graph for shrubs only. Three points in time and space were found where the rate of change in area increased nonlinearly to indicate a threshold (T1, T2, T3). The slope of each line segment between thresholds is a measure of the connectivity of shrubs over that time period. Insets show homogeneous plant cover [green in (b)] when the area is dominated by grasses. Under shrub dominance, patches of shrubs [green in (c)] are disconnected by bare interspaces that allow erosion by wind and water. Adapted from Peters et al. (2004).

grass-dominated (Fig. 12.2b) to shrub-dominated cover (Fig. 12.2c) through time. At any point in time, in general, the pattern across the landscape changes from grass-dominated in the west-southwest (left side of panels in Fig. 12.2a) to shrub-dominated in the east (right side of panels in Fig. 12.2a). Aggregating this information to the entire landscape results in a nonlinear increase in area dominated by shrubs through time (Fig. 12.2d). Three thresholds occur that are likely associated with a change in the dominant process driving dynamics across the landscape, from interspecific competition between individual plants in the early stages (prior to T1) to connections between shrubs by fine-scale water redistribution and long-distance seed dispersal (T1–T2). Recruitment and seed dispersal (T2–T3) create connections among shrub patches as infilling occurs, although at a slower rate than the previous years. At later stages (T3→), the density and spatial arrangement of shrub patches result in low connectivity among isolated shrubs. In contrast, bare soil interspaces become highly connected by wind erosion to result in deposition of soil and nutrients under shrub canopies. These positive feedbacks to shrub persistence promote the development of dune fields that further limit success of grasses (Peters et al. 2004).

Additional propositions under global change

There are two additional propositions that are unique to and characterize phenomena considered under global change (Table 12.1), and these in particular present unique challenges to existing theories (Fig. 12.3b).

Proposition 3. Human drivers of global change. The third proposition is that human activities associated with disturbance and modifications of resources are ultimately the dominant drivers of global change (P3: Fig. 12.3b top panel). The consequences of this proposition are that the dynamics and characteristics of key drivers previously recognized as governed by natural earth systems processes and feedbacks, such as atmospheric CO_2 and other greenhouse gases, climate, and nitrogen deposition, are now determined to a large extent by human activities. These activities are a product of cultural, economic, and social systems (Pickett et al. 2001b).

When combined with the widespread direct impacts of human actions on biological and ecological systems, this proposition means that many of the primary forces of change in ecology, as well as responses, interactions, and consequences of change, operate either partially or completely independent of evolutionary mechanisms, such as natural selection, that historically have been considered essential for understanding the ultimate basis of and context for ecological dynamics (Vitousek et al. 1997b; Palmer et al. 2004). For example,

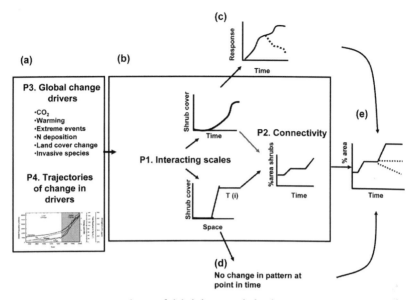

Figure 12.3 Our emerging theory of global change includes the two propositions under current conditions—variation in system properties (P1) and connectivity (P2)—combined with global change drivers (P3) that are experiencing changing trajectories through time (P4). The result is that connections among spatial units are increasing in both magnitude and frequency of occurrence primarily as a result of changes to trajectories through time rather than substantive changes in pattern.

plant communities that exist in urban ecosystems do not necessarily reflect adaptations to the local environment—instead the environment is often altered to permit the species to coexist. Thus, the spatial patterns of individuals, their population dynamics, and overall productivity are largely a function of human activities and preference driven by socioeconomic factors (Grimm et al. 2008a). Other examples include agricultural fields, water bodies devoted to aquaculture, planted "improved" pastures, and forest plantations in which the dominant species and their traits are no longer a product of evolutionary and ecological interactions, but instead are largely influenced by a human value system. Less obvious, but no less pervasive, is the attempted management and restoration (decidedly human activities) of natural areas to match environments and to achieve ecological states that may no longer exist (Hobbs et al. 2006), thus requiring significant resource inputs and human intervention (Seastedt et al. 2008).

Proposition 4. Change in trajectories of global change drivers. Our fourth proposition is that global change drivers are of historically unprecedented magnitude. As a consequence, they are leading to trajectories of ecosystem responses that differ radically from those observed in the past (Fig. 12.3b bottom panel). Increasing concentrations of atmospheric CO_2 and other greenhouse gases are primarily related to human activities, in particular fossil fuel emissions and land use change driven mainly by tropical deforestation (IPCC 2007). Atmospheric CO_2 concentrations have increased ca.100 ppm since 1750, and are currently higher than at any time in at least the past 650,000 years (Siegenthaler et al. 2005). These changes in atmospheric chemistry result in global temperature increases and regional increases or decreases in precipitation that often interact with changes in land cover to feed back to local weather (Pielke et al. 2002). Human activities also result in increases in nitrogen deposition in the form of nitrate from the combustion of fossil fuels and from ammonium, a by-product of animal metabolism and fertilization (Vitousek et al. 1997a; Fenn et al. 2003).

Recent studies predict that some future climates will have no historic analogs, and some extant climates may disappear (Fox 2007; Williams et al. 2007a). These novel climates would likely result in new species associations (Hobbs et al. 2006), and the disappearance of climates could result in species extinctions (Overpeck et al. 1991). These "no analog" communities may result in ecological surprises with unknown responses to future climates (Williams and Jackson 2007). Additional global change drivers will likely interact with novel climates to result in even more surprising dynamics (Hobbs et al. 2006). One likely result of novel climates interacting with changes in other global drivers is that ecosystems will be pushed past critical thresholds to result in irreversible ecosystem state changes. These state changes or regime shifts are increasingly recognized as important consequences of global change (Scheffer et al. 2001; Folke et al. 2004). Critical thresholds are often crossed either during or following a state change such that a return to the original state is difficult or seemingly impossible (Bestelmeyer 2006).

Combining these four propositions enables new predictions to be made about the effects of changing global drivers on ecosystem responses across scales. Because global change drivers have altered trajectories through time compared to historic dynamics, ecological responses through time at any point in space may have complex dynamics that may either continue to increase or even decrease (Fig. 12.3c). For example, shrub cover in Fig. 12.2 could continue to expand across arid and semiarid landscapes under conditions of increasing CO_2 concentrations and higher winter precipitation that favors shrub growth over grasses (Morgan et al. 2007). Alternatively, shrub cover could decrease if climatic changes favor grasses and increase fire frequencies (Briggs et al. 2005). Because the processes associated with shrub expansion (recruitment, competition, mortality) are not expected to change under global change, the general spatial pattern at any point in time is not expected to change (Fig. 12.3d). Thus, one prediction is that changes in the temporal characteristics of global change drivers will impact arid landscapes more than changes in the spatial pattern of these drivers. The combination of altered temporal dynamics and no changes in spatial patterns can generate system responses with either increases, decreases, or no changes in responses through time and space with thresholds that indicate changes in level of connectivity (Fig. 12.3e). In addition, changes in spatial pattern may result via unexpected interactions in drivers and responses, such as the emergence of extreme climatic events, pest outbreaks, and altered disturbance regimes (Running 2008), that would result in even more complex or surprising behavior.

A framework for global change

The four propositions combine to form a conceptual framework that has connectivity as its foundation (Fig. 12.4; modified from Peters et al. 2008). At the global scale, a hierarchy of interacting scales governs dynamics. Dynamics at any one location on the globe depend on both local patterns and processes at that location, and the movement of materials via transfer processes from other locations. All places on Earth are connected through a globally mixed atmosphere and regionally through a variety of biotic and abiotic mechanisms, such as human transport of propagules, toxins, and diseases as well as propagation of disturbances and changes in land use as influenced by global economics. Thus, changes in one location can have dramatic influences on both adjacent and nonadjacent areas, either at finer or broader scales. Transfer processes associated with the movement of air, water, animals, and humans provide these connections, both within and across scales (Fig. 12.2).

Disruptions in connectivity are also becoming increasingly important, often in different parts of a system that are increasing in connectivity. For example, land use practices over the past several centuries have increased the density of corn and soybean fields in the Midwestern U.S. to result in a highly connected mosaic of agricultural fields. In contrast, the plowing of tallgrass prairie for agricultural fields has resulted in disconnected remnant prairie locations throughout the region. As a result, movement of agricultural pests and disease among fields is facilitated, but the movement of plants and animals between fragmented remnant prairies is difficult because of the large distances between fragments.

Transfer processes and spatial heterogeneity can either amplify or attenuate system response to broad-scale drivers. Amplification occurs when the rate of

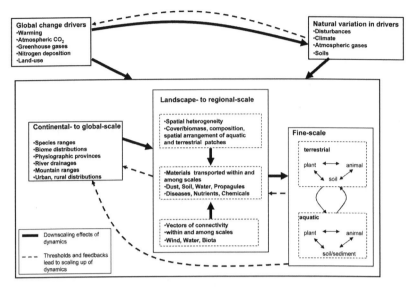

Figure 12.4 Connectivity framework. Global change drivers and natural drivers are influencing a hierarchy of scales of system properties and responses. Each scale has its characteristic patterns and processes, and scales interact to generate changing pattern-process relationships through time and across space. Downscaling occurs when broad-scale drivers overwhelm fine-scale pattern (e.g., hurricanes), and upscaling occurs when fine-scale patterns and processes propagate to influence large spatial extents (e.g., dust plumes from field-based landuse practices). Redrawn from Peters et al. (2008).

change in system properties increases. This increase can result from high spatial heterogeneity or homogeneity that promotes cascading events, such as the nonlinear spread of wildfires (Peters et al. 2004). Cascading events in which a fine-scale process propagates nonlinearly to have an extensive impact have also been documented in the climate system, in lakes, and in the invasion of perennial grasslands by woody plants (Lorenz 1964; Peters et al. 2004; Wilson and Hrabik 2006). Attenuation occurs when the rate of change decreases through time, such as the decrease in wave amplitude as the wave form associated with a tsunami increases (Merrifield et al. 2005). The result is that the greatest effects of a tsunami occur closest to the source of the seismic event, and spatial heterogeneity in land or sea features become increasingly important as distance from the seismic event increases (Fernando and McCulley 2005). The spread of wildfires also attenuates with time and with decreases in fuel load or changes in weather conditions. In addition, broad-scale drivers, such as drought, can act to overwhelm fine-scale variation in vegetation, topography, and soils to result in homogeneous responses over large areas (Albertson and Weaver 1942).

The relative importance of fine- or broad-scale pattern-process relationships can vary through time and compete as the dominant factors controlling system dynamics. For example, connectivity of larvae from coral reef fishes is more locally important and regionally more variable than previously thought based on new analyses of dispersal constraints interacting with physical oceanography (Cowen et al. 2006). Processes that connect spatial units, such as dispersal of woody plants, are important under some conditions whereas local processes, such as soil texture, dominate on other sites; in both cases, broad-scale drought can overwhelm finer-scale processes to result in similar dynamics during dry years (Yao et al. 2006).

Insights from connectivity perspective

We illustrate the importance of a connectivity-based theory of global change for addressing one specific ecological problem: the effects of hurricanes on ecological systems. We first show how our first two propositions apply to current conditions for both drivers and ecosystem responses. We then show how the two global change–specific propositions (anthropogenic origin of drivers and changes in trajectories of drivers) are needed to understand and predict the impacts of hurricanes within the context of other global change drivers in the future.

Current conditions

Drivers of hurricane activity. Although it is readily accepted that hurricanes are disturbances with major impacts on ecosystems, the drivers controlling the formation, intensity, and track of hurricanes remain poorly understood. Recent research indicates that hurricane development is affected by drivers and processes operating across a range of spatial and temporal scales (proposition 1) and that these drivers and processes interact such that spatial heterogeneity and connectivity among spatial units are important (proposition 2). Because these propositions are related, we discuss them together within the context of hurricane development.

Physical processes interacting within and among scales predominate in the development of hurricanes (Goldenberg et al. 2001); these processes are directly or indirectly affected by global change drivers. Hurricanes that affect North America most often start as small thunderstorms in the Western Sahel region of Africa and increase in spatial extent and intensity as they propagate westward across the Atlantic Ocean (Dunn 1940; Landsea 1993). Both local factors, such as sea surface temperatures within the region of hurricane development, and broad-scale factors, such as the El Niño-Southern Oscillation (ENSO) in the tropical Pacific and continental precipitation in West Africa determine whether or not an African thunderstorm develops into a hurricane (Gray 1990; Glantz et al. 1991; Landsea and Gray 1992; Saunders et al. 2000; Donnelly and Woodruff 2007).

Ecological responses to hurricanes. Hurricanes impact ecosystems across a range of spatial and temporal scales that influence ecosystem responses. Spatial pattern of damage resulting from hurricanes is scale-dependent: at the scale of individual trees and small stands, tree age and height, species composition, stand structure, and soil conditions influence amount and type of damage (Foster and Boose 1992; Ostertag et al. 2005). At broader scales, spatial variability in vegetation, land use, environmental conditions, and disturbance history are important as well as landscape- and watershed-scale factors that connect spatial units, such as wind speed and direction, precipitation intensity, and topographic gradients (Boose et al. 2001; Sherman et al. 2001). Recent studies suggest that ecosystem properties, such as stand age and condition, forest type, and aspect, and landscape-scale measures of connectivity, including distance to nearest perennial stream, are more important predictors of forest damage patterns than broad-scale drivers of wind speed and duration (Kupfer et al. 2008).

Ecosystem responses following hurricanes are also scale-dependent (reviewed in Everham and Brokaw 1996; Lugo 2008), and can include interactions across scales as a result of changes in connectivity among spatial units (Willig et al. 2007). For example, landscape reconfiguration and disruption of dispersal among patches by hurricanes can interact with local demographics of species to influence patterns in biodiversity across scales (Willig et al. 2007).

Global change conditions

Drivers of hurricanes, although incompletely understood, are being influenced by anthropogenic sources of variation (proposition 3), and the trajectories of these drivers are changing (proposition 4). We focus on both local and broad-scale drivers that are likely to change.

Local drivers. Sea surface temperatures (SSTs) have increased nonlinearly over the 20th century in the Atlantic Ocean (Trenberth 2005). This trend has been attributed to global warming and human activities (IPCC 2007). In addition, the amount of total column water vapor over the global oceans has increased 1.3% per decade (Trenberth 2005). Both higher SSTs and increased water vapor tend to increase energy available for atmospheric convection and thunderstorm production that can lead to hurricane development. There is general agreement that human-induced environmental changes occurring in hurricane regions can increase hurricane intensity and rainfall (Goldenberg et al. 2001; Emanuel 2005; Webster et al. 2005). There is less agreement on predicted effects of global warming on hurricane frequency with unclear evidence that frequencies are changing beyond the range of historic variation (Henderson-Sellers et al. 1998; Goldenberg et al. 2001).

Broad-scale drivers. Over the past 5000 years, the frequency of intense hurricane landfalls on centennial to millennial timescales was likely related to variations in ENSO and the strength of the West African monsoon (Donnelly and Woodruff 2007). Thus, reliable forecasts of remote conditions will be needed for predicting the occurrence and intensity of hurricanes in North America (Pielke and Landsea 1999). In addition, nonlinearities in the climate system that lead to threshold dynamics may make predictions based on historic trends difficult and unreliable (Rial et al. 2004).

For ENSO, a 155-year reconstruction from the tropical Pacific shows a gradual transition in the early 20th century and an abrupt shift in 1976 to new periodicities that reflected changes in the regional climate towards warmer and wetter conditions (Urban et al. 2000). The dramatic shift in 1976 coincided with a global shift in temperatures attributed to anthropogenic global warming (Graham 1995; Mann et al. 1998). Thus, global warming could further alter the frequency of ENSO cycles. However, additional factors need to be considered, such as sharp decreases in sea surface temperatures in the Atlantic Ocean interacting with dust-induced feedback processes that can moderate hurricane intensity (Lau and Kim 2007).

Factors that influence future rainfall in the Western Sahel will undoubtedly affect the number, duration, and intensity of hurricanes in North America (Webster et al. 2005). Climate projections for this region are unclear: one model predicts severe drying in the latter part of the 21st century while another predicts wet conditions throughout this time period, and a third model predicts modest drying (Cook and Vizy 2006). Clearly, better climate predictions and an understanding of the relationship between rainfall and wave formation are needed before rainfall on the continent of Africa can be used to predict hurricanes in the North Atlantic.

New insights to ecological systems

Atmospheric scientists have known since at least the 1960s that hurricanes connect the African and North American continents (Gray 1968). However, the complex cross-scale interactions determining hurricane development, intensity, and movement track have only recently been recognized and better appreciated as critical elements of a connected Earth system. A sense of urgency in understanding and prediction now predominates in the literature, in particular as our global environment continues to change and as hurricane damage increases with population density and wealth along U.S. coastlines (Pielke and Landsea 1999).

Ecological systems will continue to be influenced by hurricanes, both in obvious and subtle ways because of connections that link spatial and temporal scales (Hopkinson et al. 2008). Ecosystems in the track of hurricanes along the coast of North America are composed of species that evolved under the current hurricane regime, and it remains to be seen how different parts of these systems and different ecosystem types will respond as hurricane activity changes in the future (Michener et al. 1997). Even ecosystems located outside the direct path of hurricanes can be affected by a change in disturbance regime: deserts in southern New Mexico received within several days ca. 43% of the annual average rainfall as a result of the remnants of Hurricane Dolly in 2008. These extreme, remote events are not included in climate change projections for these systems (Seager et al. 2007), yet an increase in hurricane activity would reverse the direction of these projections from drier to wetter. Because these deserts have undergone dramatic changes from grasslands to shrublands over the past 150 years that are at least partially related to drought cycles, an increase in rainfall may provide opportunities for some landscape locations to revert to grass dominance, a state change that is considered unlikely under current climatic conditions.

Furthermore, hurricanes are not the only disturbance with local and global drivers that impact ecological systems across scales (Dale et al. 2001). Hurricanes are often associated with other disturbance events that accentuate the effects of wind and rain: drought often follows hurricanes with its own effects on surviving organisms (Covich et al. 2006). Fire can follow hurricanes with greater impacts on birds than the hurricane itself (Lynch 1991).

Research directions

Future research should include both theory development and advances in the types of studies conducted within the realm of global change. We outlined key characteristics of a developing theory of global change based on existing knowledge. As new information is obtained that increases both the depth of knowledge on specific aspects of ecosystem responses to changing drivers and the breadth of knowledge on interrelationships among components of the

Earth System, the propositions developed here will likely need to be refined, expanded, or replaced.

A consideration of drivers and responses interacting across multiple spatial and temporal scales is needed in global change studies as well as an explicit measurement of transport processes when they are important to dynamics. Five characteristics of systems have been identified to account for multiple scales of variation (reviewed in Peters et al. 2006): (1) local processes (e.g., recruitment, competition, and mortality) interacting with microsite environmental variability (e.g., climate, soils, disturbance history), (2) historic legacies that influence the local environmental conditions, the current assemblage of species, and their ability to respond, (3) current environmental drivers, (4) future environmental drivers with local- to global-scale variability, and (5) transport processes that connect spatial units across a range of scales, from the landscape to the globe.

It is the spatial scaling of characteristics 3–5 that ecologists need to consider when studying ecological problems within the context of global change, yet this aspect has received the least amount of attention to date. A consideration of variability in drivers from remote locations, such as rainfall in West Africa that influences hurricane activity in North America, is often missing from ecological studies. Although there is increasing recognition of the importance of ENSO and other climatic cycles on local rainfall patterns, the interaction of these cycles with other drivers (e.g., elevated CO_2, nitrogen deposition) has not typically been considered, although climatic cycles are related to disturbances, such as wildfire (Kitzberger et al. 2007).

Because measurement of transport processes is typically time and labor intensive, it is important that an initial step in ecological studies be to determine where, when, and how transport processes *may be* important relative to the other drivers in order to decide if sampling is justified. An important aspect of this developing global change theory and its associated framework is that transport processes need to be considered within the context of the properties of the system and the questions to be addressed, but they do not need to be explicitly sampled for all questions.

Both direct and indirect drivers, such as disturbances, have characteristic spatial and temporal scales that need to be studied as interactive effects on ecosystems. Wildfires, floods, insect outbreaks, and other episodic events are not yet included in climate change models (Running 2008), yet they often interact across scales to result in surprising ecosystem responses (e.g., Allen 2007; Ludwig et al. 2007; Young et al. 2007).

Recently, new approaches to studying continental-scale problems under global change have been presented (e.g., Crowl et al. 2008; Grimm et al.

2008b; Hopkinson et al. 2008; Marshall et al. 2008; Williamson et al. 2008). Here we summarize three key recommendations from these papers that are relevant across scales. First, existing long-term datasets can be used to compare trends across sites. Similar patterns in data through time for sites located throughout a region or in different parts of the continent can suggest the presence of a global driver determining synchronicity in these dynamics, such as observed with wildfires and climatic cycles (Kitzberger et al. 2007). Alternatively, similar internal processes may be controlling system dynamics in different locations to overwhelm variation in drivers. A major limitation to these multisite analyses has been accessibility of comparable data. Recent efforts at synthesizing long-term data and metadata from many U.S. sites are allowing these comparisons to be conducted (e.g., http://www.ecotrends .info). Second, coordinated efforts are needed to explicitly examine the importance of fine- to broad-scale drivers and transport processes to ecosystem dynamics across many sites. Existing networks of sites need to be coordinated such that comparable data are collected and compared dynamically in order to identify connections among sites, and to predict effects of cascading events as they influence adjacent and noncontiguous areas, such as the impacts of wildfires on air and water quality in burned sites and at distant locations. Third, simulation models are needed to complement experiments in order to provide more complete spatial and temporal coverage of ecosystem responses to global change drivers. Process-based models will be required to forecast a future with conditions that are unprecedented in Earth's history. In contrast, an empirical extrapolation of responses based on current or past conditions will result in large uncertainty. Multidisciplinary approaches will be needed to account for the complexity of interactions across scales and levels of organization in the ecological hierarchy.

Summary

Direct and indirect drivers of ecological systems are changing nonlinearly in response to human activities. These drivers and ecological systems interact across a range of spatial and temporal scales that often result in nonintuitive ecosystem dynamics. We outline some basic propositions to frame the development of a theory of global change based on connectivity within and among adjacent and noncontiguous spatial units. This nascent theory builds on several more mature bodies of theory developed for specific scales or levels of organization, and uses the concept of cross-scale interactions based on transport processes to link several of these theories. As our knowledge of the interacting components of the Earth System expands with improvements in sensing, measuring, and modeling technologies, we expect corresponding refinements to the theory that will improve its coherence and utility.

Acknowledgments

This study was supported by National Science Foundation awards to LTER programs at the Jornada Basin at New Mexico State University (DEB 0618210) and the Konza Prairie at Kansas State University (DEB 0218210).

Geographical
Gradients

Island
Biogeography Ecological Succession
Gradients

Niche

Metacommunity

CHAPTER 13

Ecological Gradient Theory: A Framework for Aligning Data and Models

Gordon A. Fox, Samuel M. Scheiner, and Michael R. Willig

Understanding the heterogeneous nature of species distributions is central to ecology as embodied in the first fundamental principle of its general theory (Table 1.3). As early as the 18th century, it was noted that species richness differed across the globe (von Humboldt 1808; Hawkins 2001). Today, it is widely recognized that species richness changes along a variety of gradients. Some gradients are spatial (e.g., latitude, depth, elevation; Willig et al. 2003), but may reflect underlying or correlated environmental variation (e.g., solar insolation with latitude). In this chapter, we do not discuss gradients from a purely spatial perspective, leaving such consideration to Colwell (Chapter 14). Here, we focus on gradients of species richness that pertain to environmental characteristics (e.g., disturbance, salinity, succession; Grace 1999). Gradients with respect to productivity are probably the most widely discussed of these ecological gradients (Waide et al. 1999). As productivity increases, species richness may increase, decrease, or assume a hump-shape or a U-shape; the pattern may change with geographical or ecological scale (Mittelbach et al. 2001).

Our goal is to further develop a constitutive theory of environmental gradients of species richness as first promulgated by Scheiner and Willig (2005). We expand on that effort by refining the propositions of that theory, revealing hitherto concealed assumptions, and providing a conceptual framework that further unifies seemingly disparate models. We also examine an oft-cited model in detail, show that it is interpreted incorrectly by many, and discuss approaches for revising it. Besides improved understanding of the particular theory examined here, this exercise illustrates the process of theory development, emphasizing its dynamic nature.

Domain of the theory and its models

The domain of our constitutive theory is environmental gradients in species richness. Literally, *gradient* refers to the slope of a curve; in this case the curve is richness as a function of some environmental characteristic. Slopes range from negative infinity to positive infinity. Although most thinking about these gradients has concerned continuous variation in the environment (so that richness describes a smooth curve with continuous derivatives), there is no logical, biological, or mathematical reason why this must be so. Indeed, one can imagine a limiting case: a threshold in some environmental variable x, such that locations with $x < x_{crit}$ have dramatically lower richness than locations with $x > x_{crit}$. Rather than a smooth curve, the graph will be flat except at x_{crit}, where the change in richness will be immediate and represented by a vertical line. Mathematically, this is described by a step function, which has a slope of 0 everywhere but at the step itself, where the slope is infinite. While real ecological examples are probably less extreme than this, the step function is instructive in making clear that the theory of species richness gradients should account for very sudden changes in richness in ecological space, as well as gradual changes in richness for continuously varying environments. Models used to study richness under continuous environmental variation are likely to take a different form than those used for a small number of discretely different environments; we focus on the continuous case unless stated otherwise.

Ecological gradients occur in spatial contexts, but the theory itself is not necessarily spatial; in its broadest sense, the theory refers to species richness as a function solely of some environmental characteristic. Most intuitively think of ecological gradients as occurring over space, like the gradient from drier to wetter soils that occurs along a hillside. This intuition can be misleading, as the theory encompasses environmental variation occurring in any spatial or temporal pattern, on any spatial or temporal scale. The gradient need not be spatially contiguous or arranged so that the most similar environments are nearest to each other. For example, the theory may apply to a landscape consisting of randomly distributed patches in which environmental characteristics do not show spatial autocorrelation. The pattern of spatial or temporal contiguity and autocorrelation, or lack thereof, can determine which models are appropriate for consideration in any particular situation. The models that we consider here are not spatially explicit, although spatially explicit versions are possible.

That said, particular models may be relevant only to particular spatial or temporal scales, in that it is likely that different mechanisms (e.g., competition, speciation) will dominate at particular scales. The appropriate scale of a model generally depends on the assumptions of the model itself and on the biology of the taxa under consideration, rather than on an a priori scale. A critical distinction is whether the set of sites or collections of species under consideration draws on organisms from a single pool of species (a metacommunity; Leibold Chapter 8), or from multiple pools. Again, the importance of various ecological processes will differ in these instances. Although the constitutive theory that we describe can apply to gradients at any scale, our focus in this chapter will be models with domains at the regional scale (10s to 100s of km^2) and with mechanisms operating in ecological time. These are the spatiotemporal scales for which most models of environmental gradients in species richness have been developed.

To clarify the relationship between the theory of ecological gradients and spatial issues, it is useful to consider the relationship between gradient theory and species-area theory. Both theories involve predictions of richness as a function of another variable: resource or stressor concentration in gradient theory, and area in species-area theory (Fig. 13.1). Richness is, of course, a function of both area and resources or stressors; however, we cannot yet draw a surface connecting the two-dimensional graphs in Fig. 13.1 unless we assume that there is no interaction between area and resources or stressors. Currently both of these theories only permit limited views of such a relationship: gradient theory predicts richness as a function of resources or stressors for a fixed area (along a single plane slicing the three-dimensional space perpendicular to the area axis). Changing assumptions about the landscape (how patches of different resource levels are arranged in space, relative to dispersal processes) lead to different models under gradient theory. By contrast, species-area theory predicts richness along a single plane slicing the three-dimensional space perpendicular to the resource axis, and also requires assumptions about landscape-level variation. This suggests that a complete theory of species richness may have landscape-related dimensions in addition to the area and resource or stressor dimensions. The identities and number of these axes represents a problem that has yet to be explored.

Although a number of models examine aspects of richness gradients, few have clearly defined the relevant characteristics of the species pool. Models often fail to indicate whether taxonomic or ecological attributes delimit the

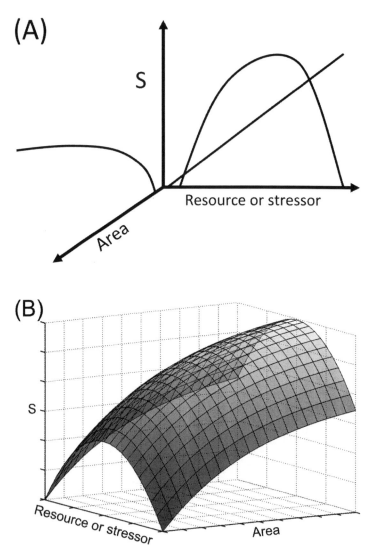

Figure 13.1 Gradient theory and species-area theory both predict species richness (S) by fixing the value of the other axis (area and resource/stressor, respectively) and making landscape-level assumptions. (A) For gradient theory the relationship can take a variety of forms; we depict three here. For species-area theory the relationship is generally assumed to be monotonically increasing or asymptotic. (B) We do not yet know how to draw the surface connecting these graphs in general, because we do not know whether area can interact with the resource or stressor axis, and in most cases would need additional axes for landscape-level variables. The surface shows what the resulting model would look like for a unimodal resource gradient in a uniform landscape without interactions between the axes.

species under consideration. For example, analyses may focus on all species within a clade and a particular level in the taxonomic hierarchy (e.g., a family), a functional guild (e.g., diurnal foliage-gleaning insectivores), an ensemble (e.g., frugivorous bats), or a trophic level (e.g., herbivores). There has been little consideration as to when a model should concern one or the other of these species pools. A species pool may also have a spatial component that is often ill-defined (e.g., biome specific or continental). A related issue is whether there is a single pool or a number of different pools, which again may be strictly a spatial phenomenon (e.g., a gradient that stretches over multiple continents) or be ecological (e.g., a gradient that involves clades that specialize on different conditions).

As we show later, extant models treat species as identical in resource requirements, dispersal ability, and extinction probability. Clearly this violates something that is probably better documented than any other fact in ecology—species differ from one another. In practice, the models make this assumption, but in the literature, the species are assumed only to be roughly equivalent. Most authors refer to these models—or to related species-abundance models—as applying to limited groups such as particular taxa or guilds. For example, herbaceous annual plants might be thought of as roughly equivalent, whereas herbaceous plants and trees are certainly not, because herbs and trees have very different mortality patterns. It is less clear whether herbaceous perennials are roughly equivalent to one another, or whether seed-eating birds are roughly equivalent to seed-eating rodents because they consume the same resources. The exact meaning of "roughly equivalent" requires exploration within particular models. Indeed, the extent to which the assumption of equivalent species can be violated remains a thorny problem for theoretical and empirical research.

Although models within the domain of this constitutive theory are often described as models of species diversity, they are more precisely models of species richness or species density (richness per unit area). Species richness is well defined. In contrast, there are many different definitions of diversity (Whittaker et al. 2001). All involve consideration of species richness, but also include the relative abundances or importances of the species. Some ecologists use the terms richness and diversity interchangeably. In almost all cases, discussions of "diversity gradients" are really discussions of "richness gradients." This is not a semantic argument, as gradients of different aspects of biodiversity (e.g., richness vs. evenness vs. diversity) can be quite different or even independent of each other (Stevens and Willig 2002; Wilsey et al. 2005; Chalcraft et al. 2009).

A theory of environmental gradients of species richness

Our theory rests on four propositions (Table 13.1), set within a conceptual framework (Fig. 13.2). All models of gradients in species richness use the first two propositions, whereas only some include one or both of the last two. These propositions are not universal statements about the world: we do not claim that all propositions hold under all circumstances. Rather, the propositions are statements about the structure of current models of ecological gradients.

Our four propositions are of different kinds. The first proposition is a

Table 13.1 The domain, background assumptions, and propositions that constitute the theory of species richness gradients. Propositions 1 and 2 are used by all models, whereas propositions 3 and 4 are used only by some.

Domain	Environmental gradients in species richness. The gradient can extend over very short spatial distances or be global, or it can extend over short or very long periods of time.
Assumptions	Systems are at equilibrium at some spatial or temporal scale. [most models]
	The species under consideration are roughly equivalent in their resource requirements, dispersal abilities, and extinction probabilities.
	Each species restricts itself more than it restricts other species.
	Local assemblages tend to be in persistent states. [local extinction models]
	The regional species pool contains only species that can coexist with one another. [random placement models]

Propositions

1. A gradient implies one or more limiting resources or conditions that differ in space or time.

2. In a uniform environment of fixed area, more individuals lead to more species.

3. Within an area of fixed size or a unit of time of fixed duration, the variance of an environmental factor increases with its mean.

4. All nonmonotonic relationships require a tradeoff in organismal, population, or species characteristics with respect to the environmental gradient.

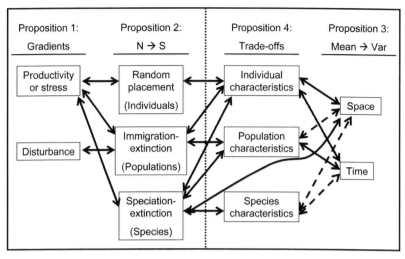

Figure 13.2 A diagram indicating how the four propositions (Table 13.1) can be assembled into different models. The vertical dotted line separates the two propositions (1 and 2) that must be included in any model from the two (3 and 4) that are optional. Solid arrows indicate propositions that have been linked in at least one model. Not all possible combinations of linkages appear in current models. Of the 43 possible models based on unique combinations of linkages, only 9 have been developed to date. Dashed arrows indicate linkages that have not been made. Additional linkages could be developed between propositions 2 and 3, but are not included in the diagram for clarity. The absent linkages between propositions 2 and 4 may not exist because of incommensurate timescales; however, we do not preclude the development of such linkages.

definition, establishing the essence of an environmental gradient in abundance. The second proposition encompasses several mechanisms that can be derived from first principles, each of which assures that the environmental gradient in abundance is also a gradient in richness. The third proposition is a description of a common empirical pattern, or is a general statement about ecological variation. The fourth proposition comprises a heterogeneous mix of mechanisms that derive from other domains and theories that influence the nature of environmental gradients in species richness.

The scale of the data or the model, including aspects of grain and extent, determine the particular mechanisms in effect for each of the propositions. Propositions 1 and 2 (Table 13.1) are functions of the extent, the range of environmental conditions encompassed by the data or being described by a model. Proposition 3 is a function of grain, the sizes of sampled patches or local communities. These scales are always determined by the biology of the species under consideration. A failure to recognize the scales within which particular mechanisms operate has led to a misapplication of a much-cited model (Wright 1983) that has been used to explain global species richness gradients (see below). Recognizing such misapplication is an example of how the process of theory formalization (as illustrated throughout this book) can provide critical insights and guide future research.

Gradients

Our first proposition is that variation characterizes a limiting environmental factor X, which affects variation in the number of individuals that can persist in a sample location of a particular size, thus creating an environmental gradient in abundance. The abundance gradient exists in space and time, although the environmental factor need not be autocorrelated spatially or temporally. Models typically consider only one aspect—time or space—rarely both.

This proposition is part of all models of gradients in species richness, but it is often implicit. The environmental factor could be the concentration of one or more resources, or some condition such as stress or disturbance. For the purposes of our presentation, the exact mechanism creating the link between number of individuals and the factor(s) X does not matter and will differ for each particular situation. Importantly, not all environmental variation creates variation in numbers of individuals, thereby constraining the domain to which our theory applies. In particular, the modifier "limiting" implies that the value of X at a site determines, at least at equilibrium, the number of individuals present, $N(X)$. Most models assume that the system has approached some sort of long-term behavior, i.e., an asymptote or a dynamic equilibrium.

Careful consideration of this proposition helps to clarify limitations regarding the scope of particular models. If variation in X leads to variation in N, then we can write $N(X)$ as a function predicting the equilibrial or asymptotic number of individuals. This requires that the model be general—it predicts the long-term number of individuals that can persist at a particular level of X, not the number or identities of species found at a particular instance of X. The individuals are therefore assumed to be identical in key ecological respects, such as physical traits like body size or demographic traits, insofar as these reflect resource use. This assumption holds only for those key characteristics. For example, models that invoke niche partitioning (Hutchinson 1959; Schoener 1974; Chesson and Huntly 1988; Leibold 1995; Rosenzweig 1995; Chase and Leibold 2003; Kelly and Bowler 2005) assume that species are equal only in resource use.

Although the mathematics in the current literature use the equality assumption, the models are universally interpreted as applying to cases in which the species under consideration are roughly similar but not identical. Under this relaxed assumption, species must be sufficiently similar to one another that, to a rough approximation, individuals in one species require the same amount of resource, or respond in a similar fashion to some condition, as do those of any other species in the species pool. This could easily be generalized to allow for an equivalence among species (e.g., 1 individual of species A equals 1.7 individuals of species B). Thus, a model might describe gradients in the richness of granivorous rodents, but cannot be expected to describe gradients for all vertebrates or even all mammals. It might be reasonable to develop a model that describes a richness gradient for herbaceous C_3 plants with respect to some environmental characteristic. But because "all plants" includes organisms with profoundly different metabolic pathways that span seven or eight orders of magnitude in dry weight, a single model likely will not describe the gradient in richness of all plants.

Model construction is simple enough when the environmental factor is a single resource, but becomes more complicated if multiple resources govern variation in abundance. If the same resource is limiting for all species at each location or time, the minimum (limiting) resource dominates (Liebig's law of the minimum; Sprengel 1839; van der Ploeg et al. 1999). If two or more resources (say, water and light availability for plants) are both limiting (either simultaneously or each one limiting at different times or sites), the combination can be quantified by the vector $\vec{X}$. If the resources affect abundance additively, then the vector $\vec{X}$ can be treated as a single resource X_c, a linear combination of the multiple resources.

Thus, gradients can be grouped into two general classes. In the first class are systems constrained by a single factor X or a combination of factors $\vec{X}$, and $N(\vec{X})$ increases monotonically with the linear combination X_c. In the second class are systems in which one or more pairs of limiting factors are negatively correlated. The negative correlation could be intrinsic (e.g., as soil water content increases, oxygen levels in the soil must decrease), or could be extrinsic to the factors themselves.

Our description of the first class as following a single constraint needs further explanation because circumstances can be more complicated. Multiple resources can interact, so that more than one is limiting at a particular time or place (Gleeson and Tilman 1992). In principle, the only difference is that instead of the number of individuals $N(\vec{X})$ being a curve, it will be a surface with the number of dimensions equal to the number of factors that are limiting at some point. Consider a system in which two resources—each limiting at different concentrations—vary in a nonlinear fashion with respect to each other.

Then it is possible for the system to switch between limitation by factor 1 and factor 2 more than once. Such is not a problem for the models, but in practice it may not be easy to determine what is limiting at each particular location. Indeed, most empirical studies do not address this concern. In practice, most studies examine only single factors (Scheiner and Willig 2005).

Finally, when gradients involve tradeoffs (proposition 4), it is generally useful to separately consider two different aspects of the environmental factors, one relating to resource attributes and the other to stress attributes. For example, consider communities of herbaceous plants arrayed from upland to wetland. Water may be a limiting resource at the higher elevations, but a stressor at lower elevations. It can be useful to model the responses of richness with the associated attributes as two separate factors because the biological responses to water as a resource and water as a stressor are different.

Many models assume that $N(\bar{X})$ is a linear function, although the critical nature of this assumption has not been explored in a comprehensive fashion. When a single factor is limiting, the important assumption is that $N(X)$ is monotonic. It is often possible to select a transformation (e.g., the log function) to linearize a monotonic pattern. However, multiple limiting factors that act singly or interactively create complications: if the factors are interactive, $N(\bar{X})$ may not be monotonic, and there may not be a transformation that will linearize the function.

One common instance of multiple factors is when richness is regulated by bottom-up and top-down interactions. If X is a resource (such as prey items) or an abiotic stressor, regulation is bottom-up; if X is predation, regulation is top-down. Because both top-down and bottom-up regulation occur in many systems, many models consider both kinds of factors in producing gradients of richness.

Individuals ∝ species

The proposition that the number of species increases with the number of individuals was developed by Fisher et al. (1943) and Preston (1962a). It has been termed the "more individuals hypothesis" (Srivastava and Lawton 1998), although it is not necessarily a hypothesis. Under random placement it is a simple sampling relationship, but if extinction or speciation mechanisms are involved in creating the richness gradient, it is indeed a hypothesis. Three mechanisms can account for this pattern: random placement, local extinction, and speciation. Random placement and local extinction are modeled using similar mathematical constructions but are distinct in their biological causations. Moreover, they operate at somewhat different scales of time and space.

Given an environmental gradient in the number of individuals, each of these mechanisms can lead to an environmental gradient in the number of species. All models of species richness gradients invoke at least one of these mechanisms, at least implicitly. In many cases, the models focus on causes of gradients in the number of individuals and assume a mechanism whereby more individuals give rise to more species.

Random placement (also called passive sampling) refers to the movement of individuals among patches or communities. It creates a relationship between the number of individuals and the number of species if local species richness is determined by random sampling of individuals from a regional species pool (Coleman 1981; Coleman et al. 1982) or metacommunity (Hubbell 2001). Biologically, this occurs as individuals move independently but tend to concentrate in areas of greatest resource or least stress. The assumptions behind random placement models are thus identical to those leading to an "ideal free distribution," an idea that has played a critical role in behavioral ecology (Fretwell and Lucas 1970; Sih Chapter 4). As the number of individuals in a local area increases, the number of species should increase because the likelihood of including a rare species increases due to chance. Under this model the species identity of each individual is random, but the number of individuals in a local assemblage is not—it is given by the function $N(X)$. That the relationship between local and regional richness is positive and monotonic does not depend on the abundance distribution locally or regionally, although those distributions determine the exact form of the relationship.

Local extinction is the mechanism invoked by Preston (1962a; 1962b) and highlighted most often as part of the MacArthur and Wilson (1967) theory of island biogeography (Sax and Gaines Chapter 10). This mechanism assumes that a local population will persist only above some minimum abundance. If an area holds more individuals, more populations can attain species-specific minimum viable sizes. Although details can differ about the exact form of the relationship between numbers of individuals and numbers of species, the core assumption is simply that the relationship is positive and monotonic.

Random placement and local extinction share some features: both operate through a balance between the entry of individuals into a site and their departure—by movement under random placement, and by death in extinction. That entry and exit could occur within the lifetimes of individuals through movement, or it could occur across generations through colonization and extinction. For convenience we divide this continuum into an individual-level mechanism (random placement) and a population-level mechanism (local extinction). Mathematically, they can be treated as equivalent for an equilibrial theory, although models will differ in detail depending on the particular set of species under consideration. In general, random sampling operates at local to landscape scales over short time periods, whereas local extinction operates at landscape to regional scales over longer time periods, with the exact meaning of these differences determined by the species' biology.

This distinction between individual- and population-level processes can define the domain of a particular model. Previously, we asked whether seed-eating birds and seed-eating rodents are roughly equivalent. If the abundance of birds at sites is determined by the movement of individuals (e.g., Coleman et al. 1982), while the abundance of rodents is determined by population growth and extinction, then application of existing models to, say, richness gradients of granivorous vertebrates would be misleading. Because the entry and departure processes are quite different for granivorous birds and rodents, involving different parameter values, one would need to model the richness of the two groups separately. If interactions between the two groups could be ignored, then the predicted richness of the combined taxa would simply be the sum of the two predicted richnesses. However, if granivorous birds and rodents interact, studying the richness of both would require modeling their interactions as well.

Speciation operates at scales of time and space that are much greater than that of random placement or local extinction. It assumes a positive relationship between the number of individuals and the net rate of speciation (i.e., speciation minus extinction; VanderMeulen et al. 2001). This mechanism most appropriately deals with species richness patterns at large spatial scales and may provide an explanation for the richness of the regional species pool.

All extant models of richness gradients make another important assumption: *they do not consider species interactions.* To see this, consider a gradient model derived from one of the versions of neutral theory (Hubbell 2001; Chave 2004; Etienne and Olff 2004; Volkov et al. 2005). A model using either random placement or local extinction can logically find the expected number of species at any location along the gradient, using only propositions 1 and 2 (Table 13.1). Now consider a gradient model that concerns niche partitioning, using character displacement, microhabitat variation, or temporal niches (e.g., Hutchinson 1959; Schoener 1974; Chesson and Huntly 1988; Leibold 1995; Rosenzweig 1995; Chase and Leibold 2003; Kelly and Bowler 2005). This model can also find the expected number of species, given one of the following assumptions:

1. Local assemblages generally have reached persistent states. By persistent we include the textbook equilibria of Lotka-Volterra competition models, as well as the more complex kinds of persistence possible with multiple species and nonlinear interactions (Armstrong and McGehee 1980). This assumption

implies that local population and community dynamics dominate, so logically this assumption might hold under a local extinction model, but not under a random placement model.

2. The regional species pool contains only those species that can coexist with one another. This assumption could hold under random placement, which implies nothing about the long-term persistence of competitors. It could also hold under the (rather unlikely) assumption that the species in the regional pool have all coevolved to coexist.

3. Competitive coexistence of particular species in a given sample unit is not necessarily guaranteed, but sampling processes still guarantee that on average there will be $S[N(X)]$ species at X resource level.

Having said this, we hasten to add that it is not logically necessary for community theory (Holt Chapter 7; Leibold Chapter 8; Pickett et al. Chapter 9) to be external to gradient theory—just that this is presently the case. It is certainly possible for local community dynamics to interact with the factors determining the existence of a richness gradient; addressing this possibility is an open theoretical question.

Mean ∝ variance

The proposition that the mean and the variance of environmental characteristics are related positively is based, in part, on the recognition that most environmental factors are bounded by 0 (i.e., have a theoretical minimum). Such a bound can lead to a positive relationship between the mean and variance, although such a relationship need not exist empirically. If the magnitude of an environmental factor is 0 or close to 0, then perforce the variance initially will increase as the mean increases. A continued rise in the mean allows for the possible continued rise in the variance, unless an upper bound also exists (e.g., water saturation of soil). Thus, this proposition is limited to those environmental variables that have a lower but not an upper bound within the range of environmental conditions of the gradient. If the upper bounds on a limiting environmental factor also restrict the number of individuals, the theory as described in this chapter can be applied to that part of the environmental gradient where the variance does increase with the mean.

This mean-variance relationship is invoked in models that focus on patch dynamics (e.g., Abrams 1988). More specifically, species richness is measured in some area within which there are multiple patches. For some models, heterogeneity is generated by interactions among individuals (e.g., Tilman 1982; Huston 1994; Currie et al. 2004). Most commonly, the invoked mean-variance relationship is spatial (e.g., wet vs. dry, good vs. bad). A meta-analysis

(Lundholm 2009) found that plant species richness or diversity frequently increases with spatial heterogeneity, but does not always do so. Some models invoke temporal heterogeneity, considering specialization on different year-types as a mechanism that promotes coexistence of multiple species (Chesson and Huntly 1988; Rosenzweig 1995; Kelly and Bowler 2005). Regardless, if species specialize on combinations of environmental characteristics that occur in patches in which they can out-compete other species, then richness should increase as the number of patches (i.e., heterogeneity) in an area increases.

This proposition is explicitly scale-dependent (Lundholm 2009) as it deals with changes in variation within the grain of a particular model, the unit for which richness is measured. This grain is always dependent on the biology of the species under consideration. Thus proposition 2 also contains a hidden assumption that the species are equivalent in their use of space or time. At the lower end, the minimal grain size is that needed to hold one individual. At the upper end, the maximal grain size is such that all possible heterogeneity or habitat types are encompassed within a single grain.

The form of the relationship between mean patch characteristics and their variance is related to theories of species-area relationships (SARs; Fig. 13.1). SARs are determined by a variety of factors: more individuals are contained in larger areas, and environmental heterogeneity increases with greater area. Clearly, models of SARs share many features with models of species richness gradients. Models of SARs are currently being developed and debated (e.g., Scheiner 2003; Tjørve 2003; Maddux 2004; Ostling et al. 2004; Adler et al. 2005; Fridley et al. 2006; Chiarucci et al. 2009). Thus, we postpone any attempt to develop formal models of SARs specific to the context of species richness gradients until the more general forms of those models have been resolved more thoroughly.

Tradeoffs and hump-shaped curves

Many models of environmental gradients in species richness posit that a tradeoff leads to a hump-shaped pattern, with the maximum value of richness at some intermediate point along the axis of an environmental factor. The models differ with regard to the nature of the invoked tradeoff. Nonetheless, they share the basic proposition that a change in the sign of the slope arises as a consequence of two mechanisms acting in concert but in an opposite fashion on each species. Commonly invoked tradeoffs are competitive ability versus a variety of other abilities (e.g., stress tolerance, colonizing ability). The tradeoffs that matter in a particular instance depend on the species and type of environmental variation, including its scale (grain and extent) in time and space.

Tradeoffs may occur at different levels in the biological hierarchy. For example, the tradeoff may involve the characteristics of individuals, such as competitive ability versus stress tolerance (Grime 1973). In other cases, the tradeoff may involve the characteristics of populations, such as the intensity of interspecific competition versus the intensity of predation (Oksanen et al. 1981). In yet other cases, the tradeoff may involve characteristics of species, such as speciation rates versus extinction rates (VanderMeulen et al. 2001). Scheiner and Willig (2005, Table 1) listed 17 different models of species richness gradients. In the conceptual scheme presented here (Fig. 13.2), we treat mechanisms that operate at the same level (i.e., individual, population or species) as mathematically equivalent. In doing so, we can unify some of those models, reducing the list of models from 17 to 9 (Table 13.2).

The maximum (or minimum) point in the curve describing an environmental gradient in species richness arises because of a change in the relative importance of factors that control the number of individuals. This tradeoff can be conceptualized as environmental variation in each of two factors that are negatively correlated. Along one portion of the environmental axis, the first factor limits the number of individuals; at some point a second factor becomes limiting. This switch results in the number of individuals increasing along one portion of the environmental axis and decreasing along another. For many models, this shift in importance is controlled by inherent properties of species. For example, Tilman (1988) theorized that in terrestrial plant communities increasing nitrogen availability causes an increase in numbers of individuals, until plant density is great enough that light becomes limiting and numbers of individuals begin to decrease. Although tradeoffs are invoked in models that produce a hump-shaped pattern, the mechanism can explain U-shaped patterns as well (Scheiner and Willig 2005).

In many models, the interacting mechanisms that determine the number of individuals are not stated explicitly. Similarly, the unique contributions of each mechanism to total abundance are rarely quantified with respect to variation in the environmental factors. As a result, the mechanistic tradeoff is neither emphasized in conceptual models nor detailed in quantitative models. The absence of mathematical or logical rigor enhances the likelihood that such concealment persists, diminishing an appreciation for the similarities of form that the details obscure. For example, various models posit tradeoffs between competition for different resources (e.g., Tilman 1982; 1988; Huston 1994) or competition vs. resistance to predation/herbivory (Leibold 1996; 1999).

Table 13.2 Models of diversity gradients and their components and mechanisms. Proposition 1: Type of gradient; Proposition 2: Mechanism linking the number of individuals and number of species; Proposition 3: Environmental heterogeneity; Proposition 4: Type of tradeoff. Previous number(s) refers to models listed in Table 1 of Scheiner and Willig (2005).

No.	Proposition 1	Proposition 2	Proposition 3	Proposition 4	Previous number(s)	Sources
1	Productivity or stress	Random placement	N/A	Individual characteristics	1	Oksanen (1996), Stevens (1999)
2	Productivity or stress	Local extinction	N/A	N/A	5	Connell (1964), Wright (1983)
3	Productivity or stress	Local extinction	N/A	Individual characteristics	11	Rosenzweig (1993), Tilman (1993)
4	Productivity or stress	Local extinction	Spatial	Individual characteristics	3, 7, 10	Tilman (1982; 1988), Abrams (1988), Huston (1994), Leibold (1996; 1999)
5	Productivity or stress	Local extinction	Temporal	Individual characteristics	4	Rosenzweig (1995)
6	Productivity or stress and disturbance	Local extinction	Temporal	Individual characteristics	6, 8	Grime (1973; 1979), Huston (1979), Huston and Smith (1987)
7	Productivity or stress	Local extinction	N/A	Population characteristics	2, 9	Rosenzweig (1971; 1995), Wollkind (1976), Oksanen et al. (1981)
8	Productivity or stress	Speciation	N/A	Individual characteristics	12, 15	Denslow (1980), Rosenzweig and Abramsky (1993), VanderMeulen et al. (2001)
9	Productivity or stress	Speciation	N/A	Species characteristics	13, 14	VanderMeulen et al. (2001)

Such models all have a similar mathematical form, but this similarity is not apparent until they are placed within a single framework.

Arguments for and against particular models often boil down to a personal preference for one tradeoff over another. We take a more catholic position by not advocating any one in particular. Rather, we embrace all of them as theoretical possibilities, although it remains to be seen whether some tradeoffs are more common than others. Perhaps most critically, the posited mechanisms often are not mutually exclusive. Tradeoffs may simultaneously exist between competition for two different resources and herbivory, for example. As with multiple environmental factors, it may be possible to model such multiple tradeoffs as an additive pair of tradeoffs. Otherwise, more complex models will be needed.

Although a specific tradeoff may exist for a particular set of species, we should not expect the same tradeoff to be ubiquitous for all species in a guild, trophic level, or community, thus limiting the scope of any particular model. It is possible that more closely related species will share a tradeoff, whereas more distantly related taxa will have different constraints, but this should not be assumed (Losos 2008). Thus, the type and form of tradeoffs sets another boundary on the conditions under which individuals of different species must be roughly equivalent. It is not known how rough this equivalence can be and still be consistent with the underlying models.

Relationship to the theory of ecology

The four propositions of the theory of environmental gradients of species richness (Table 13.1) derive from the fundamental principles of the theory of ecology (Table 1.3). Proposition 1 is a consequence of principles 4 or 5, depending on the nature of the environmental factor(s). The finite nature of resources (principle 5) creates the constraint that allows one or more resources to be limiting. The heterogeneity of environmental characteristics in space or time (principle 4) creates the potential for variation in resources or stressors. Environmental heterogeneity in time leads to the potential for variation among patches in the rate of disturbance. Proposition 2 is a consequence of principles 1, 2 or 7. The process of random sampling is one mechanism that creates the heterogeneous distribution of organisms (principle 1). Immigration-extinction balance comes about through the combination of processes that lead to heterogeneous distributions or organisms' and species' interactions (principle 2). Speciation is a suite of processes that derive from principle 7. Proposition 3 is a direct manifestation of environmental heterogeneity (principle 4). Finally, the tradeoffs embodied in proposition 4 derive from principles 6 and 7.

Applying the theory: the energy model

To see how the constitutive theory relates to current models of productivity-richness relationships, consider the energy model (Connell and Orias 1964; Wright 1983). We focus on this model for two reasons: (1) it has been very influential [we found 343 citations of Wright (1983) in the Web of Science database on February 16, 2009], and (2) it is one of the few that is written in explicit mathematical form. We follow the formal presentation of Wright (1983), which is couched in terms of the relationship of species richness and area. The model predicts the number of species in a sampling unit (Wright thought of these as islands) as a function of local energy availability. Wright considered energy input per unit area to be fixed so that his model predicts the consequences of variation in area on species richness. By contrast, the models considered in this chapter examine the consequences of variation in environmental resources or stressors among different locations while holding area constant. Thus, our explication of this model does not include terms for area as in Wright (1983).

Wright's model is $S = a(E\rho/m)^z$, where E is the amount of energy locally available for biosynthesis, ρ is an empirical constant for a given set of species describing the number of individuals that are supported per unit of available energy, and m is the population size of the smallest extant population. The terms a and z are empirical constants estimated from the data, although as we shall see, a, z, and m appear in this model because of some strong assumptions. We now examine how this model relates to our propositions, and consider some consequences of its assumptions.

Existence of a gradient and its consequences

Rewriting Wright's model in our more general terms, we begin with $N(X) = X\rho$. The use of the common term ρ means that the model describes richness when derived from a set of roughly similar species, as our explication of the general theory suggests it must. Wright further posited that species richness increases with decreasing latitude because available energy increases, a contention still advanced by many (e.g., Mittelbach et al. 2001; Hawkins et al. 2003a). This is a sensible model only if individuals of all species along the latitudinal gradient require about the same level of resources, which is certainly not true.

From individuals to species

Proposition 2 posits that more individuals lead to more species, and in Wright's model, most of the action is in proposition 2. To model species richness as a function of X, we need to model $S = f[N(X)]$, where S is the number of species present and f is some function. Wright's choice of f is $f[N] = a(N/m)^z$, which comes from Preston (1962a). This equation, with $N = X\rho$ as above, produces a positive monotonic relationship between the resource X and species richness S. The exact shape of the relationship depends on a, z, and m. The model requires the first two propositions (Table 13.1), and nothing more. The energy model does not attempt to explain the source of available energy or its relationship to climate, which is the domain of other theories (e.g., O'Brien et al. 2000).

The Wright and Preston models rely on the local extinction mechanism, delimiting the temporal and spatial scales for which the model makes predictions. In particular, this model makes predictions about the equilibrial number of species at a location with resource concentration X, when individuals are drawn from a fixed regional pool of species. The model should not be interpreted as making predictions about variation in species richness over large spatial extents (e.g., across continents) because such variation cannot result from local extinctions from a single species pool. Such large-scale gradients must involve (at least) several species pools, and likely involve speciation processes as well. Thus, by its implementation of both propositions 1 and 2, Wright's model has a far more limited interpretation than stated by Wright or many subsequent authors (e.g., Currie 1991; Mönkkönen and Viro 1997; O'Brien 1998; Gaston 2000; Allen et al. 2002; Currie et al. 2004). Although some found apparently good fits of the model for continental- to global-scale data, because those data represent an inappropriate spatial domain, it is illogical to assign meaning to estimated parameters in terms of the Wright model.

Although the power-law function used by Preston and Wright is simple and familiar to several generations of ecologists, its derivation in this case rests on a complex and rather narrow argument concerning the distribution of species abundances and how population size relates to extinction probability. In particular, the power-law function depends on Preston's assumption that species-abundance curves are described by a form of the lognormal distribution that he termed canonical.

The division by m—not explained by Wright other than his citation of Preston (1962a)—seems odd. It is natural, albeit wrong, to assume that this division (N/m) is aimed at calculating the maximum number of species. Under Preston's (1962a) canonical lognormal distribution, one specifies the shape of the species-abundance distribution with any two of three quantities: the total number of species, the standard deviation of the lognormal distribution, and the number of species in the modal octave. The quantity m is required to specify the position of this distribution along the horizontal axis (the $\log_2$ of abundance). Preston (1962a) calls m the size of the smallest population, but he also calls this a "tentative" definition (Preston 1962a, p. 190), and notes "in practice that m is less, even appreciably less, than unity, and the temporary interpretation we have given [as the size of the smallest population] then has no meaning." In other words, m is just a parameter that defines the location of the species-abundance distribution, in the same sense that statisticians speak of the mean as characterizing the location of the normal distribution.

Regardless of whether m is the size of the smallest population or an empirically estimated parameter, Wright's use of m, a, and z to defines his model links it intimately to the somewhat arbitrary assumptions of Preston's canonical lognormal distribution. Despite numerous criticisms of aspects of Preston's work (e.g., Pielou 1969; May 1975; Williamson and Gaston 2005), it has had remarkable staying power in the ecological literature: remarkable, because neither Preston nor subsequent researchers have linked the canonical lognormal to any underlying mechanisms. Preston himself (1962a) made it clear that he had none in mind. Unfortunately, ecologists are sometimes satisfied with curve-fitting exercises without concern with the underlying mechanisms. Such exercises teach us nothing beyond the narrow lesson that the particular data set is well-described by a particular curve, providing only a phenomenological description.

Most ecologists, trying to justify the use of the lognormal for species abundance distributions, do so with a vague and incorrect reference to the central limit theorem (Williamson and Gaston 2005). The central limit theorem predicts that each species' abundance will be lognormally distributed over time; unless the abundances are independently and identically distributed among species (i.e., the species are equivalent), this does not lead to a jointly lognormal distribution of abundances at a particular time. If the species are different from one another (i.e., they have a different means and variances of abundance), the joint distribution of abundances at a given time will not be lognormal. Šizling et al. (2009) proposed a more satisfying (and rigorous) explanation as to why species abundance distributions are often similar to the lognormal. Their derivation requires only that the abundance distribution be based on the combination of abundances in many nonoverlapping subplots.

None of this implies that the Wright energy model is wrong in some sense; rather, its basis is weaker than one might hope (given its influence), as it depends on the phenomenological assumption that species abundances are given by Preston's canonical lognormal. Other mechanisms could be invoked that yield the same qualitative relationship while differing in details (e.g., Hubbell 2001). Pueyo et al. (2007) showed that an infinite number of models varying between strict neutrality (all species identical) and strict idiosyncrasy (all species unique) can generate identical abundance patterns. Although Wright's model has been interpreted as a predictor of continental to global patterns, it cannot logically be so as it is restricted to a set of roughly equivalent species (implementation of proposition 1) in a single regional pool (implementation of proposition 2).

The converse is also true. Many studies have shown a positive relationship between energy and species richness on a continental to global scale (e.g., Field et al. 2005; Rodriguez et al. 2005; Buckley and Jetz 2007; Davies et al. 2007; Kalmar and Currie 2007; Kreft and Jetz 2007; Woodward and Kelly 2008), and this has often been taken as support for Wright's model. However, those studies do not attempt to directly parameterize Wright's model and test whether the model is accurately predicting those relationships. Instead, we merely have a qualitative agreement between various empirical relationships and one particular model. Our dissection of that model suggests that it cannot be used as an explanation for those relationships because the mechanisms underlying that model operate at different scales. Given the generality of the observed relationships, further work is necessary to connect the mechanisms of Wright's model with global-scale mechanisms, or to develop new models with mechanisms operating at that scale.

Our explication of Wright's model suggests that it must be interpreted on a regional spatial scale with species that are roughly equivalent. That does not preclude the possibility that one might find that the model provides a good fit to data from much larger spatial or taxonomic scales. Indeed, if a model like Wright's provides a good prediction of the number of species, given a level of resources, it may be useful to managers even if its assumptions are violated severely. The only problem here is with interpretation: a good fit of a model to data that violate its assumptions cannot be interpreted as support for the concepts embodied in the model, but only as a useful description of data. Prediction and understanding are not always on the same footing.

Prospects

Model development

Theory unification is an iterative process that includes recognition of similarities among ostensibly competing models, development of a common framework, and construction of new overarching models within that framework. Additional effort is needed in domains, such as the one we consider, in which many of the models are verbal and even the analytic models have not been examined deeply. We are encouraged that our refinement of the conceptual framework (Fig. 13.2) has led to further model unification (i.e., reducing the number of models from 17 to 9).

This is a step forward in model unification not simply because it reduces the number of models, but because it reveals their common bases, and because it points to some additional models that have not yet been studied (Fig. 13.2). The reduction in the number of models is a consequence of recognizing that the 14 different forms for proposition 4 listed in Scheiner and Willig (2005, Table 1) can be usefully placed in three categories: tradeoffs operating at the levels of individuals, populations, and species. For example, using the model numbers from Scheiner and Willig (2005, Table 1), we now treat models 3, 7, and 10 as equivalent because all assume that the gradient (proposition 1) is productivity or stress, the number of species (proposition 2) is generated by local extinction processes, heterogeneity (proposition 3) occurs over space, and tradeoffs (proposition 4) occur at the levels of individual characteristics. Similar reasoning leads to unifying other models.

Our approach has been to start with the simplest formal model, the Wright energy model, and carefully examine its assumptions and limitations. The challenge is to build a new, general and useful model that avoids the previously described limitations. The first limitation—restriction to a set of species with roughly equivalent requirements—is a hurdle only if one hopes to develop a model that explains richness in general. To the extent that progress can be made studying richness gradients of given taxa or guilds, there is no limitation. If interest lies in explaining more general gradients, however, it is not logically possible to follow the approach of first calculating the number of individuals (proposition 1) and then using a sampling argument (proposition 2)—either random placement or local extinction—to predict the number of species.

A more general model must incorporate the rules by which metacommunities are formed (Leibold Chapter 8). In other words, such a model would need our four propositions as well as propositions involving the way in which interactions among species determine numbers of individuals and species. One might argue that this is precisely what Preston (1948; 1962a; 1962b) attempted, but this is not the case. Preston's argument was couched entirely in terms of single species. We know of no persuasive models that jointly predict the numbers of species and the population sizes of multiple species.

How might we avoid the second limitation, being wed to a set of arbitrary assumptions necessary to go from $N(X)$, the number of individuals, to $S = f[N(X)]$, the number of species? Numerous models of species abundance distributions arise from quite different assumptions (Fisher et al. 1943; Preston 1948; Zipf 1965; Kempton and Taylor 1974; Pielou 1975; Mandelbrot 1977;

Engen and Lande 1996; Engen 2001; Hubbell 2001; Dewdney 2003; Lande et al. 2003; Williamson and Gaston 2005). At this point there is no basis for concluding that any particular model is either logically best or empirically most supported by available data. In the absence of such a model, assumptions about the form of $S = f[N(X)]$ are arbitrary. This does not necessarily mean that more progress in gradient theory must await developments in the theory of species abundance distributions. It is possible to make progress by using a number of different species abundance distributions and asking how the choice of distribution affects the model predictions about richness gradients. Many gradient models may be robust to such choices.

Further work is needed to relax the assumption of species equivalence. For example, for the random placement mechanism, one could substitute a distribution of body mass frequencies for the constant p. Such a model would still assume that the shape of the distribution is the same for all sites, but that is a much weaker assumption.

Linking models to data

More challenging than model development is linking models to data. Even for a model as simple as Wright's energy model, which does not invoke tradeoffs or spatial structure, the information necessary to estimate all of the parameters does not exist, as far as we are aware. When confronted by such challenges, ecologists often respond by questioning the utility of the model. Our reply is twofold. First, formalizing models makes data requirements clearer. Although many data have been gathered in the context of studying richness gradients (Mittelbach et al. 2001), those studies have not been guided by theory, thus the disconnect between the data and the models. For example, few studies collect data on richness, abundance, and the environmental variables thought to determine richness and abundance. It may be that sufficient data exist for some systems (e.g., Stiles and Scheiner 2010) and the challenge is to discover and assemble those data.

Second, only models can provide quantitative predictions. Enough may be known about processes such as herbivory or competition to permit a sufficiently constrained state-space within which a model can be explored. Given the growing urgency of understanding global change, these models, with their general parameters, may have to do while we work to collect more data. For example, our demonstration that as a mechanistic model, the Wright energy model should be restricted to local or regional gradients and limited sets of taxa or guilds suggests that it should not be combined with global change models to predict changes in global species richness, or that any such model should be sharply delimited in its taxonomic or ecological scope (e.g., Field et al. 2005). Obviously, Wright's model can still be used on these scales as a phenomenological model, so long as interpretation of the fit and parameter estimates is restricted appropriately. Similar hidden limitations may be discovered as we explore the details of other models.

Linkages to other constitutive theories

The theory of gradients of species richness has direct linkages to many of the other constitutive theories presented in this book. Geographic gradients (Colwell Chapter 14) concern spatial gradients only; the models considered here may have a spatial component, but typically do not. Not surprisingly, the theories share points of contact concerning the multiplicity of causes that determine gradients and how variation in species ranges along a gradient determines the form of the species richness relationship. Island biogeography theory (Sax and Gaines Chapter 10) is another one with shared mechanisms concerning immigration and extinction (Table 10.1, propositions 1, 4, and 6). The Wright energy model was first developed within the context of island biogeography theory as a way of explaining the relationship between area and species richness. As we have discussed, metacommunity theory (Leibold Chapter 8) may provide important tools for linking species abundance and species richness. Similarly, in order to formalize models that invoke tradeoffs in competition or predation/herbivory will require and examination of niche theory (Chase Chapter 5) and enemy-victim theory (Holt Chapter 7). Thus, the entire processes of theory formalization represented by this book will be an important guide and useful tool for further model development.

Acknowledgments

The evolution of ideas presented in the chapter arose in part as a consequence of a workshop held at the University of Connecticut and supported by the Center for Environmental Sciences and Engineering. We thank Nancy Huntly and Dov Sax for their comments and suggestions. Support to MRW was provided by the Center for Environmental Sciences and Engineering at the University of Connecticut, as well as by National Science Foundation grant DEB-0218039. GAF was partially supported by National Science Foundation grant DEB-0614468 and by a sabbatical from the University of South Florida. This manuscript is based on work done by SMS while serving at the

U.S. National Science Foundation and on sabbatical at both the Center for Environmental Sciences and Engineering at the University of Connecticut and the University of South Florida. The views expressed in this paper do not necessarily reflect those of the National Science Foundation or the United States Government.

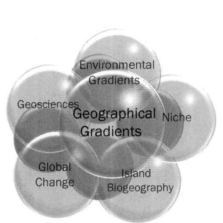

CHAPTER 14

Biogeographical Gradient Theory

Robert K. Colwell

The history and contemporary distribution of life on Earth on broad spatial scales has traditionally circumscribed the realm of biogeography. Early observers, beginning in Classical times and continuing through the 18th and 19th centuries, began by describing where the most conspicuous species were found, accumulating knowledge of vertebrates, plants, and the larger and showier arthropods on land, and of the more conspicuous, accessible, and useful macroorganisms of the seas. Meanwhile, the descriptive geography of fossils launched the twin disciplines biostratigraphy and paleogeography. The same explorations that provided the foundation for descriptive biogeography, of both extinct and living species, provided the material for taxonomists to describe species and to classify them in increasingly "natural" taxa. Knowledge of the taxonomy and geographical distribution of conspicuous species reached a critical stage in the mid-19th century, making possible the independent discovery by Wallace and Darwin of the theories of natural selection and speciation by isolation, which transformed all of biology.

The emergence of biogeographical gradient theory

In spite of insightful speculation by 19th-century observers (e.g., von Humboldt 1808; Wallace 1878) and the appearance early in the 20th century of the landscape-scale work of early plant ecologists that would eventually lay the foundations of community ecology, there was no formal theory of the geography of species and biotas on broad spatial scales. It took nearly another century of accumulated biogeographical knowledge to set the stage for the emergence of biogeographical theory, which many would trace to two pinnacles of the brief career of R. H. MacArthur: his collaboration with E. O. Wilson in *The Theory of Island Biogeography* (MacArthur and Wilson 1967), and his extension of ecological theory to broader spatial scales in *Geographical Ecology* (MacArthur 1972).

Biogeographical gradients—spatial patterns in the distribution of taxa—were crucial to both of these developments. The theory of island biogeography emerged as a joint explanation for two biogeographical gradients: the pattern of increase in species richness on islands with increasing island *size* and with decreasing *isolation* from source areas. MacArthur's fascination with the most striking biogeographical gradient on the planet, the increase in species richness with decreasing *latitude*, began with a series of papers proposing theoretical links between this gradient and community ecology (Klopfer and MacArthur 1960; 1961; MacArthur 1965; 1969) and culminated in pivotal chapters of *Geographical Ecology* that also brought to bear the key concepts of gradients in area and isolation, derived from island biogeography. Although not all of MacArthur's proposals have stood the test of time, these works challenged biogeographers to think theoretically and abstractly about biogeographical gradients.

With the further development of biogeographical databases, rapidly growing computational power, and the development of inexpensive GIS software for personal computers in the 1980s, the stage was set for two key advances in the theory of biogeographical gradients. The emergence of macroecology (Brown and Maurer 1989; Brown 1995; Gaston and Blackburn 2000) as a discipline formalized the intersection, already under way, between broad spatial scales and ecological mechanisms, inspired and supported by inferences from large empirical datasets (e.g., Currie and Paquin 1987; Currie 1991). Meanwhile, explicitly stochastic models of species distributions on biogeographical scales (Pielou 1977; Colwell and Winkler 1984; Colwell and Hurtt 1994; Pineda and Caswell 1998; Willig and Lyons 1998; Bokma et al. 2001; Hubbell 2001) stimulated new approaches to simulation modeling. In addition to bringing stochastic processes into focus, a key outgrowth of these approaches has been a shift towards modeling biogeographical patterns at the level of species ranges or populations, treating species richness as a function of the overlap of geographical ranges in geographical space, rather than a black-box emergent property.

Domain of the theory and models

This chapter aims to review the current role of biogeographical gradients within the realm of ecological theory, recognizing that biogeography has broad and deep connections with the disciplines of evolutionary biology, paleontology, and earth history. The domain of biogeographical gradient theory treated in this chapter encompasses the *characteristics and causes of spatial gradients in the occurrence and cooccurrence of species on geographical scales*. In this context, "geographical scales" refers to spatial scales from the level of ecoregions, biomes, and geographical regions to continents and ocean basins. Fox et al. (Chapter 13) treat the related but distinct concept of ecological gradients, which focus on conditions and resources, whether physical, chemical, or biological in nature. The key distinctions lie in the explicit, broad-scale spatial context of biogeographical gradients, and the restriction of focus, here, to theories and patterns of gradients in species occurrence. At the outset, however, it must be said that it would be impossible to make sense of biogeographical gradients without understanding the role of ecological gradients on broad spatial scales. Because Sax and Gaines (Chapter 10) treat the important special case of island biogeography, I will focus here on general models of biogeographical gradients (Table 14.1).

Table 14.1 The propositions of the theory of biogeographical gradients.

1. Biogeographical gradients arise from demographic and evolutionary processes acting at the level of populations, including migration, adaptation, speciation, and extinction.

2. Biogeographical gradients are composite manifestations of the location and overlap of geographic ranges.

3. Biogeographical gradients have multiple, interacting causes.

4. Biogeographical gradients characterize the physical world either as spatially explicit patterns or as ordered elements of a biogeographical mosaic.

5. Biogeographical gradients are the consequence of ecological and evolutionary responses of species and lineages to environmental gradients and mosaics.

6. Biogeographical gradients are shaped by the contingent facts of both earth history and human history.

7. Because biogeographical gradients are realized in spatially bounded domains, constraints of geometry may affect them, independent of biological and historical influences.

Gradients of species richness

Geographical patterns of species richness have seized the attention of naturalists since the earliest days of European natural history exploration, particularly the latitudinal gradient in richness (Hillebrand 2004) that characterizes most taxa (e.g., von Humboldt 1808; Wallace 1878). The remarkable richness of tropical biotas, from rainforests to reefs, contrasted with the comparative biotic simplicity of temperate and boreal biotas has inspired a wealth of proposed explanations and theories (e.g., Wallace 1878; Dobzhansky 1950; MacArthur 1965; Pianka 1966; Huston and DeAngelis 1994; Rosenzweig 1995), most recently reviewed by Willig et al. (2003). Geographical gradients of increasing richness from dryer to wetter climates, doubtless familiar for millennia to travelers, offer a second repeated pattern in need of explanation (O'Brien 1998; Hawkins et al. 2003a). Changes in richness with elevation on land (Rahbek 1995), and with depth in the oceans (Pineda and Caswell 1998; Levin et al. 2001), although even more complex and varied than the latitudinal or dry-wet gradients, provide additional examples of repeated biogeographical gradients that have been the focus of intensive study.

Two key concepts in island biogeography theory (see Sax and Gaines Chapter 10) are the decrease in species richness with isolation from biotic source areas and the increase in species richness with island area. The isolation effect also plays a role in continental biotas when topographic or climatic factors isolate regions from sources of colonists (Lomolino et al. 1989) or when distance per se plays a role in preventing species from reaching suitable locations. In a more general sense, isolation is an expression of dispersal limitation (Ehrlen and Eriksson 2000; Shurin 2000), which is a fundamental characteristic of living things. The increase of richness with area is equally general and enjoys an enormous literature of it own in under the rubric of species-area relations (SARs; Rosenzweig 1995; He and Legendre 1996; Scheiner 2003). In this chapter I will touch on the role of area as it is thought to affect biogeographical gradients.

Theories and models of geographic gradients of species richness

The transformation of ecology into a self-consciously experimental science in the last several decades of the twentieth century (e.g., Hairston 1989), with its attendant enthusiasm for careful design and rigorous analysis, raised the bar for what counts as explanation in the discipline. Theories and the models that embody them count too, of course, but they are traditionally viewed as being on a different plane from experiments, and some of the most admired work combines analytical models with experimental field studies (e.g., Simberloff and Wilson 1969). Unfortunately, the spatial and temporal scales on which physically manipulative experiments are feasible (and the relatively constrained circumstances under which they are ethical) place such experiments out of reach for many compelling and longstanding questions, including the causes of species richness patterns on geographical scales. Models offer the way forward for the investigation of such questions, and certain kinds of models permit experimentation, as I will explain later.

Statistical models

Statistical models have been widely used to study the correlates, and where possible to suggest the multiple causes (proposition 3), of geographical gradients of species richness (e.g., Currie et al. 2004). This approach can be conceived as treating biogeographical gradients as ordered mosaics (proposition 4) of spatial conditions and represents a clear point of contact between biogeographical gradients and environmental gradients (Fox et al. Chapter 13). For example, a continental-scale map of actual evapotranspiration (AET) might be gridded at a specified resolution (e.g. 2° latitude by 2° longitude cells, or equal-area cells). A scatterplot of the number of bird species (for example) recorded from each cell, plotted against AET for each cell, amounts to ordering the cells (along the abscissa) by AET: an ordered mosaic (e.g., Hawkins et al. 2003b, Fig. 1). Multivariate environmental correlates have often been modeled as a multidimensional, ordered mosaic, with species richness (at some specified spatial scale) treated as the response variable. From such statistical models emerge linear or more complex functions fitted to the data, from which the richness of individual map cells may deviate positively or negatively.

Because nearby map cells are more likely to share similar conditions and similar levels of species richness than are distant map cells, spatial autocorrelation, if not taken into account, can inflate the apparent sample size and Type 1 error. Regardless of spatial autocorrelation in the environment or in richness, however, it is the regression *residuals* that matter: if they show substantial spatial autocorrelation (by Moran's I or other measures), then spatial regression methods must be used to account for any unexplained spatial structure remaining in the regression residuals. On the other hand, if the model residuals show little or no spatial autocorrelation in the best models (that is, the models explain the data very well), then spatial regression methods are unnecessary (Diniz-Filho et al. 2003; Rangel et al. 2010).

Using statistical models applied to ordered multivariate mosaics, a substantial body of work has demonstrated highly significant correlations between species richness and environmental variables, particularly (for terrestrial habitats) measures of solar energy, available water, productivity, and topographic heterogeneity (e.g., Wright 1983; Currie and Paquin 1987; Currie 1991; O'Brien 1998; Rahbek and Graves 2001; Jetz and Rahbek 2002; Hawkins et al. 2003b; Graves and Rahbek 2005; Storch et al. 2006; Davies et al. 2007). These statistical models do not, in themselves, constitute causal theories, but have inspired mechanistic theories that make additional testable predictions (Huston 1994; Rosenzweig 1995; Willig et al. 2003; Currie et al. 2004). These ideas are discussed and evaluated by Fox et al. (Chapter 13).

At present, however, the widely documented correlation between productivity-related climatic variables and richness still lacks a well-supported mechanistic explanation (Currie et al. 2004). Topographic heterogeneity, on the other hand, as a regional level promoter of isolation and speciation (propositions 1 and 5), has behind it a solid body of natural history, phylogeographic, and biogeographical evidence (e.g., Graves 1985; Fjeldså 1994; Rahbek and Graves 2001; Graves and Rahbek 2005; Hughes and Eastwood 2006).

Simulation models and experiments

Biogeographical gradients cannot generally be studied experimentally, but statistical analyses and simulations can help reveal their causes. As an alternative to statistical models, spatially explicit models can be designed to predict biogeographical gradients of species richness (in one or more spatial dimensions) by modeling the underlying geographical ranges, or even the populations of individual species (propositions 1 and 2). Richness emerges from the model as the overlap of ranges as expressed by joint occupancy of map cells. Ecologists and statisticians (Peck 2004; Grimm et al. 2005; Clark and Gelfand 2006) have recently followed philosophers of science (Winsberg 1999; 2001; 2003), physicists, climate modelers, and others in arguing that experimental methods need not be limited to concrete systems but can be legitimately applied to computer simulations carefully constructed to reflect known underlying processes.

Experiments on simulated systems, just like experiments on real ones, are constantly in dialogue with theory and abstract models on the design side, and with statistics on the results side. Astrophysicists, for example, were interested in the unusual convective structure of giant red stars (Winsberg 2001), on which manipulative experiments are not feasible. Although internal convective patterns are known and the laws of fluid dynamics that govern convection are well understood, modeling the complex internal turbulence of such a star proved analytically intractable. Instead, a spatially explicit, discrete-time simulation model was constructed, simplifying where possible (ignoring the internal dynamics of the core and treating it as a simple heat source). Parameters were varied experimentally to discover which factors most influenced the convective patterns and which parameter values produced simulated results that best matched observed ones.

Biogeographical simulation modeling calls upon the same principles of strategic simplification, experimental exploration of parameters, and statistical analysis of results as this example from astrophysics (a field in which manipulation of the physical objects of study is even more impractical than in biogeography) to assess mechanistic hypotheses about biogeographical patterns (proposition 3).

The Neutral Model: no niches, with and without boundaries

The most fundamental of biogeographical simulation models are the spatially explicit, individual-based "neutral models" of Hubbell (2001) and Bell (2000), in which speciation, extinction, and migration take place on a grid, but the spatial domain is unbounded, with no niche differentiation—or, viewed another way, all species share an identical niche. Beta diversity—change in species composition with distance—develops, but only because of dispersal limitation, not because of environmental gradients, of which there are none. Alpha diversity—the number of species occurring in a unit area—does not vary spatially, nor does mean population size or geographic range size, averaged over species (proposition 1). Realistically considered, a classic neutral model is appropriate baseline for a reasonably homogeneous ecoregion, but not for explicitly environmentally heterogeneous regions or gradients (Hubbell 2005). Local communities are imbedded in broader "metacommunities," but the spatial domain is effectively unbounded, and "edge effects" are avoided or intentionally ignored (e.g., Chave and Leigh 2002; Solé et al. 2004; Uriarte et al. 2004; de Aguiar et al. 2009).

Imposing a spatial boundary on a neutral model of community structure is a simple step towards added realism as a biogeographical model (proposition 7). Implementing Hubbell's (2001) model in a bounded lattice [using the genealogical version of the neutral model of Etienne and Olff (2004)], Rangel and Diniz-Filho (2005b) showed that, with dispersal limitation (reasonably short migration distances in relation to domain size), steady-state species richness is no longer uniform across the domain, as in classic neutral models, but instead peaks in mid-domain: a simple (perhaps the simplest possible) biogeographical gradient in species richness (Fig. 14.1). The decline in richness towards the boundary is caused by increasingly asymmetric migration, mak-

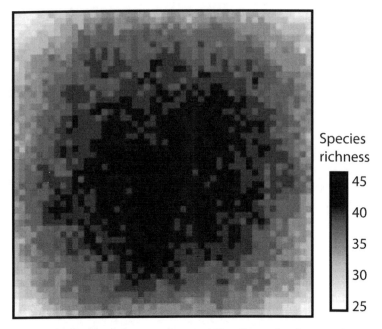

Figure 14.1 Spatial patterns of species richness in a 50 × 50 virtual grid system, after 3000 time steps, averaged among 10 replicates, when species have short-distance migration capacity. From Rangel and Diniz-Filho (2005a).

ing species loss more frequent nearer the boundaries after local disappearance following disturbance events (proposition 1). In Rangel and Diniz-Filho's (2005b) bounded neutral model, the range for each species can be defined the same way that ranges are specified on real-world maps: as minimum polygons surrounding all the occurrences of each species. When patterns from the bounded neutral model are analyzed in this way, the mid-domain peak of richness remains (proposition 2). Arita and Vazquez-Dominguez (2008) explored the temporal and spatial dynamics of a one-dimensional, bounded neutral model, with speciation and extinction. They found that the mid-domain peak of species richness was associated with higher speciation and extinction rates.

Range-based models: still no niches

Modeling at the species (range) level, directly, rather than at the individual level, greatly simplifies the prospect of building simulation models that add realism by integrating niche-driven processes and historical contingencies (proposition 2). The simplest range-based stochastic models of species richness gradients, however, incorporate neither niche differentiation nor history, and model range location on a one-dimensional spatial domain. Colwell and Hurtt (1994) first showed that random placement of one-dimensional "ranges" (line segments) within a one-dimensional "geographical domain" (a bounded line) produces a peak of range overlap (richness) in the middle of the line segment, a pattern that Colwell and Lees (2000) later named the *mid-domain effect* (MDE) (proposition 7). Willig and Lyons (1998) developed a simple mathematical model for one-dimensional MDE models, for hypothetical ranges, and Pineda and Caswell (1998) introduced the key idea of shuffling empirical ranges at random within a bounded, one-dimensional domain. Lees et al. (1999) developed an analytical method for this *range-shuffling* approach, which they applied to fauna of the Madagascan rainforest biome, strongly implicating MDE as the primary driver of both latitudinal and elevational mid-domain richness peaks (see also Kerr et al. 2006; Currie and Kerr 2007; Lees and Colwell 2007).

Two-dimensional stochastic, range-based models also produce mid-domain richness peaks in a homogenous domain, as initially shown by Bokma et al. (2001) and Jetz and Rahbek (2001) (proposition 7). In the latter study, cohesive (continuous) hypothetical ranges were created within a bounded domain (the map of Sub-Saharan Africa), sampling without replacement from the empirical range size frequency distribution (RSFD) for Sub-Saharan African birds. The ranges were simulated by an algorithm inspired by a verbal description by Gotelli and Graves (1996, p. 256), which later came to be known as the *spreading dye algorithm* (Connolly 2005). In this algorithm, *n* hypothetical ranges are randomly placed within the domain, each matching in size (but not in location) one of the *n* empirical ranges for some group of real species endemic to the domain. For each range, a map cell is chosen randomly, then contiguous cells are added one at a time to the range, until its size equals that of the empirical range it matches.

Whether in one or two dimensions, the mid-domain effect arises from geometric constraints on the location of species ranges within a bounded domain (proposition 7). The larger a range, the more constrained its location within the domain, such that larger ranges tend to overlap towards the center of the domain. For a unit-line domain, the midpoint of a range of length r ($0 < r \leq 1$) is geometrically constrained to be located in a region of length $1 - r$ in the middle of the domain. For this reason, in a domain with environmental gradients, a key prediction of MDE theory is that the location of smaller ranges on a domain is expected to be influenced more by the environment and less by geometric constraints (proposition 3) than larger ranges (Colwell et al. 2004; 2005; Dunn et al. 2007).

MDE models were originally intended as null models (Colwell and Hurtt 1994; Willig and Lyons 1998; Lees et al. 1999; Colwell and Lees 2000), from which deviations could be interpreted as caused by environmental or historical gradients (e.g., Connolly et al. 2003). From both a statistical and mechanistic point of view, however, geometric constraints are better treated as explanatory factors, in a multivariate context, on a par with environmental gradients and historical effects (Jetz and Rahbek 2002; Colwell et al. 2004, 2005) (proposition 3). To this end, the effects of constraints are modeled, with the "niche-shuffling" algorithm of Pineda and Caswell (1998) on one-dimensional transects, or with the spreading dye algorithm for two dimensions. This approach has been controversial (Hawkins and Diniz-Filho 2002; Zapata et al. 2003, 2005; Hawkins et al. 2005; Sandel and McKone 2006; Currie and Kerr 2008), in part because of divergent approaches to inference. Some authors (most recently Currie and Kerr 2008) appear committed to a strictly Popperian approach, when it comes to MDE. They view MDE as a "null hypothesis," subject to rejection unless empirical patterns conform within statistical limits to the predictions of an MDE model, while, paradoxically, treating interacting environmental and historical factors as jointly explanatory in a model-selection approach (proposition 3).

Niche-based models on a bounded environmental gradient

Species interact with their environment in neutral models of community assembly only by undirected drift. As soon as even the simplest environmental gradient is introduced and niche differentiation among species is permitted in a model, the rules change. Although classic demographic models of the evolution of range size on gradients have played a key role in our understanding the interaction of selection and migration on gradients (Kirkpatrick and Barton 1997; Case and Taper 2000), modeling biographical gradients of species richness at the population level has not yet been attempted, probably more because of computational limitations than conceptual ones. Instead, models have been constructed at the level of the niches and geographical ranges of species, rather than individuals. Ranges are represented by occupied cells on a gridded domain, with range expansion or contraction as a rough proxy for population dynamics (propositions 1 and 2).

The gradient model of Rangel and Diniz-Filho (2005b) is perhaps the simplest possible dynamic niche-based evolutionary model of biogeographical gradients of species richness. This model sets up a monotonic environmental "suitability" gradient on a bounded, linear domain (in effect, a transect, represented as a row of adjacent cells). The "best" environment is at one end of the domain, where initial species establishment and subsequent speciation is stochastically favored, to a degree controlled by the slope of the linear environmental suitability gradient. Each species is limited in its geographic range (number of occupied cells) by an interaction between the steepness of the environmental gradient and a species-specific, environmental tolerance—its niche breadth (following Kirkpatrick and Barton 1997). Niche shifts (adaptation) occur instantaneously at the time of speciation: each new species assumes as its niche optimum the gradient value of its root cell, with a stochastically assigned niche breadth (propositions 1 and 5). Species do not interact.

The conventional prediction—that the peak of richness will appear at the more "suitable" end of the domain after a period of random speciation and extinction (e.g., Currie 1991; Currie et al. 2004)—is realized only when the environmental gradient is very strong, forcing species to have small ranges, given their environmental tolerances (Fig. 14.2a). With weaker gradients, ranges are larger, given the same tolerances, and a richness peak appears toward the

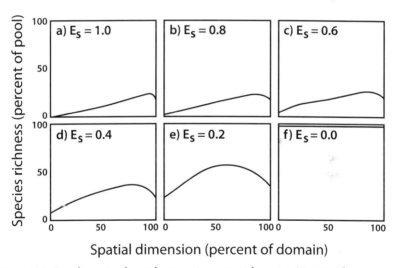

Figure 14.2 One-dimensional spatial patterns in species richness in a 30 × 1 grid, representing a bounded domain with a simple environmental gradient. Panels (a) to (f) illustrate model predictions for an environmental "suitability" gradient of decreasing strength (E$_s$, which ranges from 1 to 0). As the gradient weakens, ranges expand and control over the pattern gradually shifts from the environmental gradient to geometric constraints. From Rangel and Diniz-Filho (2005b).

"suitable" end of the gradient. With successively weakened gradients, the richness peak shifts toward the center of the domain (Fig. 14.2b–e). As the environmental gradient weakens, ranges become larger (Kirkpatrick and Barton
1997), and the influence of geometric constraints becomes stronger relative
to direct effects of the environmental gradient on richness patterns. The mid-
domain peak disappears entirely, and the species richness pattern becomes flat,
only at precisely 0 gradient strength (Fig. 14.2f). Similar results emerge from
Connolly's (2005) process-based analytical models.

   This simulation model demonstrates the interaction between simple niche
structuring and adaptation on gradients and the stochastic effects of traditional neutral model processes (speciation, extinction, and implicit range
dynamics) within the geometric constraints imposed by a bounded domain
(propositions 1, 2, 3, and 5). Models like this one and its two-dimensional analogues (next section) offer an opportunity to explore a challenging and often
contentious issue in quantitative biogeography: the detection of the relative
influence of candidate explanatory factors for species richness, which can be
investigated rigorously and experimentally in a model system, such as this one,
where the causes of pattern are known with certainty.

Range-based models in a bounded, environmentally heterogeneous domain

Models in this category add realism—and complexity—by replacing the linear
gradient of the simplest niche-based gradient models (e.g., Fig. 14.2) with a
bounded heterogeneous environmental mosaic in two dimensions (proposition 5). Although hypothetical landscapes would offer more control of environmental patterns, models in this category, to date, have been based on gridded maps of real continents or regions (*environmental maps*), allowing them
to be realistically complex spatially, and environmentally multivariate.

   Two distinct (and incompatible) approaches have been taken to modeling species distributions, and thus species richness, on environmental maps
(Gotelli et al. 2009). In the first, an empirical range size frequency distribution (RSFD) for some particular taxon for the modeled domain is used as the
basis for the modeled distributions. This approach has its roots in the work of
Pineda and Caswell (1998) and Lees et al. (1999). For example, to investigate
the drivers of gradients of bird species richness worldwide, Storch et al. (2006)
used the empirical RSFD for all land birds and a worldwide terrestrial map of
actual evapotranspiration (AET). Independently, Rahbek et al. (2007) used
the empirical RSFD for South American land birds to model avian species
richness for multifactor environmental maps of the continent. In both studies,
the list of actual range sizes was used, one by one, to map a hypothetical range

of the same area on the map following a modified form of Jetz and Rahbek's (2001) spreading dye algorithm in which stochastic range location is guided by the magnitude of one or more mapped environmental variables. Davies et al. (2007) followed a similar approach. Although in different ways, each of these studies considered MDE as a potential explanatory factor in driving spatial patterns of species richness.

The second approach explicitly models evolutionary processes of speciation, range expansion, range shift with environmental change, and extinction, at the species level. In this approach, not only patterns of species richness, but also the range size frequency distribution, arise from the biogeographical dynamics of the model itself (propositions 1 and 5). In this class of models, there is no intended one-to-one correspondence with any particular empirical species, but simulated species are instead viewed as collectively representative of the particular empirical taxon being modeled. Bokma et al. (2001) developed a predecessor of this approach—a cellular automaton model in a domain without environmental gradients. The first model in this class to simulate evolutionary processes on an environmentally textured domain was developed by Brayard et al. (2005), for foraminifera in the Atlantic Ocean. These authors showed that an interaction between sea surface temperature, currents, and MDE—for both the shape of the domain and for temperature—can produce the twin peaks of richness, north and south of the Equator, that characterize the empirical pattern of richness for these organisms (proposition 3).

Species ranges and the biogeographical gradients that arise from them in physical space can be mapped and modeled on abstract environmental gradients in niche space. This reciprocal correspondence between the cells of an n-factor set of environmental maps (in GIS terminology, a map with n environmental layers) and the corresponding n-dimensional niche space has been called *Hutchinson's duality* (Colwell and Rangel 2009). This conceptualization not only informs species distribution modeling (e.g., Guisan and Zimmermann 2000; Elith et al. 2006), but it also lies at the heart of dynamic, mechanistic models of biogeographical gradients and relates them conceptually to ecological gradients (Fox et al. Chap. 13). Models relying on this duality have begun to incorporate temporal variation in the environment, both stochastic and periodic (proposition 6). In the model of Rangel et al. (2007), empirical environmental maps undergo simultaneous sinusoidal variation in environmental factors. Tolerance for and adaptation to these shifting environmental factors is modeled directly in multivariate niche space (Hutchinson 1957; 1978; see Chase Chapter 5), whereas range expansion, range fragmentation, speciation, and extinction are modeled as projections from niche space onto geographic space (Pulliam 2000).

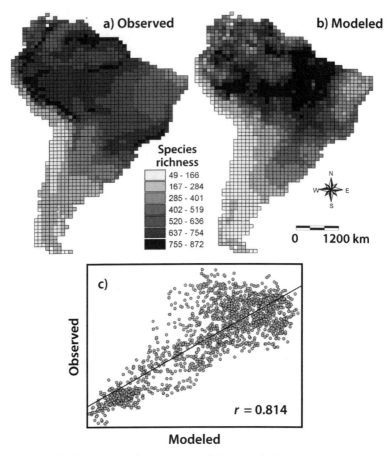

Figure 14.3 (A) Observed spatial patterns in South American bird species richness.
(B) Simulated spatial patterns in South American bird species richness for the best-fitting
model. (C) Relationship and OLS (ordinary least squares) fit between the patterns in maps
(A) and (B). From Rangel et al.(2007).

In the Rangel et al. (2007) model, for example, "life" originates in a single,
randomly chosen starting cell in the geographical space, the gridded environmental map of South America. Based on its n environmental values (layers), this initial map cell corresponds to a single point in the corresponding
n-dimensional environmental niche space, in which the point is declared the
niche center for this founder species. A niche breadth for each factor is assigned independently and stochastically for each environmental axis in niche
space. The niche hypervolume thus defined is projected back onto map space,

defining the geographical range of the seed species as all cells on the map that are *contiguous* with the founder species' seed cell, and that lie within its niche in niche space. As the environmental factors fluctuate, the range of the founder species becomes fragmented, interrupted by areas of the map that no longer correspond to areas within its niche. Smaller fragments face stochastic extinction; larger ones become daughter species. The model allows each surviving fragment to adapt to its new conditions and expand its new niche in niche space around its own niche center, to a degree controlled by "niche conservatism" parameters in the model. The dynamic processes of range fragmentation, extinction, speciation, and constrained adaptation continue until some specified number of species ranges populate the geographical space (propositions 1 and 5). The model thus produces a phylogeny of niches in niche space, mapped by Hutchinson's duality into geographical space. With the empirical species richness map for South American birds as a criterion (Fig. 14.3), Rangel et al. (2007) treated their exploration of model parameter space as a multifactorial, multilevel experiment, concluding that a relatively low extinction rate, a substantial level of niche conservatism (Wiens and Donoghue 2004), and an equatorial latitude for the founder species' seed cell are key conditions for a good fit between the observed and modeled geographical patterns of species richness of birds (proposition 3).

Biogeographical gradients of range size

Biogeographical gradients in range size and species richness are difficult to disentangle, and the quest to understand their interactions has motivated the history of quantitative biogeography for decades. Moreover, from a theoretical viewpoint, it makes little sense to consider static biogeographical patterns of range size without considering the roles of adaptation, speciation, and extinction in relation to range size (propositions 1 and 5), as discussed in the previous section. These are complex topics, encompassing a large literature. I offer here only an outline of the theoretical underpinnings.

Rapoport's rule

In *Aerografía*, his ground-breaking but long-underappreciated monograph, Eduardo Rapoport (1975, English translation in 1982) reported that latitudinal range size tends to be smaller for tropical subspecies than for temperate subspecies, within species of New World mammals and three orders of birds in Asia. As applied to species, rather than subspecies (the difference not always being obvious, in any case), Stevens (1989) canonized this pattern as *Rapoport's rule*. Stevens presented several examples of the same qualitative pattern, all for north temperate groups (latitude 25° N and beyond), which he documented by regressing mean range size against latitude for latitudinal bands (proposition 2).

Two decades after Stevens' (1989) paper and more than three after Rapoport's (1975) far-sighted explorations of geographical ranges, we can now look back and appreciate their profound effect on the theory of biogeographical gradients, primarily by making us all try to think more clearly about how the geography of ranges produces the geography of richness (proposition 2). On the methodological side, the seemingly straightforward concept that Stevens put forward quickly ran into trouble, coming as it did at a time when we ecologists were coming to our senses regarding spatial autocorrelation (Pagel et al. 1991; Legendre 1993), pseudoreplication (Hurlbert 1984), and phylogenetic nonindependence (Harvey and Pagel 1991). Because Stevens' approach counts the same ranges repeatedly (in proportion to their size), the variables for Stevens' regressions are not statistically independent, and degrees of freedom are thus inflated. Colwell and Hurtt (1994), using an approach introduced by Graves (1985), suggested, instead, a "midpoint plot," in which each species is plotted once, in a scattergram of range size vs. latitudinal midpoint (Colwell and Hurtt 1994, Fig. 9). The midpoint plot made clear that range size and range midpoint are inherently nonindependent: a range with a high-latitude midpoint cannot be as large a range with a midpoint nearer the equator (proposition 7). Exploration of this geometric constraint, in the context of Rapoport's rule, was the key to the discovery of the mid-domain effect, independently, by both Colwell and Hurtt (1994) and by Willig and Lyons (1998).

Rohde et al. (1993) compared Stevens' method (range size mean for species occurring in each latitudinal band) to what they called the "midpoint method" (range size mean for only those species whose midpoint occurs in each latitudinal band). They not only found no support for Rapoport's rule for marine teleost fishes, by either method, but showed that species with tropical range midpoints have broader ranges than species with midpoints at temperate latitudes. Although these two methods measure different things (Colwell and Hurtt 1994; Connolly 2009), they often produce qualitatively concordant results for empirical data.

While avoiding statistical nonindependence and spatial autocorrelation due to repeated contributions from larger ranges in the Stevens method, the midpoint method does nothing to deal with phylogenetic nonindependence. In an assessment of Rapoport's rule for New World endemic landbirds, Blackburn and Gaston (1996) attempted to account for this problem by using phylogenetically independent contrasts. They compared results of this method with three others: Stevens' method, Colwell and Hurtt's (1994) midpoint scatterplot method (which Blackburn and Gaston named the "across species method"), and Rohde et al.'s (1993) midpoint (band mean) method. Independent of method, they reported that range size indeed reaches a minimum in the tropics, but at about 12° N latitude (Nicaragua), not at the equator, a pattern that the authors attribute to biogeographical history (proposition 6), rather than the mechanisms that Stevens conjectured to be driving Rapoport's rule (discussed below).

However, the coarse data quality (published range maps, rather than primary data) and the very large spatial grain of the analysis (10° cells) used by Blackburn and Gaston (1996) and in numerous other studies blinds the analysis to the spatial pattern of small-ranged species. For example, nearly 70% of the 241 species of South American hummingbirds, the majority Andean, have ranges smaller than a single 10° quadrat (Rahbek and Graves 2000). This source of error is not a random one, but would tend to bias against extending the pattern of decreasing range size (Rapoport's rule) into the equatorial tropics. On the other hand, primary survey and collections data for most taxa, unless corrected for it, are likely to display the opposite bias (underestimation of range size in the tropics), because of undersampling in rich communities (Colwell and Hurtt 1994).

Colwell and Hurtt (1994) had shown that, depending upon the range size frequency distribution and random placement algorithm, a reverse Rapoport effect (larger ranges in the tropics, smaller ranges at higher latitudes) may appear simply as a result of geometric constraints (proposition 7), using Steven's method of analysis. Willig and Lyons (1998) took the nonindependence of midpoint and range into full account through simulation, and concluded, for midpoint plots, that New World bats and marsupials both support Rapoport's rule.

The conflicting results from these studies are typical of what was already quite a substantial literature on the subject when reviewed a decade ago (Gaston et al. 1998). Published empirical support for the "rule" turns out to be variable, not only taxonomically but also geographically, with land areas north of the Tropic of Cancer offering the strongest evidence for it. Existing studies covering tropical latitudes offer considerably less support, but most of these suffer from data quality and methodological issues, suggesting that the "epitaph" famously called for by Gaston et al. (1998) may in the long run proved to be a case of premature burial (Poe 1844). On the conceptual side, Stevens' (1989) seminal paper appears to have survived the challenges and controversies surrounding the prevalence and detection of the empirical pattern.

Range size and richness: latitude

Stevens went beyond describing Rapoport's rule as an empirical pattern to conjecture that the pattern was a key to understanding latitudinal (Stevens 1989) and, later, elevational (Stevens 1992) and depth (Stevens 1996) gradients of species richness. His idea was founded on the landmark "mountain passes" paper of Janzen (1967), who speculated that tropical species, because they have evolved in climates (particularly temperature regimes) that vary little seasonally compared with seasonal patterns at higher latitudes, would tend to have narrower climatic tolerances than species at higher latitudes (proposition 5). Stevens proposed that these narrow tolerances (particularly temperature, but also precipitation) would restrict tropical species to smaller geographical ranges than would the wider climatic tolerances expected for species living at higher latitudes—a mechanism that he suggested might explain Rapoport's findings for vertebrate subspecies. If geographical ranges are small, Stevens argued, then demographic sink areas, lying outside the range limits for positive fitness, would overlap broadly among many species, yielding areas of high local species density. This hypothesis, which is logically quite separate from the range-size pattern described by Rapoport's rule, has come to be known as the "Rapoport rescue effect" (Stevens 1992), after Brown and Kodric-Brown's (1977) idea that demographic sink populations are continually "rescued" from extinction by immigration (proposition 1).

In a reverse approach, Taylor and Gaines (1999) used stochastic range simulations on a spherical domain to force a "classic" Rapoport effect, with or without demographic sink perimeters on species ranges. They found that the resulting pattern of species richness was opposite the empirical one—the poles were richer than the equator—calling into question Stevens' (1989; 1992) conjecture that the latitudinal gradient in species richness might be a consequence of Rapoport's rule.

For the latitudinal gradient of species richness, Gaston and Chown (1999) pointed out that a fundamental empirical problem undermines Stevens' (1989) appealing line of reasoning. Stevens' argument implicitly assumes that the latitudinal gradient in mean annual temperature, familiar to anyone who has gone south to Mexico or Spain, or north to Queensland, Mombasa, or Rio for a winter holiday, continues all the way to the Equator. In fact, as pointed out by Terborgh (1973b) and discussed by Rosenzweig (1995), Gaston and Chown (1999), and Colwell et al. (2008), the latitudinal gradient in mean annual temperature, which rises almost linearly from the poles and to the Tropics of Cancer and Capricorn for lowland continental stations, levels off to a

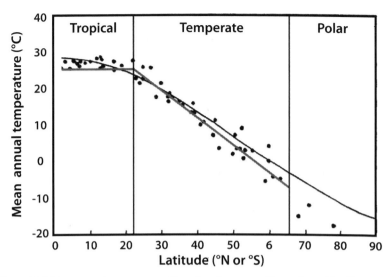

Figure 14.4 Mean annual temperature of low elevation, non-arid, continental localities over a wide range of latitudes. The smooth curve is a theoretical expectation. The trend lines show the tropical plateau and temperate linear decline in mean temperature. From Terborgh (1973b).

broad plateau within the tropics (Fig. 14.4). For this reason, the difference in the scope of thermal tolerance between tropical and temperate species anticipated under Janzen's (1967) hypothesis would not be expected to produce a latitudinal Rapoport pattern in the first place, at least within the tropics, in the absence of other limitations to species ranges, such as topography or simply dispersal limitation. Indeed, in principle, a tropical species with a narrow temperature tolerance might well have a broader latitudinal range than a temperate species with a larger temperature tolerance, given the broad thermal plateau between 20° S and 20° N (Terborgh 1973b).

The key to understanding this issue is an accurate mapping between niche space (thermal niche breadth, in this case) and environmental map space. At the very least, based on mean annual temperatures, latitudinal ranges of species restricted to the tropics cannot be assumed to be correlated in any simple way with their temperature tolerances. On the other hand, Stevens' reasoning makes more sense for temperate and boreal species, along the nearly linear portion of the latitudinal temperature gradient. Perhaps it is no coincidence that many of the best cases for Rapoport's rule (Gaston et al. 1998), including every one of Stevens' (1989) examples, covered latitudes beyond 25° N.

A definitive answer, however, requires more and better data and analyses that account for geometric constraints, spatial autocorrelation, and phylogenetic nonindependence.

Range size and richness: elevation

In contrast with the latitudinal gradient in temperature (for a given elevation), which is mostly a temperate and arctic pattern, the lapse rate (decline in temperature with elevation) differs little with latitude, on a mean annual basis (Colwell et al. 2008). Whether you walk up a tropical mountain or temperate one, the temperature declines roughly the same amount, about 5° C for every 1 km of elevation. However, the fact that seasonal temperature variation at all elevations is greater at temperate latitudes than within the tropics suggested to Janzen (1967) that elevational ranges of tropical species should be narrower than elevational ranges for related temperate species (proposition 5). Surprisingly, it was not until 40 years later that McCain (2009) carried out a test of this prediction, using a meta-analysis of datasets published for other reasons, albeit of varying quality and completeness (undersampling in rich, tropical biotas remains a hazard). Even in the most stringently restricted subset of these data, however, the results of this analysis support Janzen's prediction for vertebrates. Assuming this latitudinal gradient in elevational range sizes is correct, we may expect that the same topographical gradient would yield smaller geographical (mapped) ranges for tropical species than for temperate species. This expectation puts a different spin on Rapoport's rule, if it is to include tropical species: the rule would be expected to apply only to the degree that elevational, rather than latitudinal, temperature gradients are driving latitudinal gradients of range size. Clearly, the degree to which this conjecture is correct depends upon geographical patterns of topography, not on climate alone (Rahbek and Graves 2001; Graves and Rahbek 2005).

What about Stevens' (1992) application of Rapoport's rule and the Rapoport rescue hypothesis to elevational gradients? In the supposed parallel between latitude and elevation upon which Stephens based his argument, the tropical lowlands were assumed to represent the highest richness and smallest mean range size (both latitudinal and elevational) on Earth, with parallel declines in richness and increases in range size with both latitude and elevation. Under this model, elevational gradients at temperate and boreal latitudes would also show a monotonic decrease in richness and a monotonic increase in elevational range size, but scaled to a higher mean range size because of greater seasonality.

There are at least four serious problems with any parallel application of Rapoport's rule to latitudinal and elevational gradients. First, as discussed above, based solely on the evolution of temperature tolerances, there is no reason to expect (and so far, little evidence to show) that latitudinal ranges are generally smaller in the tropics than at higher latitudes, at a given elevation and for similar topography. Second, it is unclear whether range size increases routinely with elevation, particularly in the tropics, although a thorough meta-analysis of worldwide datasets is needed. The evidence Stevens (1992) presents for increasing range size with elevation in the tropics is almost certainly biased by undersampling, which creates a spurious negative correlation between richness and range size (Colwell and Hurtt 1994). Despite this almost inevitable sampling bias, recent quantitative sampling for some 2000 species of plants and insects on a tropical elevational gradient in Costa Rica revealed no conspicuous increase in elevational range size with elevation (Colwell et al. 2008). The third problem, discussed earlier, is that species richness more often peaks at mid-elevations than in the lowlands, regardless of latitude (Colwell and Hurtt 1994; Rahbek 1995; 1997; 2005; Nogués-Bravo and Araujo 2008). The fourth issue is that the absolute spatial scale encompassed for a given amount of environmental change on a latitudinal gradient is much greater than on an elevational gradient, with important demographic and genetic consequences (Rahbek 2005). For temperature, the elevational gradient (per km elevation) at subtropical to high temperate latitudes is nearly 1000 times greater than the latitudinal gradient (per km poleward). (On the ground, the temperature gradient is 100 times greater, even on a 1% slope.) At tropical latitudes, and there is virtually no latitudinal gradient in mean temperature, so the contrast is even more extreme (Colwell et al. 2008).

In summary, biogeographical gradients in range size very well may be driven, as Stevens' (1989) proposed, by adaptive tolerances to seasonal fluctuations (proposition 5), but the expected translation from thermal tolerance limits in niche space to geographical ranges in map space (Hutchinson's duality) requires an accurate environmental map. Stevens' idea that tropical richness could be the product of wide "sink" margins around small geographical ranges is not in accord with the wide geographical scope of tropical climates, at any constant elevation. On the other hand, on steep elevational gradients, local richness may well be enhanced by source-sink dynamics [mass effect (Schmida and Wilson 1985) from physically nearby source areas at lower and higher elevations], particularly in the tropics (Rahbek 1997; Kessler 2000; Grytnes 2003).

Conclusion

As delimited in this chapter, the theory of biogeographical gradients comprises the patterns and causes of geographic variation in the size and location species ranges and their overlap, which we express as species richness. I have focused on concepts and models, attempting to work from the simpler to the more complex. The underlying demographic, ecological, and evolutionary processes that ultimately determine all biogeographical patterns in nature have played key roles in this exploration, particularly in applying Hutchinson's duality to translate between the evolution of environmental tolerances, which is best modeled in niche space, and the realization of distributions on the planet, as expressed in geographical space.

A promising development in the study of biogeographical gradient theory is the spatial analysis of another duality: the reciprocal relationship between the spatial distribution of geographical ranges that overlap at a point (the dispersion field of Graves and Rahbek 2005) and the scope of species richness at sites encompassed by a single species' range (the diversity field of Arita et al. 2008; Villalobos and Arita 2009). On the temporal side, although increasingly sophisticated models have begun to incorporate the role of deep time in biogeography, the role of earth history in shaping biogeographical gradients—through plate tectonics, mountain-building, changing connections between land masses and seas, and climate history—represents an important modeling frontier for theoretical biogeography. I have not attempted to discuss the role of the rapidly growing body of knowledge of phylogenetic history and phylogeographical studies in biogeography, but these data also promise to enrich modeling and inform theory in the study of biogeographical gradients. Finally, biogeographical models can neither hope to be meaningful, nor can they be rigorously assessed without accurate, carefully compiled data on the distributions of organisms. There is no substitute for good data.

Acknowledgments

For helpful comments on the manuscript, I am grateful to Carsten Rahbek, Thiago Rangel, an anonymous reviewer, and the editors. This work was supported by National Science Foundation Grants DEB-0639979 and DBI 0851245 and by the Center for Macroecology, Evolution and Climate of the University of Copenhagen.

SYNTHESIS

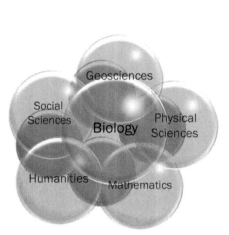

CHAPTER 15

The State of Theory in Ecology

Michael R. Willig and Samuel M. Scheiner

There is a need to subject current theory to stringent empirical test, but ecology can never have too much theory.

MacArthur 1972

An article in *Wired Magazine* proclaimed in its title "The End of Theory: The Data Deluge Makes the Scientific Method Obsolete" (Anderson 2008). The basic premise of that essay was that evolving computational capabilities will allow large and heterogeneous datasets to be mined efficiently and effectively. The result would be the production of pattern without the need of hypothesis formation and testing, so much so that correlation would supersede causation. Science would "advance without coherent models, unified theories, or really any mechanistic explanation at all." Although advanced cyberinfrastructure will revolutionize much of the scientific enterprise as it relates to data collection and visualization, the overarching thesis of the article invites replies on many levels. Rather than do so here, we use that statement as a springboard from which to emphasize the unsophisticated view held by many that there is but one scientific method and that the accuracy and utility of models are the essential characteristics of theory. Indeed, perusal of the preceding chapters in this book or the influential tome by Pickett et al. (2007) suggests that Anderson's (2008) conceptualization of theory is flawed from many perspectives. Mark Twain, upon reading his obituary in a New York journal, is quoted as having replied, "The reports of my death are greatly exaggerated." So too, this is our response to Anderson's contention. Theory in general, and theory in ecology in particular, are alive and flourishing, providing much impetus to deepen and broader our understanding of the natural world. Our goal in this chapter is to briefly summarize where the discipline now stands with regard to that understanding.

The hierarchy of ecology

The domain of a theory defines its central focus. In the case of a general theory, that domain also circumscribes a scientific discipline. In Chapter 1, we defined the domain of the theory of ecology as the spatial and temporal patterns of the distribution and abundance of organisms, including their causes and consequences. Nonetheless, this definition requires additional exposition to understand the nature of the patterns and processes under consideration (Kolasa and Pickett 1989). The brief definition of ecology's domain does not address the nature of the interactions that define the levels of the ecological hierarchy, and how that hierarchy fits within the relationships and interactions that define the rest of the domain of biology (Fig. 15.1; Scheiner 2010).

Our hierarchical perspective makes clear that ecological theory is directed at understanding biological entities at or above the level of individuals. The birth, death, growth, and movement of individual organisms give rise to the complex spatial and temporal tapestry of life that is the focus of ecological studies, and these basic attributes arise from the dynamics involved in the acquisition of energy and nutrients from the environment. Most of the chapters in this book focus on the middle of that hierarchy: populations (Hastings Chapter 6; Holt Chapter 7), single communities (Chase Chapter 5; Pickett et al. Chapter 9), or collections of communities (Leibold Chapter 8; Sax and Gaines Chapter 10; Fox et al. Chapter 13; Colwell Chapter 14).

A few chapters examine theory associated with the ends of the hierarchy. At the level of individuals, the domain of the theory of ecology intersects with the domain of the theory of organisms (Scheiner 2010; Zamer and Scheiner in prep.), resulting in such disciplines as physiological ecology. The theories that define those disciplines are examples of how the domains of a constitutive theory can overlap the domains of more general theories. In this book, the individual-level perspective is represented by foraging theory (Sih Chapter 4), which is representative of the broader domain of behavioral ecology.

At the other end of the hierarchy, the theory of ecology overlaps with theories from the geological sciences. Ecological processes have a dramatic effect on the distribution of biologically important chemicals (e.g., C, N, P, O). Over billions of years, ecological interactions have transformed the planet from an

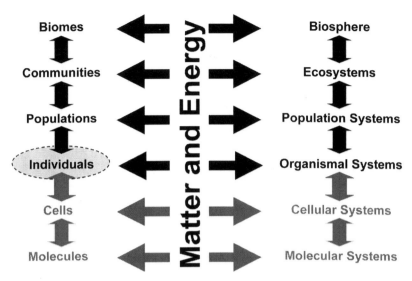

Figure 15.1 One way of organizing living systems is as a hierarchy that extends from molecules to biomes. At each level, biological entities (e.g., cells, individuals, communities) interact with matter and energy (double-headed horizontal arrows) to form living systems. The domain of ecology is defined by interactions at the level of individuals and higher (signified by black rather than gray lettering), and is characterized by an organismal perspective (signified by the shaded ellipse at the individual level). Each level in the biological hierarchy is associated with other levels (e.g., populations comprise individuals of the same species and populations of different species comprise communities) in the hierarchy (represented by vertical arrows on the left). Additional connectivity among levels occurs because the matter and energy that fuel the activities of all biological entities flow and cycle, respectively, in biological systems regardless of level in the hierarchy (indicted by vertical arrows on the right). Moreover, then nature of the ecological entities and their interactions changes over time as a consequence of evolution, resulting in complex dynamics and multiple feedbacks. (Modified from Odum 1971.)

environment with high ultraviolet radiation, low availability of oxidizing and reducing ion pairs, and few energy sources to an environment with low ultraviolet radiation, high availability of oxidizing and reducing ion pairs, and a diversity of energy sources (Burke and Lauenroth Chapter 11). Moreover, the interplay between matter and energy with the biotic portions of the environment creates dynamic interacting systems at all levels of the ecological hierarchy (e.g., organismal systems, population systems, ecosystems) that can play out over global scales (Peters et al. Chapter 12). This framework makes clear that the study of ecosystems (i.e., community systems) is an integral part of the domain of ecology. Ecosystems arise from the interactions of communities with matter and energy, and the resultant cycles, flows, and pools or standing stocks are consequences, to at least some extent, of the activities of the biota.

Implicit in our definition of ecology is an emphasis on spatial and temporal perspectives. As elucidated in many of chapters in this book, ecological relationships can vary as a function of the scales of space or time, and environmental drivers that strongly affect variation at one scale may be markedly different from those at other scales. Consequently, ecological understanding, especially predictive understanding, is a challenge when the form or parameterization of a relationship may differ across scales, or when the identity of the dominant driver of a pattern changes with scale. One of the central challenges in ecology is the development of theories and models that integrate across levels in the biological hierarchy (Fig. 15.1). In this book, the only theories that explicitly integrate across the hierarchy are metacommunity theory (Leibold Chapter 8), succession theory (Pickett et al. Chapter 9), and island biogeography theory (Sax and Gaines Chapter 10), each of which integrates population- and community-level processes, and global change theory (Peters et al. Chapter 12), which integrates from individuals to the biosphere. Other chapters (e.g., Fox et al. Chapter 13; Colwell Chapter 14) hint at such integration, but do not explicitly model it.

The concept of hierarchical levels as applied to ecology (Odum 1971) has long been recognized in many realms of the natural and social sciences, along with important philosophical considerations (e.g., Novikoff 1945; Feibleman 1954; Greenberg 1988). In addition to facilitating communication and classification in ecology, the integrative levels of organization in that hierarchy suggest that moving from individuals to communities involves increases in complexity, and that properties at higher levels can emerge from lower levels. At each level in the ecological hierarchy, emergent characteristics manifest that cannot be predicted or fully understood based on just the patterns and processes at lower levels. Moreover, interactions are horizontal, among entities at the same level (e.g., different species interact within a community) as well as vertical. In addition, influences are fully complementary, in that processes at higher levels can affect properties at lower levels (e.g., communities affect populations). For these reasons, both reductionist and system approaches to ecological understanding are by themselves insufficient and sometimes misleading. In ecology, an inability to clearly and unambiguously identify the spatiotemporal limits of entities at each hierarchical level (e.g., individual, population, community) may conspire to further challenge the development of predictive understanding. For example, we may find it easier or less arbitrary to distinguish unitary individuals than to distinguish unitary communities, making it progressively more complicated to understand the linkage between pattern and process as we traverse up the ecological hierarchy.

The development of theory

The chapters in this book represent a diversity of ecological theories that differ greatly in content and scope, as well as in their degree of theoretical maturation. The chapters also differ in the extent to which their syntheses summarize, clarify, amplify, integrate, or unify theoretical constructs to the advancement of ecological understanding. In Chapter 1 we presented a hierarchical view of theory consisting of three tiers: general theories, constitutive theories, and models (Table 1.1). We noted, however, that this division into three tiers was arbitrary. The hierarchy is really a continuum. About half of the chapters primarily focus on the model end of this continuum because they provide explicit directions for the building of models: foraging theory (Sih Chapter 4), niche theory (Chase Chapter 5), population dynamics theory (Hastings Chapter 6), enemy-victim theory (Holt Chapter 7), island biogeography theory (Sax and Gaines Chapter 10), and ecological gradient theory (Fox et al. Chapter 14). The others—metacommunity theory (Leibold Chapter 8), succession theory (Pickett et al. Chapter 9), ecosystem theory (Burke and Lauenroth Chapter 11), global change theory (Peters et al. Chapter 12), and biogeographical gradient theory (Colwell Chapter 14)—are much more like general theories because their propositions are closer in nature to fundamental principles and define the domain of their models rather than provide explicit rules for model building. So, even for the constitutive theories presented in this book, additional constitutive theories could be developed that are either narrower in scope and act to unify some particular set of models or are broader in scope and aim at uniting other constitutive theories.

To some extent the tendency of a chapter to be at one end or the other of that continuum from general to specific is a function of the maturity of the theory. Foraging theory, population dynamic theory, enemy-victim theory, and island biogeography theory are all quite mature and the authors of those chapters focused on models. In contrast, metacommunity theory, global change theory, and biogeographical gradient theory are less mature and those authors presented more general views of their topics. In some cases, the very maturity of the theory in combination with the complexity of the domain led to a chapter that was more general in focus (succession theory and ecosystem theory). In other cases, the theories arose out of attempts to synthesize across competing models, leading to a more model-focused approach of a less mature theory (niche theory and ecological gradient theory).

Theories may assume a number of different roles (R. Creath, unpublished ms.). They represent generalizations that extend the scope of the particular data that espoused those generalizations. They generate concepts that extend beyond what can be expressed in observation alone. In these two capacities, theories are primarily descriptors of the world. In addition, theories are a framework for guiding and evaluating research paradigms sensu Kuhn (1962) or research traditions sensu Laudan (1977). To a greater or lesser extent, all of the constitutive theories in this book play each of these roles.

Many controversies arise within a domain because of a failure to differentiate between a core concept that is neutral and broad on the one hand, and various incarnations of that concept that may be narrow and specific on the other. The differences among particular models allow the more general theory to be broadly applicable, depending on circumstances defined by the distinguishing assumptions of those models. Understanding the features that favor one model over another thereafter becomes a unifying strength of the theory to account for myriad empirical observations. The framework for understanding disturbance and succession (Pickett et al. Chapter 9) exemplifies this process. The frameworks for niche theory (Chase Chapter 5), enemy-victim theory (Holt Chapter 7), and metacommunity theory (Leibold Chapter 8) perform similar unifications.

Roles of theory in ecology

The body of theory in ecology accounts for our observations about the natural world and gives us our predictive understanding through the use of models. It organizes those models into constitutive domains that provide a robust intellectual infrastructure. That organizational framework provides a blueprint of the strengths and weaknesses of our understanding, motivating future empirical and theoretical work and catalyzing research agendas. Refinements of theory can identify the mechanistic bases of patterns and processes about which we have considerable confirmation, as well as distinguish the ideas and relationships that are in flux or about which there is considerable uncertainty.

In the title to his chapter, Kolasa (Chapter 2) emphasizes that "theory makes ecology evolve." Using a historical perspective, he identifies the genesis of the ideas that formed the bases of the fundamental principles of the domain of ecology (Table 1.3). He forcefully argues that theory and empiricism are inextricably intertwined, not mutually exclusive undertakings, and that theory is a vehicle for sharing knowledge across domains as well as for targeting efforts to fruitfully deepen or broaden the scope of ecological understanding. This too is the broad perspective that can be gleaned from other chapters in this book. It is a basis for our contention that the proclamation about the death of theory (Anderson 2008) is fatally flawed, even in an age of cyberinformatics. Given the deluge of data, theory helps winnow out data that are irrelevant to a particular domain of interest while highlighting that which advances understanding. Theory helps to organize multiple lines of evidence in an efficient manner. Theory provides connections among ideas and concepts within and among domains. Theory provides insights into new data requirements needed to distinguish among or resolve differences among competing views of the world.

Odenbaugh (Chapter 3) clarifies the nature of a unifying theory, and reinforces Kolasa's contention that models alone, no matter how mathematically elegant or predictive, are incomplete aspects of a mature theory. Odenbaugh challenges ecologists and evolutionary biologists to further integrate disciplinary understanding with a goal of exposing the spatiotemporal interdependence of ecological and evolutionary processes: current ecological processes are in play because of past evolutionary processes and current evolutionary processes are in play because of past ecological processes. He illustrates past efforts in this area by focusing on the work of MacArthur and his collaborators. Odenbaugh argues that although their ideas were formative and stimulatory to generations of ecologists, they did not succeed in unifying ecology, despite their intentions. Rather, their work provided the discipline with natural selection thinking, a focus on model building, and a strategy aimed at predictive understanding and generality, instead of only descriptive understanding, as in natural history. This same tension between predictive understanding and descriptive natural history helped shape the origins of ecology as a discipline at the beginning of the 20th century (Hagen 1992). As evidenced by the chapters in our book, this striving towards predictive understanding continues.

Multicausality

Ecological systems have a critical property—multicausality—that affects the structure and evaluation of ecological theories (Pickett et al. 2007). In general, multicausality (Fig. 15.2A) occurs when more than one driving factor (Xs in figure) effects an outcome (Y in figure). For heuristic purposes, we distinguish a number of general types of multicausality. First are instances where variation in a particular characteristic arises as a consequence of variation in only a subset of the possibilities driving factors (Fig. 15.2B). For example, each of three factors (e.g., X_1, X_2, or X_3) could affect an outcome, but they do not all do so in concert in all circumstances. In some circumstances, only X_1 and X_2 might effect the outcome whereas in other circumstances, only X_1 might do so. From

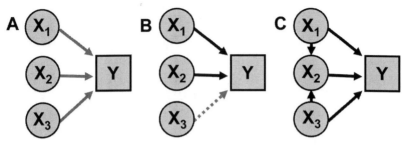

Figure 15.2 These diagrams represent various types of multicausality, a situation in which more than one driving factor (represented by Xs) effects variation in focal characteristic (represented by Y). (A) A general model that defines the candidate driving factors (solid grey lines) and the responding characteristic of interest. (B) Only a subset of the possible driving factors have an effect depending on particular circumstances (represented by solid arrows). In this case, the model is additive and multivariate, but under other circumstances it could be univariate. (C) All of the candidate driving factors have an effect on the responding characteristic of interest. In this case the effects of the factors are interactive. Even more complex situations can arise, where only subsets of the candidate driving factors come into effect in particular circumstances, and the factors interact in a non-additive way, including direct and indirect effects.

an analytical perspective, this results in multivariate causation in the former situation and univariate causation in the latter situation. Second are instances where all of the driving factors in concert effect an outcome (Fig. 15.2C). The multiple driving factors could act together in an additive manner or could do so in complex ways that are nonlinear and include direct and indirect effects.

These aspects of multicausality are important for the structure of ecological models and their evaluation. If a model includes all of the multiple causes, it will provide robust predictions or explanations. For multicausal models of the first type (Fig. 15.2B), it is necessary that the model include only the particular driving factors acting in a particular situation, but the causes that are included in the model may differ from situation to situation. Thus, it is not necessary to know all of the possible causes in all situations, just the ones that are important in the situation under consideration. For models of second type (Fig. 15.2C), it is necessary to know all of the driving factors in advance of model construction.

For all types of multicausality, if a model does not include all causes, the utility of the model depends on whether those causes have additive or nonadditive (i.e., interactive) effects on the outcome. When effects are additive (Fig. 15.2B), conclusions about the relative magnitudes of the processes included in the model are robust. The excluded factors may affect absolute predictions of a model, but not relative ones. On the other hand, if the causes interact (Fig. 15.2C), then the magnitudes and rank orders associated with one causative mechanism may depend on those of another. At minimum, it is necessary to acknowledge that this is an inherent assumption in particular models.

In this book, we find both types of multicausal models. Models that deal with independent, additive causes are exemplified by niche theory (Chase Chapter 5), population dynamics theory (Hastings Chapter 6), enemy-victim theory (Holt Chapter 7), island biogeography theory (Sax and Gaines Chapter 10), and ecological gradient theory (Fox et al. Chapter 13). Interactive causes are notable in foraging theory (Sih Chapter 4), metacommunity theory (Leibold Chapter 8), ecosystem theory (Burke and Lauenroth Chapter 11), global change theory (Peters et al. Chapter 12), and biogeographic gradient theory (Colwell Chapter 14). Perhaps the most extreme version of such interactions is found in succession theory (Pickett et al. Chapter 9).

When evaluating models, the two types of multicausality and the details of their interactions have important implications for how an experiment would be designed. For the first type of multicausality with additive effects, rigorously holding constant all factors other than those under investigation would be most informative. Even with nonadditive effects, an experiment would manipulate just a few factors. Most laboratory and greenhouse experiments are of these types. For the second type of multicausality, unless one knew all of the necessary causes and their interactions, a field experiment would be more informative. Although one or a few factors might be deliberately manipulated, other necessary factors would also be free to also contribute. Importantly, statistical techniques such as structural equation modeling, which are capable of identifying causal factors and of incorporating direct and indirect effects (Grace 2006; Grace et al. 2010), could be employed with nonexperimental data.

Because of the second type of multicausality, some philosophers of science conclude that we can never determine the true explanation of a phenomenon because multiple alternative explanations always exist (Suppe 1977). In practice, ecologists must often use multiple lines of evidence to discern the relative roles of ecological processes in producing patterns (e.g., Carpenter 1998). See Scheiner (2004a) for a more complete discussion of the use of total evidence in ecology.

Another aspect of multicausality is that some causal processes are proximate and others ultimate. Consider the question: Why are male lions larger than female lions? A proximate explanation involves development and food intake during growth. A more ultimate explanation involves sexual selection: larger males are better able to monopolize a group of females. Beyond those processes may be phylogenetic effects involving all felids or carnivores. These alternative explanations often derive from different general theories, so a given constitutive theory needs to either draw on those multiple general theories or acknowledge the limitations of its explanatory scope.

Spatial variation, temporal variation, and scale

Environmental heterogeneity, both abiotic and biotic, is core to ecological processes, as shown by its prominence in the theory of ecology (Table 1.3, principles 3, 5, and 6). This heterogeneity creates a central role for the importance of scale in ecological theories.

Geographic space and ecological space are intimately intertwined. This intersection can be seen most clearly in two theories. Biogeographic gradient theory (Colwell Chapter 14) presents a synthetic framework for the creation of a theory of spatial gradients (e.g., latitude, elevation, depth) that operate at broad geographic scales. Broad-scale patterns of species richness and range size are an emergent property arising from the sum of species-specific responses. Metacommunity theory (Leibold Chapter 8) bridges local and regional scales (i.e., mesoscale ecology) by considering the extent to which local filters and dispersal determine the composition and species richness of sets of communities.

Interactions of ecological processes can change over space and time. Within a single community their relationships change as a result of disturbance and succession (Pickett et al. Chapter 9). Those interactions are now mostly understood as befits a theory that has been developing since the origins of ecology in the late 19th century (Cowles 1899; Clements 1916). At the other end of the spatial and temporal scale are those global changes initiated by human activities (Peters et al. Chapter 12). Borrowing concepts from hierarchy theory (Allen and Starr 1982) and landscape ecology (e.g., Peters et al. 2006; Peters et al. 2008), global change theory addresses issues associated with the consequences of large-scale human-initiated disturbances such as global warming, urbanization, and agricultural intensification. This theory is implicitly scale-sensitive, suggesting that fine-scale relationships between pattern and process interact with broad-scale relationships, resulting in spatial heterogeneity and differential connectivity among spatial units.

Conservation, management, and policy

Ecology as a discipline and ecologists as scientists have changed greatly from the middle of the last century, when the Nature Conservancy was formed. That organization was founded by a group of ecologists who were frustrated with their inability to get the leadership of the Ecological Society of America (ESA) to address the practical and policy implications of their science. In contrast, today the ESA has taken a leadership role in translating science into policy.

It is telling, though, that such linkage is mostly absent from this book, despite our instructions to authors that they should address those issues. Only two chapters do so explicitly: island biogeography theory (Sax and Gaines Chapter 10) and global change theory (Peters et al. Chapter 12). It is not surprising that these chapters address those concerns. Island biogeography theory has long been entwined in efforts to determine the best design for nature reserves (Burgman et al. 2005), especially the SLOSS ("single large or several small") debate of the 1980s. Today, global change has become a central focus of both science and public policy. The magnitude and rate of change are both great, and these anthropogenically induced changes will likely affect all levels in the ecological hierarchy, often in dramatic ways and likely over broad spatial extents.

That is not to say that the other theories in this book are not also relevant to applied issues. For example, population dynamics theory (Hastings Chapter 6) is used extensively for population viability analyses. Similarly, enemy-victim theory (Holt Chapter 7) is useful in understanding pathogen-host interactions in agricultural settings as well as the dynamics of infectious diseases as they relate to public health. Rather, we ecologists tend to separate theory development from theory application. The drive for theory development often comes from basic research questions, with application and additional refinement of theory coming later. Global change theory is a notable exception. Its impetus arises from current concerns about where our planet is headed as a consequence of anthropogenic contributions to greenhouse gases and expansive modifications of landscape structure and configuration throughout the world.

Much of the application of theory to questions of management has focused on optimization issues (i.e., maximum sustained yield) related to production of particular agricultural crops or harvests of particular species of wildlife for human consumption or use. A more holistic approach that considers management from an integrated, multispecies ecosystem perspective is gaining ground because of its ability to include both direct and indirect effects on targeted species, the species with which they interact, and the ecosystem services that they provide to humans (Peterson 2005). In many ways, this heralds the emergence of a new scientific discipline—socioecology—at the intersection of the social sciences, environmental sciences, and engineering.

This new discipline explicitly considers coupled human and natural sys-

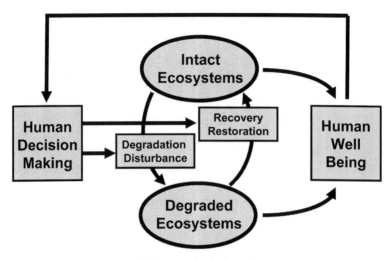

Figure 15.3 This conceptual model illustrates key linkages between natural and human systems that together constitute a socioecological system. It recognizes (1) that the functionality of natural systems varies along a continuum from intact to degraded, with each providing inputs (both positive or negative) to human well-being; (2) that human decisions affect ecosystems positively, via recovery, restoration, or reclamation, or negatively via degradation; and (3) that real or perceived well-being of humans should directly feed back on decision making (policy) so as to adaptively and sustainably manage natural systems.

tems as the domain of interest (Fig. 15.3). At its foundation is the theory of disturbance and succession, with human social systems as both the drivers of and respondents to change. These dynamic feedbacks must be used in policy decisions if they are to lead to adaptive management with a goal of enhancing resilience and long-term sustainability. Such a theory would focus on cycles of disturbance and recovery (succession) within the socioecological system (including its biotic and abiotic constituents), and would rely on an understanding of successional dynamics (Pickett et al. Chapter 9) and ecosystem function (Burke and Lauenroth Chapter 11). In so doing, it considers humans as ecological engineers or drivers of change (disturbance agents) that affect landscape configurations of local ecosystems, each with positive or negative consequences to human well-being. Moreover, it considers human well-being as providing feedback to human actions via policy and management. Perhaps the greatest challenge to face society and science in the 21st century will be developing a predictive understanding of coupled human and natural systems—socioecological systems—so that policy and management can be responsive to long-term goals of sustainability. The further development of theory across all of ecology will play a critical role in the ultimate success of such an endeavor.

Integration and unification: the future of ecology

Despite the different levels in the biological hierarchy that the chapters in this book address, as well as particular interactions that form the focus of their expositions, they share a number of features. Each chapter defines a suite of basic propositions within a particular domain, and combines different state variables or parameters in alternative ways to provide understanding or prediction about central ecological phenomena. Each links the propositions associated with its domain back to the fundamental principles of ecology (Table 1.3). The chapters identify central models within their domains; some are conceptual while others are more precise and mathematical in nature. Thus, these chapters consolidate the state of understanding and accelerate the process of theory unification. In addition, each chapter clarifies connections between its focal domain and the domains of other chapters or subdisciplines of ecology, enhancing integration. Various chapters illustrate how different assumptions lead to different models. A failure to substantiate a particular model does not necessarily mean that the more general theory with which it is associated is wrong or useless. Rather, much of ecology deals with understanding the conditions that favor one model over another, and how these conditions relate to the formalized assumptions of each model.

During discussions at the workshop that preceded this book and during the process of articulating the various constitutive theories, a common claim was that one person's theory was central to all of ecology and that all other domains could be viewed as ancillary to her or his domain. Such viewpoints are to be expected as we attempt to build a set of integrated constitutive theories. Our general theory tells us that the constitutive theories must be linked to each other. As noted in the various chapters, each constitutive theory links directly with numerous other theories. In some cases the theories share similar propositions. Some of this sharing is expected and obvious [e.g., population dynamics theory (Hastings Chapter 6) and enemy-victim theory (Holt Chapter 7); ecological gradient theory (Fox et al. Chapter 13) and biogeographical gradient theory (Colwell Chapter 14)]. In other cases, overlaps become more apparent after propositions are formalized, for example the role of connectivity in metacommunity theory (Leibold Chapter 8) and global change theory (Peters et al. Chapter 12). In yet other cases, the propositions of one theory can point to ways that other theories can be modified, for example the role of species interactions in niche theory (Chase Chapter 5) as a guide to adding such interactions to island biogeography theory (Sax and Gaines Chapter 10) or ecological gradient theory (Fox et al. Chapter 13). No single constitutive theory is at the center of ecology. Rather all are connected and overlap to some degree and together define the science of ecology.

A more comprehensive unification and integration of ecology would be advanced by applying these approaches to domains of ecology beyond those considered in this book (e.g., competition, mutualism, food webs, and landscapes). For example, the metabolic theory of ecology (West et al. 1997; Brown et al. 2004) currently consists of a single model that has been applied to a variety of questions (e.g., Allen and Gillooly 2009). Progress would be furthered through the articulation of the propositions that underlie that model coupled with an attempt to develop alternative models derived from those same propositions. Such alternative models would help to clarify the debate around this theory (Hawkins et al. 2007; del Rio 2008). It would lead to the testing of alternative hypotheses, going beyond the current practice of simply fitting data to a single model, as is frequently done across all of ecology.

We do not mean to imply that ecology will advance or mature only via a single approach, such as that advocated in this book. Indeed, understanding within a domain evolves via a variety of activities. Such a diversity of approaches can lead to robust formulations of the intellectual framework—the theory—that distinguishes ecology, integrates its components, and identifies lacunae in understanding or shortcomings in empirical validation.

The past 50 years in ecology have seen the development of two contrasting approaches to model development. One is the ecosystem approach, exemplified by the work of Odum and his collaborators (Odum 1971), which attempts to build models that are highly complex and specific. The other is the evolutionary ecology approach, exemplified by the work of MacArthur and his colleagues (Odenbaugh Chapter 3), which aims to build very simple and general models. Although often seen as antithetical (Odenbaugh 2003), the theoretical framework presented in this book can encompass both modeling approaches (e.g., Holt Chapter 7, Burke and Lauenroth Chapter 11). The challenge for all modeling approaches is to determine the underlying propositions that provide the theoretical framework for a set of models. For simple, general models moving to more general propositions is relatively straightforward, although still not a trivial exercise, as was discovered by the contributors to this book. For complex, specific models, deriving general propositions is less straightforward. Recent advances in structural equation modeling (Grace et al. 2010) provide one avenue by which such models can be united within a general framework.

In summary, the process of unification and integration is well under way within the various domains of the constituent theories of ecology, as well as at the level of the general theory, including its integration with the rest of biology (Scheiner 2010). An uber-theory, in the sense of an all-encompassing model or mathematical formulation, is unlikely to characterize ecology in its full diversity of content based on the hierarchy of interacting systems.

We are hopeful that from these modest beginnings, advances in ecological understanding will be accelerated by a faithful and consistent application of integrative and unifying approaches to the development of theory, such as those considered in this book. We trust that these chapters will strengthen the foundations of ecological understanding and help to herald a time of an intensified interest in ecological theory. We are not viewing the death of theory. Borrowing from Winston Churchill (10 November 1942), "his is not the end. It is not even the beginning of the end. But it is, perhaps, the end of the beginning" of a revitalization in the advancement of theory as a vehicle for promoting deep understanding of ecological systems.

Acknowledgments

Ecological Understanding: The Nature of Theory and Theory of Nature, by Pickett, Kolasa, and Jones (2007), was seminal to the completion of this chapter. In addition, many of the ideas we develop arose from interactions among the chapter authors, who were participants in a workshop on the Theory of Ecology, supported by the University of Connecticut through the Center for Environmental Sciences and Engineering and the Office of the Vice Provost for Research and Graduate Education. Support to MRW was provided in part by the National Science Foundation via grant DEB-0614468. This work was done while SMS was serving at the U.S. National Science Foundation and on sabbatical at the Center for Environmental Sciences and Engineering at the University of Connecticut. The views expressed in this chapter do not necessarily reflect those of the National Science Foundation or the United States Government.

ACKNOWLEDGMENTS

We thank the Center for Environmental Sciences and Engineering in cooperation with the Office of the Vice Provost for Research and Graduate Education at the University of Connecticut for sponsoring a workshop that brought the chapter authors and others together, as well as all workshop participants. We thank Kayla Scheiner for producing the graphic figures that head each chapter and the many people who reviewed all or parts of the book. The book could not have been produced without the help of the people at the University of Chicago Press, including Abby Collier, Michael Koplow, Amy Krynak, and especially Christie Henry.

CONTRIBUTORS

Brandon T. Bestelmeyer
USDA Agricultural Research Service
  Jornada Experimental Range

Ingrid C. Burke
Department of Botany
University of Wyoming

Mary L. Cadenasso
Department of Plant Sciences
University of California–Davis

Jonathan M. Chase
Department of Biology
Washington University

James P. Collins
School of Life Sciences
Arizona State University

Robert K. Colwell
Department of Ecology and Evolu-
  tionary Biology
University of Connecticut

Gordon A. Fox
Department of Integrative Biology
University of South Florida

Steven D. Gaines
Bren School of Environmental Science
  and Management
University of California–Santa
  Barbara

Alan Hastings
Department of Environmental Science
  and Policy
University of California–Davis

Robert D. Holt
Department of Biology
University of Florida

Alan K. Knapp
Department of Biology
Colorado State University

Jurek Kolasa
Department of Biology
McMaster University

William K. Lauenroth
Department of Botany
University of Wyoming

Mathew A. Leibold
Section of Integrative Biology
University of Texas–Austin

Scott J. Meiners
Department of Biological Sciences
Eastern Illinois University

Jay Odenbaugh
Department of Philosophy
Lewis & Clark College

Debra P. C. Peters
USDA Agricultural Research Service
    Jornada Experimental Range

Steward T. A. Pickett
Institute for Ecosystem Studies

Dov F. Sax
Department of Ecology and Evolu-
    tionary Biology
Brown University

Samuel M. Scheiner
Arlington, VA

Andrew Sih
Department of Environmental Science
    and Policy
University of California–Davis

Michael R. Willig
Center for Environmental Sciences
    and Engineering
University of Connecticut

BIBLIOGRAPHY

Aber, J. D., and J. M. Melillo. 2001. Terrestrial Ecosystems. New York, Harcourt Academic Press.

Abrahams, M. V. 1986. Patch choice under perceptual constraints: a case for departures from an ideal free distribution. Behavioral Ecology and Sociobiology 19:409–415.

Abrams, P. A. 1983. The theory of limiting similarity. Annual Review of Ecology and Systematics 4:359–376.

———. 1988. Resource productivity-consumer species diversity: simple models of competition in spatially heterogeneous environments. Ecology 69:1418–1433.

———. 1992. Predators that benefit prey and prey that harm predators: unusual effects of interacting foraging adaptations. American Naturalist 140:573–600.

———. 1993. Why predation rate should not be proportional to predator density. Ecology 74:726–733.

———. 1994. Should prey overestimate the risk of predation? American Naturalist 144:317–328.

———. 2000. The evolution of predator-prey interactions: theory and evidence. Annual Review of Ecology and Systematics 31:79–105.

———. 2007. Habitat choice in predator-prey systems: spatial instability due to interacting adaptive movements. American Naturalist 169:581–594.

Abrams, P. A., and L. R. Ginzburg. 2000. The nature of predation: prey dependent, ratio dependent, or neither? Trends in Ecology and Evolution 15:337–341.

Abrams, P. A., and R. D. Holt. 2002. The impact of consumer-resource cycles on the coexistence of competing consumers. Theoretical Population Biology 62:281–295.

Ackakaya, H. R., R. Arditi, and L. R. Ginzburg. 1995. Ratio-dependent predation: an abstraction that works. Ecology 76:995–1004.

Adler, P. B., E. P. White, W. K. Lauenroth, D. M. Kaufman, A. Rassweiler, and J. A. Rusak. 2005. Evidence for a general species-time-area relationship. Ecology 86:2032–2039.

Adler, P. B., J. HilleRisLambers, and J. M. Levine. 2007. A niche for neutrality. Ecology Letters 10:95–104.

Ahl, V., and T. F. H. Allen. 1996. Hierarchy Theory: A Vision, Vocabulary, and Epistemology. New York, Columbia University Press.

Albertson, F. W., and J. E. Weaver. 1942. History of the native vegetation of western Kansas during seven years of continuous drought. Ecological Monographs 12:23–51.

Alig, R. J., J. D. Kline, and M. Lichtenstein. 2004. Urbanization on the U.S. landscape: looking ahead in the 21st century. Landscape and Urban Planning 69:219–234.

Allen, A. P., and J. F. Gillooly. 2009. Towards an integration of ecological stoichiometry and the metabolic theory of ecology to better understand nutrient cycling. Ecology Letters 12:369–384.

Allen, A. P., J. H. Brown, and J. F. Gillooly. 2002. Global biodiversity, biochemical kinetics, and the energetic-equivalence rule. Science 297:1545–1548.

Allen, A. P., J. F. Gillooly, V. M. Savage, and J. H. Brown. 2006. Kinetic effects of temperature on rates of genetic divergence and speciation. Proceedings of the National Academy of Sciences 103:9130–9135.

Allen, C. D. 2007. Interactions across spatial scales among forest dieback, fire, and erosion in northern New Mexico landscapes. Ecosystems 10:797–808.

Allen, T. F. H., and T. B. Starr. 1982. Hierarchy: Perspectives for Ecological Complexity. Chicago, University of Chicago Press.

Alonzo, S. H. 2002. State-dependent habitat selection games between predators and prey: the importance of behavioural interactions and expected lifetime reproductive success. Evolutionary Ecology Research 4:759–778.

Amarasekare, P. 2008. Spatial dynamics of foodwebs. Annual Review of Ecology, Evolution, and Systematics 39:479–500.

Amarasekare, P., and R. M. Nisbet. 2001. Spatial heterogeneity, source–sink dynamics, and the local coexistence of competing species. American Naturalist 158:572–584.

Anderson, C. 2008. The end of theory: the data deluge makes the scientific method obsolete, Wired Magazine, 16(7). [http://www.wired.com/science/discoveries/magazine/16-07/pb_theory]

Anderson, R. M., and R. M. May. 1991. Infectious Diseases of Humans. Oxford, Oxford University Press.

Andrewartha, H. G., and L. C. Birch. 1954. The Distribution and Abundance of Animals. Chicago, University of Chicago Press.

Arditi, R., and L. R. Ginzburg. 1989. Coupling in predator-prey dynamics: ratio-dependence. Journal of Theoretical Biology 139:311–326.

Arditi, R., J.-M. Callois, Y. Tyutyunov, and C. Jost. 2004. Does mutual interference always stabilize predator-prey dynamics? A comparison of models. Comptes Rendus: Biologies 327:1037–1057.

Arita, H., and E. Vazquez-Dominguez. 2008. The tropics: cradle, museum or casino? A dynamic null model for latitudinal gradients of species diversity. Ecology Letters 11:653–663.

Arita, H., J. Christen, P. Rodrìguez, and J. Soberón. 2008. Species diversity and distribution in presence-absence matrices: mathematical relationships and biological implications. American Naturalist 172:519–532.

Armesto, J. J., and S. T. A. Pickett. 1986. Removal experiments to test mechanisms of plant succession in oldfields. Vegetatio 66:85–93.

Armstrong, R., and R. McGehee. 1976. Coexistence of two competitors on one resource. Journal of Theoretical Biology 56:449–502.

———. 1980. Competitive exclusion. American Naturalist 115:151–170.

Arnold, W., T. Ruf, and R. Kuntz. 2006. Seasonal adjustment of energy budget in a large wild mammal, the Przewalski horse (Equus ferus przewalskii) II. Energy expenditure. Journal of Experimental Biology 209:4566–4573.

Attwill, P. M., and M. Adams. 1993. Tansley Review No. 50: Nutrient cycling in forests. New Phytologist 124:561–582.

Auclair, A. N. D., and F. G. Goff. 1971. Diversity relations of upland forests in the western Great Lakes area. American Naturalist 105:499–528.

Austin, M. P. 2005. Vegetation and environment: discontinuities and continuities, Pages 52–84 in E. van der Maarel, ed. Vegetation Ecology. Malden, MA, Blackwell Science.

Baker, H. G., and G. L. Stebbins. 1965. The Genetics of Colonizing Species. New York, Academic Press.

Barot, S., M. Blouin, S. Fontaine, P. Jouquet, J. C. Lata, and J. Mathieu. 2007. A tale of four stories: soil ecology, theory, evolution and the publication system. PLoS ONE 2:1248.

Bazzaz, F. A. 1979. The physiological ecology of plant succession. Annual Review of Ecology and Systematics 10:351–371.

———. 1983. Characteristics of populations in relation to disturbance in natural and man-modified ecosystems, Pages 259–275 in H. A. Mooney, and M. Godron, eds. Disturbance and Ecosystems: Components of Responses. New York, Springer-Verlag.

———. 1986. Life history of colonizing plants: some demographic, genetic, and physiological features, Pages 96–110 in M. A. Mooney, and J. A. Drake, eds. Ecology of Biological Invasions of North America and Hawaii. New York, Springer-Verlag.

Bazzaz, F. A., and T. W. Sipe. 1987. Physiological ecology, disturbance, and ecosystem recovery, Pages 203–227 in E. D. Schulze, ed. Potentials and Limitations of Ecosystems Analysis. New York, Springer-Verlag.

Beatty, J. 1997. Why do biologists argue like they do? Philosophy of Science 64:S432–S443.

Beddington, J. R. 1975. Mutual interference between parasites or predators and its effect on searching efficiency. Journal of Animal Ecology 44:331–340.

Bednekoff, P. A. 2007. Foraging in the face of danger, Pages 305–330 in D. W. Stephens, J. S. Brown, and R. C. Ydenberg, eds. Foraging. Behavior and Ecology. Chicago, University of Chicago Press.

Begon, M., J. L. Harper, and C. R. Townsend. 1996a. Ecology: Individuals, Populations, and Communities. Oxford, Blackwells.

Begon, M., M. Mortimer, and D. J. Thompson. 1996b. Population Ecology: A Unified Study of Animals and Plants. 3rd ed. Oxford, Blackwell Science.

Begon, M., C. R. Townsend, and J. L. Harper. 2006. Ecology: From Individuals to Ecosystems. Oxford, Blackwell Publishing.

Beisner, B. E., P. R. Peres, E. S. Lindstrom, A. Barnett, and M. L. Longhi. 2006. The role of environmental and spatial processes in structuring lake communities from bacteria to fish. Ecology 87:2985–2991.

Belgrano, A., and J. H. Brown. 2002. Ecology: oceans under the macroscope. Nature 419:128–129.

Bell, G. 2000. The distribution of abundance in neutral communities. American Naturalist 155:606–617.

———. 2001. Neutral macroecology. Science 293:2413–2418.

———. 2003. The interpretation of biological surveys. Proceedings of the Royal Society B: Biological Sciences 270:2531–2542.

Belovsky, G. E. 1978. Diet optimization in a generalist herbivore: the moose. Theoretical Population Biology 14:105–134.

Benkman, C. W., W. C. Holimon, and J. W. Smith. 2001. The influence of a competitor on the geographic mosaic of coevolution between crossbills and lodgepole pine. Evolution 55:282–294.

Berryman, A. A. 2003. On principles, laws and theory in population ecology. Oikos 103:695–701.

Bestelmeyer, B. T. 2006. Threshold concepts and their use in rangeland management and restoration: the good, the bad, and the insidious. Restoration Ecology 14:325–329.

Bever, J. D. 2003. Soil community feedback and the coexistence of competitors: conceptual frameworks and empirical tests. New Phytologist 157:465–473.

Biox, D., X. D. Quintana, and R. Moreno-Amich. 2004. Succession of the animal community in a mediterranean temporary pond. Journal of the North American Benthological Society 23:29–49.

Blackburn, T., and K. Gaston. 1996. Spatial patterns in the geographic range sizes of bird species in the New World. Philosophical Transactions of the Royal Society B: Biological Sciences 351:897–912.

Blackburn, T. M., and K. J. Gaston. 2003. Macroecology: Concepts and Consequences: the 43rd Annual Symposium of the British Ecological Society. Malden, Blackwell Publisher.

Bohannan, B. J. M., and R. E. Lenski. 1997. Effect of resource enrichment on a chemostat community of bacteria and bacteriophage. Ecology 78:2303–2315.

Bokma, F., J. Bokma, and M. Mönkkönen. 2001. Random processes and geographic species richness patterns: why so few species in the north. Ecography 24:43–49.

Bolker, B., M. Holyoak, V. Krivan, L. Rowe, and O. Schmitz. 2003. Connecting theoretical and empirical studies of trait-mediated interactions. Ecology 84:1101–1114.

Bonan, G. B. 2008. Forests and climate change: forcings, feedbacks, and the climate benefits of forests. Science 320:1444–1449.

Boose, E. R., K. E. Chamberlin, and D. R. Foster. 2001. Landscape and regional impacts of hurricanes in New England. Ecological Monographs 71:27–48.

Booth, B. D., and D. W. Larson. 1999. Impact of language, history, and choice of system on the study of assembly rules, Pages 206–229 in E. Weiher and P. A. Keddy, eds. Ecological Assembly Rules. Cambridge, Cambridge University Press.

Borer, E. T., C. J. Briggs, and R. D. Holt. 2007. Predators, parasitoids, and pathogens: a cross-cutting examination of intraguild predation theory. Ecology 88:2681–2688.

Bormann, F. H., and G. E. Likens. 1979. Patterns and Processes in a Forested Ecosystem. New York, Wiley and Sons.

Botkin, D. B., and M. J. Sobel. 1975. Stability in time-varying ecosystems. American Naturalist 109:625–646.

Boucher, D., L. Jardillier, and D. Debroas. 2005. Succession of bacterial community composition over two consequitive years in two aquatic systems: a natural lake and a lake-reservoir. FEMS Microbial Ecology 55:79–97.

Bouskila, A., and D. T. Blumstein. 1992. Rules of thumb for predation hazard assessment: predictions from a dynamic model. American Naturalist 139:161–176.

Bowers, M. A. 1997. Influence of deer and other factors on an old-field plant community, Pages 310–326 in W. J. McShea, H. B. Underwood, and J. H. Rappole, eds. The Science of Overabundance. Deer Ecology and Population Management. Washington, D.C., Smithsonian Institution Press.

Brand, T., and V. T. Parker. 1995. Scale and general laws of vegetation dynamics. Oikos 73:375–380.

Brandon, R. N. 1997. Does biology have laws? The experimental evidence. Philosophy of Science 64:S444–S457.

Brayard, A., G. Escarguel, and H. Bucher. 2005. Latitudinal gradient of taxonomic richness: combined outcome of temperature and geographic mid-domains effects? Journal of Zoological Systematics & Evolutionary Research 43:178–188.

Briggs, C. J., and M. F. Hoopes. 2004. Stabilizing effects in spatial parasitoid-host and predator-prey models: a review. Theoretical Population Biology 65:299–315.

Briggs, J. M., A. K. Knapp, J. M. Blair, J. L. Heisler, G. A. Hoch, M. S. Lett, and J. K. McCarron. 2005. An ecosystem in transition: causes and consequences of the conversion of mesic grassland to shrubland. BioScience 55:243–254.

Brodin, A., and C. W. Clark. 2007. Energy storage and expenditure, Pages 221–269 *in* D. W. Stephens, J. S. Brown, and R. C. Ydenberg, eds. Foraging. Behavior and Ecology. Chicago, University of Chicago Press.

Brook, B. W., N. S. Sodhi, and P. K. L. Ng. 2003. Catastrophic extinctions follow deforestation in Singapore. Nature 424:420–423.

Brooks, C. P., J. Antonovics, and T. H. Keitt. 2008. Spatial and temporal heterogeneity explain disease dynamics in a spatially explicit network model. American Naturalist 172:149–159.

Brovkin, V., S. Sitch, W. von Bloh, M. Claussen, E. Bauer, and W. Cramer. 2004. Role of land cover changes for atmospheric CO_2 increase and climate change during the last 150 years. Global Change Biology 10:1253–1266.

Brown, J. H. 1971. Mammals on mountaintops: nonequilibrium insular biogeography. American Naturalist 105:467–478.

———. 1995. Macroecology. Chicago, University of Chicago Press.

Brown, J. H., S. K. M. Ernest, J. M. Parody, and J. P. Haskell. 2001. Regulation of diversity: maintenance of species richness in changing environments. Oecologia 126:321–332.

Brown, J. H., J. E. Gillooly, A. P. Allen, V. M. Savage, and G. B. West. 2004. Toward a metabolic theory of ecology. Ecology 85:1771–1789.

Brown, J. H., and A. Kodric-Brown. 1977. Turnover rates in insular biogeography: effects of immigration on extinction. Ecology 58:445–449.

Brown, J. H., and B. A. Maurer. 1989. Macroecology: the division of food and space among species on continents. Science 243:1145–1150.

Brown, J. S. 1988. Patch use as an indicator of habitat preference, predation risk, and competition. Behavioral Ecology and Sociobiology 22:37–47.

Brown, J. S., and B. P. Kotler. 2007. Foraging and the ecology of fear, Pages 437–480 *in* D. W. Stephens, J. S. Brown, and R. C. Ydenberg, eds. Foraging. Behavior and Ecology. Chicago, University of Chicago Press.

Brown, W. L., and E. O. Wilson. 1956. Character displacement. Systematic Zoology 5:49–65.

Bruno, J. F., J. D. Fridley, K. D. Bromberg, and M. D. Bertness. 2005. Insights into biotic interactions from studies of species invasions, Pages 13–40 *in* D. F. Sax, S. D. Gaines, and J. J. Stachowicz, eds. Species Invasions: Insights into Ecology, Evolution, and Biogeography. Sunderland, Sinauer Associates.

Buckley, L. B., and W. Jetz. 2007. Environmental and historical constraints on global patterns of amphibian richness. Proceedings of the Royal Society B: Biological Sciences 274:1167–1173.

Burgman, M. A., D. B. Lindenmayer, and J. Elith. 2005. Managing landscapes for conservation under uncertainty. Ecology 86:2007–2017.

Burke, I. C., A. R. Mosier, P. B. Hook, D. G. Milchunas, J. E. Barrett, M. A. Vinton, R. L. McCulley et al. 2008. Soil organic matter and nutrient dynamics of shortgrass steppe ecosystems *in* W. K. Lauenroth and I. C. Burke, eds. Ecology of the Shortgrass Steppe: A Long Term Perspective. New York, Oxford University Press.

Burnham, K. P., and D. R. Anderson. 2002. Model Selection and Multi-model Inference. New York, Springer-Verlag.

Cadenasso, M. L., and S. T. A. Pickett. 2000. Linking forest edge structure to edge function: mediation of herbivore damage. Journal of Ecology 88:31–44.

Cadenasso, M. L., S. T. A. Pickett, K. C. Weathers, S. S. Bell, T. L. Benning, M. M. Carreiro, and T. E. Dawson. 2003. An interdisciplinary and synthetic approach to ecological boundaries. BioScience 53:717–722.

Cadotte, M. W. 2006a. Dispersal and species diversity: a meta-analysis. American Naturalist 167:913–924.

———. 2006b. Metacommunity influences on community richness at multiple spatial scales: a microcosm experiment. Ecology 87:1008–1016.

Camerano, L. 1880. Dell'equilibrio dei viventi merce la reciproca distruzione. Atti della Reale Accademia delle Scienze di Torino 15:393–414.

Canadell, J. G., D. E. Paraki, and L. F. Pitelka. eds. 2007. Terrestrial Ecosystems in a Changing World. New York, Springer.

Canham, C. D., J. J. Cole, and W. K. Lauenroth. eds. 2003. Models in Ecosystem Science. Princeton, Princeton University Press.

Carbone, C., and J. L. Gittleman. 2002. A common rule for the scaling of carnivore density. Science 295:2273–2276.

Caro, T. 2005. Antipredator Defenses in Birds and Mammals. Chicago, University of Chicago Press.

Carpenter, S. R. 1998. The need for large-scale experiments to assess and predict the response of ecosystems to perturbation, Pages 287–312 in M. L. Pace and P. M. Groffman, eds. Successes, Limitations, and Frontiers in Ecosystem Science. New York, Springer.

———. 2008. Phosphorus control is critical to mitigating eutrophication. Proceedings of the National Academy of Sciences 105:11039–11040.

Carpenter, S. R., R. C. Lathrop, and A. Munoz-del-Rio. 1993. Comparison of dynamic models for edible phytoplankton. Canadian Journal of Fisheries and Aquatic Science 50:1757–1767.

Carpenter, S. R., and M. G. Turner. 2000. Hares and tortoises: interactions of fast and slow variables in ecosystems. Ecosystems 3:495–497.

Carpenter, S. R., B. Walker, J. M. Anderies, and N. Abel. 2001. From metaphor to measurement: resilience of what to what? Ecosystems 4:765–781.

Case, T. J. 2000. An Illustrated Guide to Theoretical Ecology. Oxford, Oxford University Press.

Case, T. J., and M. L. Taper. 2000. Interspecific competition, environmental gradients, gene flow, and the coevolution of species' borders. American Naturalist 155:583–605.

Caswell, H. 1976. Community structure: neutral model analysis. Ecological Monographs 46:327–354.

———. 2001. Matrix Population Models. Sunderland, MA, Sinauer.

Caughley, G., and J. H. Lawton. 1981. Plant-herbivore systems, Pages 132–166 in R. M. May, ed. Theoretical Ecology. 2nd ed. Sunderland, MA, Sinauer Associates.

Chalcraft, D. R., S. B. Cox, C. Clark, E. E. Cleland, K. N. Suding, E. Weiher, and D. Pennington. 2008. Scale-dependent responses of plant biodiversity to nitrogen enrichment. Ecology 89:2165–2171.

Chalcraft, D. R., B. J. Wilsey, C. Bowles, and M. R. Willig. 2009. The relationship between productivity and multiple aspects of biodiversity in grassland communities. Biodiversity and Conservation 18:91–104.

Chapin, F. S., III, P. A. Matson, and H. A. Mooney. 2002. Principles of Terrestrial Ecosystem Ecology. New York Springer-Verlag.

Chapin, F. S., III, G. Woodwell, J. Randerson, E. Rastetter, G. Lovett, D. Baldocchi, D. Clark et al. 2006. Reconciling carbon-cycle concepts, terminology, and methods. Ecosystems 9:1041–1050.

Charnov, E. L. 1976a. Optimal foraging: attack strategy of a mantid. American Naturalist 110:141–151.

———. 1976b. Optimal foraging: the marginal value theorem. Theoretical Population Biology 9:129–136.

Charnov, E. L., G. H. Orians, and K. Hyatt. 1976. Ecological implications of resource depression. American Naturalist 110:247–259.

Chase, J. M. 1999. Food web effects of prey size-refugia: variable interactions and alternative stable equilibria. American Naturalist 154:559–570.

———. 2003. Community assembly: when does history matter? Oecologia 136:489–498.

———. 2005. Towards a really unified theory for metacommunities. Functional Ecology 19:182–186.

———. 2007. Drought mediates the importance of stochastic assembly. Proceedings of the National Academy of Sciences 104:17430–17434.

Chase, J. M., P. A. Abrams, J. P. Grover, S. Diehl, P. Chesson, R. D. Holt, S. A. Richards et al. 2002. The interaction between predation and competition: a review and synthesis. Ecology Letters 5:302–315.

Chase, J. M., E. G. Biro, W. A. Ryberg, and K. G. Smith. 2009. Predators temper the relative importance of stochastic processes in the assembly of prey communities. Ecology Letters 12:1210–1218.

Chase, J. M., and M. A. Leibold. 2002. Spatial scale dictates the productivity-biodiversity relationship. Nature 416:427–430.

———. 2003. Ecological Niches: Linking Classical and Contemporary Approaches. Chicago, University of Chicago Press.

Chase, J. M., M. A. Leibold, A. L. Downing, and J. B. Shurin. 2000. The effects of productivity, herbivory, and plant compositional turnover in grassland food webs. Ecology 81:2485–2497.

Chave, J. 2004. Neutral theory and community ecology. Ecology Letters 7:241–253.

Chave, J., and E. G. Leigh. 2002. A spatially explicit neutral model of β-diversity in tropical forests. Theoretical Population Biology 62:153–168.

Chesson, P. 1979. Predator-prey theory and variability. Annual Review of Ecology and Systematics 9:323–347.

———. 2000a. Mechanisms of maintenance of species diversity. Annual Review of Ecology and Systematics 31:343–366.

Chesson, P., M. J. Donahue, B. A. Melbourne, and A. L. W. Sears. 2005. Scale transition theory for understanding mechanisms in metacommunities, Pages 279–306 in M. Holyoak, M. A. Leibold, and R. D. Holt, eds. Metacommunities: Spatial Dynamics and Ecological Communities. Chicago, University of Chicago Press.

Chesson, P. L. 1991. Stochastic population models, Pages 123–143 in J. Kolasa and S. T. A. Pickett, eds. Ecological Heterogeneity. New York, Springer-Verlag.

———. 2000b. Mechanisms of maintenance of species diversity. Annual Review of Ecology and Systematics 31:343–366.

Chesson, P. L., and N. Huntly. 1988. Community consequences of life history traits in a variable environment. Annales Zoologici Fennici 25:5–16.

Chesson, P. L., and R. R. Warner. 1981. Environmental variability promotes coexistence in lottery competitive systems. American Naturalist 117:923–942.

Chiarucci, A., G. Bacaro, D. Rocchini, C. Ricotta, M. W. Palmer, and S. M. Scheiner. 2009. Spatially Constrained Rarefaction: incorporating the autocorrelated structure of biological communities in sample-based rarefaction. Community Ecology 10:209–214.

Clark, C. W. 1990. Mathematical Bioeconomics: The Optimal Management of Renewable Resources. Pure and Applied Mathematics Series, v. 386. New York, Wiley.

———. 1994. Antipredator behavior and the asset protection principle. Behavioral Ecology 5:159–170.

Clark, J. S. 2007. Models for Ecological Data: An Introduction. Princeton, NJ, Princeton University Press.

———. 2010. Individuals and the variation needed for high species diversity in forest trees. Science 327:1129–1132.

Clark, J. S., and A. E. Gelfand. 2006. A future for models and data in environmental science. Trends in Ecology and Evolution 21:375–380.

Clark, J. S., and J. S. McLachlan. 2003. Stability of forest biodiversity. Nature 423:635–638.

Clements, F. E. 1916. Plant Succession. Washington, D.C., Carnegie Institute of Washington.

———. 1937. Nature and structure of the climax. Journal of Ecology 24:252–284.

Clements, F. E., and V. E. Shelford. 1939. Bio-ecology. New York, Wiley.

Cockell, C., and A. R. Blaustein. 2001. Ecosystems, Evolution and Ultraviolet Radiation. New York, Springer.

Cody, M. L. 2006. Plants on Islands: Diversity and Dynamics on a Continental Archipelago. Berkeley, University of California Press.

Cohen, J. E. 1978. Food Webs and Niche Space. Princeton, Princeton University Press.

Coleman, B. D. 1981. On random placement and species-area relations. Mathematical Biosciences 54:191–215.

Coleman, B. D., M. A. Mares, M. R. Willig, and Y.-H. Hsieh. 1982. Randomness, area, and species richness. Ecology 63:1121–1133.

Collins, J. P. 1986. Evolutionary ecology and the use of natural selection in ecological theory. Journal of the History of Biology 19:257–288.

Colwell, R. K., G. Brehm, C. Cardelús, A. C. Gilman, and J. T. Longino. 2008. Global warming, elevational range shifts, and lowland biotic attrition in the wet tropics. Science 322:258–261.

Colwell, R. K., and G. C. Hurtt. 1994. Nonbiological gradients in species richness and a spurious Rapoport effect. American Naturalist 144:570–595.

Colwell, R. K., and D. C. Lees. 2000. The mid-domain effect: geometric constraints on the geography of species richness. Trends in Ecology and Evolution 15:70–76.

Colwell, R. K., C. Rahbek, and N. Gotelli. 2004. The mid-domain effect and species richness patterns: what have we learned so far? American Naturalist 163:E1–E23.

———. 2005. The mid-domain effect: there's a baby in the bathwater. American Naturalist 166:E149–E154.

Colwell, R. K., and T. F. Rangel. 2009. Hutchinson's duality: the once and future niche. Proceedings of the National Academy of Sciences 106:19651–19658.

Colwell, R. K., and D. W. Winkler. 1984. A null model for null models in biogeography, Pages 344–359 in D. R. Strong, Jr., D. Simberloff, L. G. Abele, and A. B. Thistle, eds. Ecological Communities: Conceptual Issues and the Evidence. Princeton, NJ, Princeton University Press.

Conley, D. J., J. Carstensen, R. Vaquer-Sunyer, and C. M. Duarte. 2009. Ecosystem thresholds with hypoxia. Hydrobiologia 629:21–29.

Connell, J. H. 1972. Community interactions on marine rocky intertidal shores. Annual Review of Ecology and Systematics 3:169–192.

Connell, J. H., I. R. Noble, and R. O. Slatyer. 1987. On the mechanisms producing successional change. Oikos 50:136–137.

Connell, J. H., and E. Orias. 1964. The ecological regulation of species diversity. American Naturalist 98:399–414.

Connell, J. H., and R. O. Slatyer. 1977. Mechanisms of succession in natural communities and their role in community stability and organization. American Naturalist 111:1119–1144.

Connolly, S. R. 2005. Process-based models of species distributions and the mid-domain effect. American Naturalist 166:1–11.

———. 2009. Macroecological theory and the analysis of species richness gradients, Pages 279–309 *in* J. D. Witman and K. Roy, eds. Marine Macroecology. Chicago, University of Chicago Press.

Connolly, S. R., D. R. Bellwood, and T. P. Hughes. 2003. Indo-Pacific biodiversity of coral reefs: deviations from a mid-domain model. Ecology 84:2178–2190.

Connor, E. F., and D. Simberloff. 1979. Assembly of species communities: chance or competition? Ecology 60:1132–1140.

Cook, K. H., and E. K. Vizy. 2006. Coupled model simulations of the West African monsoon system: 20th and 21st century simulations. Journal of Climate 19:3681–3703.

Cooper, G. 1993. The competition controversy in community ecology. Biology and Philosophy 8:359–384.

———. 2003. The Science of the Struggle for Existence: On the Foundations of Ecology. Cambridge, Cambridge University Press.

Cooper, W. S. 1926. The fundamentals of vegetation change. Ecology 7:391–413.

Cornell, H. V., and J. H. Lawton. 1992. Species interactions, local and regional processes, and limits to the richness of ecological communities: a theoretical perspective. Journal of Animal Ecology 61:1–12.

Cottenie, K. 2005. Integrating environmental and spatial processes in ecological community dynamics. Ecology Letters 8:1175–1182.

Cottenie, K., and L. De Meester. 2005. Local interactions and local dispersal in a zooplankton metacommunity. Metacommunities: Spatial Dynamics and Ecological Communities:189–211.

Covich, A. P., T. A. Crowl, and T. Heartsill Scalley. 2006. Effects of drought and hurricane disturbances on headwater distributions of palaemonid river shrimp (*Macrobrachium* spp.) in the Luquillo Mountains, Puerto Rico. Journal North American Benthological Society 25:99–107.

Cowen, R. K., C. B. Paris, and A. Srinivasan. 2006. Scaling of connectivity in marine populations. Science 311:522–527.

Cowles, H. C. 1899. The ecological relations of the vegetation on the sand dunes of Lake Michigan. Botanical Gazette 27:95–117, 167–202, 281–308, 361–391.

Cox, J. G., and S. L. Lima. 2006. Naivete and an aquatic-terrestrial dichotomy in the effects of introduced predators. Trends in Ecology and Evolution 21:674–680.

Craine, J. M. 2007. Plant strategy theories: replies to Grime and Tilman. Journal of Ecology 95:235–240.

Crawley, M. J. 1983. Herbivory: The Dynamics of Animal-plant Interactions. Oxford, Blackwell.

———. ed. 1992. Natural Enemies: The Population Biology of Predators, Parasites and Diseases. Oxford, Blackwell.

Creel, S., and D. Christianson. 2008. Relationships between direct predation and risk effects. Trends in Ecology and Evolution 23:194–201.

Crowl, T. A., T. O. Crist, R. R. Parmenter, G. Belovsky, and A. E. Lugo. 2008. The spread of invasive species and infectious disease as drivers of ecosystem change. Frontiers in Ecology and the Environment 6:238–246.

Crowley, P. H., S. E. Travers, M. C. Linton, S. L. Cohen, A. Sih, and R. C. Sargent. 1991. Mate density, predation risk and the seasonal sequence of mate choices: a dynamic game. American Naturalist 137:567–596.

Currie, D., and V. Paquin. 1987. Large-scale biogeographical patterns of species richness of trees. Nature 329:326–327.

Currie, D. J. 1991. Energy and large-scale patterns of animal- and plant-species richness. American Naturalist 137:27–49.

Currie, D. J., and J. T. Kerr. 2007. Testing, as opposed to supporting, the Mid-domain Hypothesis: a response to Lees and Colwell (2007). Ecology Letters 10:E9–E10.

———. 2008. Tests of the mid-domain hypothesis: a review of the evidence. Ecological Monographs 78:3–18.

Currie, D. J., G. G. Mittelbach, H. V. Cornell, R. Field, J.-F. Guégan, B. A. Hawkins, D. M. Kaufman et al. 2004. Predictions and tests of climate-based hypotheses of broad-scale variation in taxonomic richness. Ecology Letters 7:1121–1134.

D'Antonio, C. M., and P. M. Vitousek. 1992. Biological invasions by exotic grasses, the grass/fire cycle, and global change. Annual Review of Ecology and Systematics 23:63–87.

Dale, V. H., L. A. Joyce, S. McNulty, R. P. Neilson, M. P. Ayres, M. D. Flannigan, P. J. Hanson et al. 2001. Climate change and forest disturbances. BioScience 51:723–734.

Dall, S. R. X., L. A. Giraldeau, O. Olsson, J. M. McNamara, and D. W. Stephens. 2005. Information and its use in evolutionary ecology. Trends in Ecology and Evolution 20:187–193.

Dammermann, K. W. 1948. The fauna of Krakatau, 1883–1933. Koninklijke Nederlandsche Akademie Wetenschappen Verhandelingen 44:1–594.

Darlington, P. J., Jr. 1957. Zoogeography: The Geographical Distribution of Animals. New York, John Wiley and Sons.

Darwin, C. 1859. On the Origin of Species by Means of Natural Selection. London, Murray.

Davidson, D. W. 1993. The effects of herbivory and granivory on terrestrial plant succession. Oikos 68:23–35.

Davies, R. G., C. D. L. Orme, D. Storch, V. A. Olson, G. H. Thomas, S. G. Ross, T. S. Ding et al. 2007. Topography, energy and the global distribution of bird species richness. Proceedings of the Royal Society B: Biological Sciences 274:1189–1197.

Davis, M. A., J. Pergl, A.-M. Truscott, J. Kollmann, J. P. Bakker, R. Domenech, K. Prach et al. 2005. Vegetation change: a reunifying concept in plant ecology. Perspectives in Plant Ecology, Evolution and Systematics 7:69–76.

Day, J. W., C. A. Hall, A. Yanez-Arancibia, D. Pimentel, C. I. Marti, and W. J. Mitsch. 2009. Ecology in times of scarcity. BioScience 59:321–331.

Dayton, P. K. 1975. Experimental evaluation of ecological dominance in a rocky intertidal algal community. Ecological Monographs 45:137–159.

de Aguiar, M. A. M., M. Baranger, E. M. Baptestini, L. Kaufman, and Y. Bar-Yam. 2009. Global patterns of speciation and diversity. Nature 460:384–387.

DeAngelis, D. L., R. A. Goldstein, and R. V. O'Neill. 1975. A model for trophic interaction. Ecology 56:881–892.

Defries, R. S., M. C. Hansen, J. R. G. Townshead, A. C. Janetos, and T. R. Loveland. 2000. A new global 1-km dataset of percentage tree cover derived from remote sensing. Global Change Biology 6:247–254.

DeGraaf, R. M., and R. I. Miller. 1997. The importance of disturbance and land-use change in New England: implications for forested landscapes and wildlife conservation, Pages 3–35 in R. M. DeGraaf and R. I. Miller, eds. Conservation of Faunal Diversity in Forested Landscapes. New York, Chapman and Hall.

DeGroot, M. H. 1970. Optimal Statistical Decisions. New York, McGraw-Hill.

del Moral, R. 1993. Mechanisms of primary succession on volcanoes: a view from Mount St. Helens, Pages 79–100 *in* J. Miles and D. W. H. Walton, eds. Primary Succession on Land. Oxford, Blackwell Scientific Publications.

del Rio, C. M. 2008. Metabolic theory or metabolic models? Trends in Ecology and Evolution 23:256–260.

De Meester, L., G. Louette, C. Duvivier, C. Van Darnme, and E. Michels. 2007. Genetic composition of resident populations influences establishment success of immigrant species. Oecologia 153:431–440.

Deng, B., S. Jessie, G. Ledder, A. Rand, and S. Srodulski. 2007. Biological control does not imply paradox. Mathematical Biosciences 208:26–32.

Denslow, J. S. 1980. Gap partitioning among tropical rainforest trees. Biotropica (Supplement) 12:47–55.

de Roos, A. M., L. Persson, and H. R. Thieme. 2003. Emergent Allee effects in top predators feeding on structured prey populations. Proceedings of the Royal Society B: Biological Sciences 270:611–618.

de Roos, A. M., O. Diekmann, and J. A. J. Metz. 1992. Studying the dynamics of structured population models: a versatile technique and its application to daphnia. American Naturalist 139:123–147.

Dewdney, A. K. 2003. The stochastic community and the logistic-J distribution. Acta Oecologia 24:221–229.

DeWitt, T. J., and S. M. Scheiner. eds. 2003. Phenotypic Plasticity: Functional and Conceptual Approaches. New York, Oxford University Press.

Diamond, J. M. 1969. Avifaunal equilibria and species turnover rates on the Channel Islands of California. Proceedings of the National Academy of Sciences 64:57–63.

———. 1975a. Assembly of species communities, Pages 342–444 *in* M. L. Cody and J. M. Diamond, eds. Ecology and Evolution of Communities. Cambridge, Harvard University Press.

———. 1975b. The island dilemma: lessons of modern biogeographic studies for the design of natural preserves. Biological Conservation 7:129–146.

Diaz, R. J., and R. Rosenberg. 2008. Spreading dead zones and consequences for marine ecosystems. Science 321:926–929.

Dietz, T., E. A. Rosa, and R. York. 2007. Driving the human ecological footprint. Frontiers in Ecology and the Environment 5:13–18.

Diniz-Filho, J. A. F., L. M. Bini, and B. A. Hawkins. 2003. Spatial autocorrelation and red herrings in geographical ecology. Global Ecology and Biogeography 12:53–64.

Dobson, A., and M. Crawley. 1994. Pathogens and the structure of plant communities. Trends in Ecology and Evolution 9:393–398.

Dobzhansky, T. 1950. Evolution in the tropics. American Scientist 38:209–221.

Donnelly, J. P., and J. D. Woodruff. 2007. Intense hurricane activity over the past 5,000 years controlled by El Niño and the West African monsoon. Nature 447:465–468.

Drake, J. A. 1991. Community-assembly mechanics and the structure of an experimental species ensemble. American Naturalist 137:1–26.

Drake, J. A., G. R. Huxel, and C. L. Hewitt. 1996. Microcosms as models for generating and testing community theory. Ecology 77:670–677.

Drossel, B., P. G. Higgs, and A. J. McKane. 2001. The influence of predator-prey population dynamics on the long-term evolution of food web structure. Journal of Theoretical Biology 208:91–107.

Dunn, G. E. 1940. Cyclogenesis in the tropical Atlantic. Bulletin of the American Meteorological Society 21:215–229.

Dunn, R. R., C. M. McCain, and N. Sanders. 2007. When does diversity fit null model predictions? Scale and range size mediate the mid-domain effect. Global Ecology and Biogeography 3:305–312.

Dwyer, G., J. Dushoff, and S. H. Yee. 2004. The combined effects of pathogens and predators on insect outbreaks. Nature 430:341–345.

Egler, F. E. 1954. Vegetation science concepts. I. Initial floristic composition: a factor in old-field vegetation development. Vegetatio 4:412–417.

Ehrlen, J., and O. Eriksson. 2000. Dispersal limitation and patch occupancy in forest herbs. Ecology 81:1667–1674.

Ejrnaes, R., D. N. Hansen, and E. Aude. 2003. Changing course of secondary succession in abandoned sandy fields. Biological Conservation 109:343–350.

Eldredge, L. G. 1985. Unfinished Synthesis: Biological Hierarchies and Modern Evolutionary Thought. Oxford, Oxford University Press.

Eldredge, L. G., and S. E. Miller. 1995. How many species are there in Hawaii? Bishop Museum Occasional Papers 41:3–17.

Eliot, C. 2007. Method and metaphysics in Clements's and Gleason's ecological explanations. Studies in History and Philosophy of Biological and Biomedical Sciences 38:85–109.

Elith, J., C. H. Graham, R. P. Anderson, M. Dudik, S. Ferrier, A. Guisan, R. J. Hijmans et al. 2006. Novel methods improve prediction of species' distributions from occurrence data. Ecography 29:129–151.

Elser, J. J., R. W. Sterner, E. Gorokhova, W. F. Fagan, T. A. Markow, J. B. Cotnew, J. F. Harrison et al. 2000. Biological stoichiometry from genes to ecosystems. Ecology Letters 3:540–550.

Elton, C. S. 1927. Animal Ecology. London, Sidgwick and Jackson.

Emanuel, K. A. 2005. Increasing destructiveness of tropical cyclones over the past 30 years. Nature 436:686–688.

Emlen, J. M. 1966. The role of time and energy in food preference. American Naturalist 100:611–617.

Engen, S. 2001. A dynamic and spatial model with migration generating the log-Gaussian field of population densities. Mathematical Biosciences 173.

Engen, S., and R. Lande. 1996. Population dynamic models generating species abundance distributions of the gamma type. Journal of Theoretical Biology 178:325–331.

Englund, G., and K. Leonardsson. 2008. Scaling up the functional response for spatially heterogeneous systems. Ecology Letters 11:440–449.

Enquist, B. J., J. H. Brown, and G. B. West. 1998. Allometric scaling of plant energetics and population density. Nature 395:163–165.

Ernest, S. K. M., B. J. Enquist, J. H. Brown, E. L. Charnov, J. F. Gillooly, V. M. Savage, E. P. White, et al. 2003. Thermodynamic and metabolic effects on the scaling of production and population energy use. Ecology Letters 6:990–995.

Essington, T. E., and S. Hanson. 2004. Predator-dependent functional responses and interaction strengths in a natural food web. Canadian Journal of Fisheries and Aquatic Science 61:2215–2226.

Etienne, R. S., and H. Olff. 2004. A novel genealogical approach to neutral biodiversity theory. Ecology Letters 7:170–175.

Everham, E. M., III, and N. V. L. Brokaw. 1996. Forest damage and recovery from catastrophic wind. Botanical Review 62:113–185.

Feibleman, J. K. 1954. Theory of integrative levels. British Journal for the Philosophy of Science 5:59–66.

Fenn, M. E., R. Haeuber, G. S. Tonnesen, J. S. Baron, S. Grossman-Clarke, D. Hope, D. A. Jaffe et al. 2003. Nitrogen emissions, deposition, and monitoring in the western United States. BioScience 53:391–403.

Fernando, H. J. S., and J. L. McCulley. 2005. Coral poaching worsens tsunami destruction in Sri Lanka. Eos 86:301–304.

Ferrari, M. C. O., A. Sih, and D. P. Chivers. 2009. The paradox of risk allocation: a review and prospectus. Animal Behaviour 78:579–585.

Field, C. B., D. B. Lobell, H. A. Peters, and N. R. Chiariello. 2007. Feedbacks of terrestrial ecosystems to climate change. Annual Review of Environment and Resources 32:1–29.

Field, R., E. M. O'Brien, and R. J. Whittaker. 2005. Global models for predicting woody plant richness from climate: development and evaluation. Ecology 86:2263–2277.

Fisher, R. A., A. S. Corbet, and C. B. Williams. 1943. The relation between the number of species and the number of individuals in a random sample of an animal population. Journal of Animal Ecology 12:42–58.

Fitz, H. C., E. B. DeBellevue, R. Constanza, R. Boumans, T. Maxwell, L. Wainger, and F. H. Sklar. 1996. Development of a general ecosystem model for a range of scales and ecosystems. Ecological Modelling 88:263–295.

Fjeldså, J. 1994. Geographical patterns for relict and young species of birds in Africa and South America and implications for conservation priorities. Biodiversity and Conservation 3:207–226.

Folke, C., S. R. Carpenter, B. Walker, M. Scheffer, T. Elmquist, L. Gunderson, and C. S. Holling. 2004. Regime shifts, resilience, and biodiversity in ecosystem management. Annual Review of Ecology and Systematics 35:557–581.

Forbes, S. 1887. The lake as a microcosm. Illinois Natural History Survey Bulletin 15:537–550.

Forster, J. R. 1778. Observations Made During a Voyage Round the World, on Physical Geography, Natural History, and Ethnic Philosophy. London, G. Robinson.

Foster, D. R., and E. R. Boose. 1992. Patterns of forest damage resulting from catastrophic wind in central New England, USA. Journal of Ecology 80:79–98.

Fox, D. 2007. Back to the no-analog future? Science 316:823–825.

Freckleton, R. P., and A. R. Watkinson. 2002. Large-scale spatial dynamics of plants: metapopulations, regional ensembles and patchy populations. Journal of Ecology 90:419–434.

Free, C. A., J. R. Beddington, and J. H. Lawton. 1977. On the inadequacy of simple models of mutual interference for parasitism and predation. Journal of Animal Ecology 36:375–389.

Fretwell, S. D. 1975. The impact of Robert MacArthur on ecology. Annual Review of Ecology and Systematics 6:1–13.

Fretwell, S. J., and H. J. Lucas, Jr. 1970. On territorial behaviour and other factors influencing habitat distribution in birds. Acta Biotheoretica 19:16–36.

Fridley, J. D., R. K. Peet, E. van der Maarel, and J. H. Willems. 2006. Integration of local and regional species-area relationships from space-time species accumulation. American Naturalist 168:133–143.

Friedman, M. 1974. Explanation and scientific understanding. Journal of Philosophy 71:5–19.

Fryxell, J. M., A. Mosser, A. R. E. Sinclair, and C. Packer. 2007. Group formation stabilizes predator-prey dynamics. Nature 449:1041–1043.

Fukami, T. 2004. Assembly history interacts with ecosystem size to influence species diversity. Ecology 85:3234–3242.

Funk, V. A., and W. L. Wagner. 1995. Hawaiian Biogeography: Evolution on a Hot Spot Archipelago. Washington, D.C., Smithsonian Institution Press.

Fussmann, G. R., G. Weithoff, and T. Yoshida. 2007. A direct, experimental test of resource vs. consumer dependence: reply. Ecology 88:1603–1604.

Gaston, K., and S. Chown. 1999. Why Rapoport's rule does not generalise. Oikos 84:309.

Gaston, K. J. 2000. Global patterns in biodiversity. Nature 405:220–227.

Gaston, K. J., and T. M. Blackburn. 2000. Pattern and Process in Macroecology. Oxford, Blackwell Science.

Gaston, K. J., T. M. Blackburn, and J. I. Spicer. 1998. Rapoport's rule: time for an epitaph? Trends in Ecology and Evolution 13:70–74.

Gause, G. F. 1934. The Struggle for Existence. Baltimore, Williams & Wilkins Co.

Gavrilets, S. 2003. Perspective: models of speciation: what have we learned in 40 years? Evolution 57:2197–2215.

Gelb, I. J. 1967. Growth of a herd of cattle in ten years. Journal of Cuneiform Studies 21:64–69.

Getty, T., A. C. Kamil, and P. G. Real. 1987. Signal detection theory and foraging for cryptic and mimetic prey, Pages 525–548 in A. C. Kamil, J. R. Krebs, and H. R. Pulliam, eds. Foraging Behavior. New York, Plenum Press.

Getz, W. M. 1984. Population dynamics: a resource per-capita approach. Journal of Theoretical Biology 108:623–644.

———. 1993. Metaphysiological and evolutionary dynamics of populations exploiting constant and interactive resources: r-K selection revisited. Evolutionary Ecology 7:287–305.

———. 1999. Population and evolutionary dynamics of consumer-resource systems, Pages 194–231 in J. McGlade, ed. Advanced Ecological Theory: Principles and Applications. Oxford, Blackwell.

Getz, W. M., and J. Pickering. 1983. Epidemic models: thresholds and population regulation. American Naturalist 121:892–898.

Giere, R. N. 1988. Explaining Science: A Cognitive Approach, Chicago, University of Chicago Press.

Gillespie, R. 2004. Community assembly through adaptive radiation in Hawaiian spiders. Science 303:356–359.

Ginzburg, L. R. 1998. Assuming reproduction to be a function of consumption raises doubts about some population predator-prey models. Journal of Animal Ecology 67:325–327.

Ginzburg, L. R., and M. Colyvan. 2004. Ecological Orbits: How Planets Move and Populations Grow. Oxford, Oxford University Press.

Giraldeau, L. A., and T. Caraco. 2000. Social Foraging Theory. Princeton, NJ, Princeton University Press.

Glantz, M. H., R. W. Katz, and N. Nicholls. 1991. Teleconnections Linking Worldwide Climate Anomalies: Scientific Basis and Societal Impact. Cambridge, Cambridge University Press.

Gleason, H. A. 1917. The structure and development of the plant association. Bulletin of the Torrey Botanical Club 44:463–481.

———. 1926. The individualistic concept of the plant association. Bulletin of the Torrey Botanical Club 53:7–26.

Gleeson, S. K., and D. Tilman. 1992. Plant allocation and the multiple limitation hypothesis. American Naturalist 139:1322–1343.

Gleeson, S. K., and D. S. Wilson. 1986. Optimal foraging and prey coexistence. Oikos 46:139–144.

Glenn-Lewin, D. C. 1980. The individualistic nature of plant community development. Vegetatio 43:141–146.

Glenn-Lewin, D. C., R. K. Peet, and T. T. Veblen. eds. 1992. Plant Succession: Theory and Prediction. New York, Chapman and Hall.

Goldenberg, S. B., C. W. Landsea, A. M. Mestas-Nuñez, and W. M. Gray. 2001. The recent increase in Atlantic hurricane activity: causes and implications. Science 293:474–479.

Golley, F. B. 1993. A History of the Ecosystem Concept in Ecology. New Haven, Yale University Press.

Gorham, E. 1961. Factors influencing supply of major ions to inland waters with special reference to the atmosphere. Geology Society of America Bulletin 72:795–840.

Gorham, E., P. M. Vitousek, and W. A. Reiners. 1979. The regulation of chemical budgets over the course of terrestrial ecosystem succession. Annual Review of Ecology and Systematics 10:53–84.

Gotelli, N. J. 2004. Assembly rules, Pages 1027–1121 in M. V. Lomolino, D. F. Sax, and J. H. Brown, eds. Foundations of Biogeography. Chicago, University of Chicago Press.

Gotelli, N. J., and G. R. Graves. 1996. Null Models in Ecology. Washington, D. C., Smithsonian Institution Press.

Gotelli, N. J., M. J. Anderson, H. T. Arita, A. Chao, R. K. Colwell, S. R. Connolly, D. J. Currie et al. 2009. Patterns and causes of species richness: a general simulation model for macroecology. Ecology Letters 12:873–886.

Gouhier, T., F. Guichard, and A. Gonzalez. 2010. Synchrony and stability of food webs in metacommunities. American Naturalist 175:E16–E34.

Gould, S. J. 1979. An allometric interpretation of species-area curves: the meaning of the coefficient. American Naturalist 114:335–343.

Grace, J. B. 1999. The factors controlling species density in herbaceous plant communities: an assessment. Perspectives in Plant Ecology, Evolution and Systematics 2:1–28.

———. 2006. Structural Equation Modeling and Natural Systems. Cambridge, Cambridge University Press.

Grace, J. B., T. M. Anderson, H. Olff, and S. M. Scheiner. 2010. Improving the connection between biological data and theory through the use of structural equation meta-models. Ecological Monographs 80:67–87.

Graham, N. E. 1995. Simulation of recent global temperature trends. Science 267:686–691.

Grand, T. C., and L. M. Dill. 1999. Predation risk, unequal competitors and the ideal free distribution. Evolutionary Ecology Research 1.

Grant, B. 2007. The powers that might be. Scientist 21:42–42.

Grassly, N. C., and C. Fraser. 2008. Mathematical models of infectious disease transmission. Nature Reviews Microbiology 6:234–244.

Gravel, D., C. D. Canham, M. Beaudet, and C. Messier. 2006. Reconciling niche and neutrality: the continuum hypothesis. Ecology Letters 9:399–409.

Graves, G., and C. Rahbek. 2005. Source pool geometry and the assembly of continental avifaunas. Proceedings of the National Academy of Sciences 102:7871–7876.

Graves, G. R. 1985. Elevational correlates of speciation and intraspecific geographic variation in plumage in Andean forest birds. Auk 102:556–579.

Gray, W. M. 1968. Global view of the origin of tropical disturbances and storms. Monthly Weather Review 96:669–700.

———. 1990. Strong association between West African rainfall and U.S. landfall of intense hurricanes. Science 249:1251–1256.

Greenberg, G. 1988. Levels of social behavior in G. Greenberg and E. Tobach, eds. Evolution of Social Behavior and Integrative Levels. Hillsdale, NJ, Erlbaum.

Grime, J. P. 1973. Competitive exclusion in herbaceous vegetation. Nature 242:344–347.

———. 1979. Plant Strategies and Vegetation Processes. New York, John Wiley and Sons.

———. 2007. Plant strategy theories: a comment on Craine (2005). Journal of Ecology 95:227–230.

Grime, J. P., and J. G. Hodgson. 1987. Botanical contributions to contemporary ecological theory. New Phytologist 106:283–295.

Grimm, N. B., S. H. Faeth, N. E. Golubiewski, C. L. Redman, J. Wu, X. Bai, and J. M. Briggs. 2008a. Global change and the ecology of cities. Science 319:756–760.

Grimm, N. B., D. Foster, P. Groffman, J. M. Grove, C. S. Hopkinson, K. Nadelhoffer, D. E. Paraki et al. 2008b. The changing landscape: ecosystem responses to urbanization and pollution across climatic and societal gradients. Frontiers in Ecology and the Environment 5:264–272.

Grimm, N. B., J. M. Grove, S. T. A. Pickett, and C. L. Redman. 2000. Integrated approaches to long-term studies of urban ecological systems. BioScience 50:571–584.

Grimm, V., and S. F. Railsback. 2005. Individual-Based Modeling And Ecology. Princeton, NJ, Princeton University Press.

Grimm, V., E. Revilla, U. Berger, F. Jeltsch, W. M. Mooij, S. F. Railsback, H. H. Thulke et al. 2005. Pattern-oriented modeling of agent-based complex systems: lessons from ecology. Science 310:987.

Grinnell, J. 1917. The niche-relationships of the California thrasher. Auk 34:427–433.

———. 1919. The English house sparrow has arrived in Death Valley: an experiment in nature. American Naturalist 53:468–472.

Grisebach, A. H. R. 1839. Genera et species Gentianearum. Stuttgart and Tübingen, J. G. Cotta.

Groffman, P., J. Baron, T. Blett, A. Gold, I. Goodman, L. Gunderson, B. Levinson et al. 2006. Ecological thresholds: the key to successful environmental management or an important concept with no practical application? Ecosystems 9:1–13.

Grover, J. P. 1997. Resource Competition. London, Chapman and Hall.

Grytnes, J. 2003. Ecological interpretations of the mid-domain effect. Ecology Letters 6:883–888.

Guisan, A., and N. Zimmermann. 2000. Predictive habitat distribution models in ecology. Ecological Modelling 135:147–186.

Gunderson, L. H. 2000. Ecological resilience: in theory and application. Annual Review of Ecology and Systematics 31:425–439.

Gunderson, L. H., and C. S. Holling. eds. 2002. Panarchy: Understanding Transformations in Human and Natural Systems. Washington, D.C., Island Press.

Gutierrez, A. P. 1992. Physiological basis of ratio-dependent predator-prey theory: the metabolic pool model as a paradigm. Ecology 73:1552–1563.

Gyllenberg, M., and I. Hanski. 1997. Habitat deterioration, habitat destruction, and metapopulation persistence in a heterogeneous landscape. Theoretical Population Biology 52:198–215.

Haberl, H., H. Erb, F. Krausmann, V. Gaube, A. Bondeau, C. Plutzar, S. Gingrich et al. 2007. Quantifying and mapping the human appropriation of net primary production in earth's terrestrial ecosystems. Proceedings of the National Academy of Sciences 104:12942–12945.

Haeckel, E. 1904. Kunstformen der Natur. Leipzig and Vienna, Verlag des Bibliographischen Instituts.

Hagen, J. B. 1989. Research perspectives and the anomalous status of modern ecology. Biology and Philosophy 4:443–455.

———. 1992. An Entangled Bank: The Origins of Ecosystem Ecology. New Brunswick, NJ, Rutgers University Press.

Hairston, N. 1989. Ecological Experiments: Purpose, Design, and Execution. Cambridge, Cambridge University Press.

Hall, J., and E. O'Connell. 2007. Earth systems engineering: turning vision into action. Proceedings of the Institution of Civil Engineers-Civil Engineering 160:114–122.

Hall, S. R., C. J. Knight, C. R. Becker, M. A. Duffy, A. J. Tessier, and C. E. Caceres. 2009. Quality matters: resource quality for hosts and the timing of epidemics. Ecology Letters 12:118–128.

Hall, S. R., K. D. Lafferty, J. M. Brown, C. E. Cáceres, J. M. Chase, A. P. Dobson, R. D. Holt et al. 2008. Is infectious disease just another type of predator-prey interaction?, Pages 223–241 in R. S. Ostfeld, F. Keesing, and V. T. Eviner, eds. Infectious Disease Ecology: Effects of Ecosystems on Disease and of Disease on Ecosystems. Princeton, NJ, Princeton University Press.

Hallam, S. J., N. Putnam, C. M. Preston, J. C. Detter, D. Rokhsar, P. M. Richardson, and E. F. DeLong. 2004. Reverse methanogenesis: testing the hypothesis with environmental genomics. Science 305:1457–1462.

Hammond, J. L., B. T. Luttbeg, and A. Sih. 2007. Predator and prey space use: dragonflies and tadpoles in an interactive game. Ecology 88:1525–1535.

Hanski, I. 1991. Single-species metapopulation dynamics: concepts, models and observations. Biological Journal of the Linnean Society 42:14–38.

———. 1994. A practical model of metapopulation dynamics. Journal of Animal Ecology 63:151–162.

———. 1995. Effects of landscape pattern on competitive interactions, Pages 203–224 in L. Hansson, L. Fahrig, and G. Merriam, eds. Mosaic Landscapes and Ecological Processes. New York, Chapman and Hall.

———. 1999. Metapopulation Ecology. Oxford, Oxford University Press.

Hanski, I., and M. Gilpin. 1991. Metapopulation dynamics: brief history and conceptual domain. Biological Journal of the Linnean Society 42:3–16.

———. 1997. Metapopulation Biology: Ecology, Genetics, and Evolution. New York, Academic Press.

Hansson, L., and P. Angelstam. 1991. Landscape ecology as a theoretical basis for nature conservation. Landscape Ecology 5:191–201.

Harper, J. L. 1967. A Darwinian approach to plant ecology. Journal of Ecology 55:247–270.

Harte, J., T. Zillio, E. Conlisk, and A. B. Smith. 2008. Maximum entropy and the state-variable approach to macroecology. Ecology 89:2700–2711.

Harvey, P. H., and M. D. Pagel. 1991. The Comparative Method in Evolutionary Biology. Oxford, Oxford University Press.

Hassell, M. P. 1978. The Dynamics of Arthropod Predator-Prey Systems. Princeton, NJ, Princeton University Press.

———. 2000. The Spatial and Temporal Dynamics of Host-Parasitoid Interactions. Oxford, Oxford University Press.

Hassell, M. P., and G. C. Varley. 1969. New inductive population model for insect parasites and its bearing on biological control. Nature 223:1133–1137.

Hastings, A. 1980. Disturbance, coexistence, history and competition for space. Theoretical Population Biology 18:363–373.

———. 1991. Structured models of metapopulations. Biological Journal of the Linnean Society 42:57–71.

———. 2004. Transients: the key to long-term ecological understanding? Trends in Ecology and Evolution 19:39–45.

Hastings, A., and C. L. Wolin. 1989. Within-patch dynamics in a metapopulation. Ecology 70:1261–1266.

Hawkins, B. A. 2001. Ecology's oldest pattern? Trends in Ecology and Evolution 16:470.

Hawkins, B. A., F. S. Albuquerque, M. B. Araújo, J. Beck, L. M. Bini, F. J. Cabrero-Sanudo, I. Castro-Parga et al. 2007. A global evaluation of metabolic theory as an explanation for terrestrial species richness gradients. Ecology 88:1877–1888.

Hawkins, B. A., and J. A. F. Diniz-Filho. 2002. The mid-domain effect cannot explain the diversity gradient of Nearctic birds. Global Ecology and Biogeography 11:419–426.

Hawkins, B. A., J. A. F. Diniz-Filho, and A. E. Weis. 2005. The mid-domain effect and diversity gradients: is there anything to learn? American Naturalist 166:E140–E143.

Hawkins, B. A., R. Field, H. V. Cornell, D. J. Currie, J.-F. Guégan, D. M. Kaufman, J. T. Kerr et al. 2003a. Energy, water and broad-scale geographic patterns of species richness. Ecology 84:3105–3117.

Hawkins, B. A., E. E. Porter, and J. A. F. Diniz-Filho. 2003b. Productivity and history as predictors of the latitudinal diversity gradient of terrestrial birds. Ecology 84:1608–1623.

He, F. 2005. Deriving a neutral model of species abundance from fundamental mechanisms of population dynamics. Functional Ecology 19:187–193.

He, F., and P. Legendre. 1996. On species-area relations. American Naturalist 148:719.

Heaney, L. R. 2000. Dynamic disequilibrium: a long-term, large-scale perspective on the equilibrium model of island biogeography. Global Ecology and Biogeography 9:59–74.

———. 2007. Is a new paradigm emerging for oceanic island biogeography? Journal of Biogeography 34:753–757.

Heinz Center. 2002. The State of the Nation's Ecosystems: Measuring the Lands, Waters, and Living Resources of the United States. Cambridge, Cambridge University Press.

Heithaus, M. R. 2001. Habitat selection by predators and prey in communities with asymmetrical intraguild predation. Oikos 92:542–554.

Hempel, C. 1966. Philosophy of Natural History. Englewood Cliffs, NJ, Prentice Hall.

Henderson-Sellers, A., H. Zhang, K. Emanuel, W. Gray, C. Landsea, G. Holland, J. Lighthill et al. 1998. Tropical cyclones and global climate change: a post-IPCC assessment. Bulletin of the American Meteorological Society 79:19–38.

Henderson, J. M., and R. E. Quandt. 1971. Microeconomic Theory: A Mathematical Approach. New York, McGraw Hill.

Hillebrand, H. 2004. On the generality of the latitudinal diversity gradient. American Naturalist 163:192–211.

Hils, M. H., and J. L. Vankat. 1982. Species removals from a first-year old-field plant community. Ecology 63:705–711.

Hobbs, R. J., S. Arico, J. Aronson, J. S. Baron, P. Bridgewater, V. A. Cramer, P. R. Epstein et al. 2006. Novel ecosystems: theoretical and management aspects of the new ecological world order. Global Ecology and Biogeography 15:1–7.

Hobbs, R. J., and V. A. Cramer. 2007. Old field dynamics: regional and local differences, and lessons for ecology and restoration, Pages 309–318 *in* V. A. Cramer and R. J. Hobbs, eds. Old fields: Dynamics and Restoration of Abandoned Farmland. Washington, D.C., Island Press.

Hochberg, M. E. 1991. Non-linear transmission rates and the dynamics of infectious disease. Journal of Theoretical Biology 153:301–321.

Hochberg, M. E., R. Gomulkiewicz, R. D. Holt, and J. N. Thompson. 2000. Weak sinks could cradle mutualistic symbioses: strong sources should harbor parasitic symbioses. Journal of Evolutionary Biology 13:213–222.

Hochberg, M. E., and J. H. Lawton. 1990. Spatial heterogeneities in parasitism and population dynamics. Oikos 59:9–14.

Holdridge, L. 1947. Determination of world plant formations from simple climatic data. Science 105:367–368.

Holling, C. S. 1959. The components of predation as revealed by a study of small mammal predation of the European pine sawfly. Canadian Entomologist 91:293–320.

———. 1973. Resilience and stability of ecological systems. Annual Review of Ecology and Systematics 4:1–23.

———. 1978. Adaptive Environmental Assessment and Management. Chichester, Wiley.

———. 1992. Cross-scale morphology, geometry, and dynamics of ecosystems. Ecological Monographs 62:447–452.

Holt, R. D. 1977. Predation, apparent competition, and the structure of prey communities. Theoretical Population Biology 12:197–229.

———. 1983. Optimal foraging and the form of the predator isocline. American Naturalist 122:521–541.

———. 1984. Spatial heterogeneity, indirect interactions, and the coexistence of prey species. American Naturalist 124:377–406.

———. 1993. Ecology at the mesoscale: the influence of regional processes on local communities, Pages 77–88 in R. E. Ricklefs, and D. Schluter, eds. Species Diversity in Ecological Communities: Historical and Geographical Perspectives. Chicago, University of Chicago Press.

———. 1997a. Community modules, Pages 333–350 in A. C. Gange and V. K. Brown, eds. Multitrophic Interactions in Terrestrial Systems. Oxford, Blackwell Science.

———. 1997b. From metapopulation dynamics to community structure: some consequences of spatial heterogeneity, Pages 149–164 in I. Hanski and M. E. Gilpin, eds. Metapopulation Biology. San Diego, CA, Academic Press.

———. 2002. Food webs in space: on the interplay of dynamic instability and spatial processes. Ecological Research 17:261–273.

———. 2009. Darwin, Malthus, and movement: a hidden assumption in the demographic foundations of evolution. Israel Journal of Ecology and Evolution 55:189–198.

Holt, R. D., J. P. Grover, and D. Tilman. 1994. Simple rules for interspecific dominance in systems with exploitative and apparent competition. American Naturalist 144:741–771.

Holt, R. D., and M. E. Hochberg. 1998. The coexistence of competing parasites. II. Hyperparasitism and food chain dynamics. Journal of Theoretical Biology 193:485–495.

Holt, R. D., M. Holyoak, and M. A. Leibold. 2005. Future directions in metacommunity ecology. Metacommunities: Spatial Dynamics and Ecological Communities:465–489.

Holt, R. D., and T. Kimbrell. 2007. Foraging and population dynamics, Pages 365–395 in D. W. Stephens, J. S. Brown, and R. C. Ydenberg, eds. Foraging. Behavior and Ecology. Chicago, University of Chicago Press.

Holt, R. D., and B. P. Kotler. 1987. Short-term apparent competition. American Naturalist 142:623–645.

Holt, R. D., and J. Pickering. 1985. Infectious disease and species coexistence: a model of Lotka-Volterra form. American Naturalist 126:196–211.

Holyoak, M., M. A. Leibold, and R. D. Holt. 2005. Metacommunities: Spatial Dynamics and Ecological Communities. Chicago, University of Chicago Press.

Hopkinson, C. S., A. E. Lugo, M. Alber, A. P. Covich, and S. J. Van Bloem. 2008. Forecasting effects of sea-level rise and windstorms on coastal and inland ecosystems. Frontiers in Ecology and the Environment 6:255–263.

Horn, H. S. 1974. The ecology of secondary succession. Annual Review of Ecology and Systematics 5:25–37.

———. 1975. Markovian properties of forest succession, Pages 196–211 *in* M. L. Cody and J. M. Diamond, eds. Ecology and Evolution of Communities. Cambridge, MA, Harvard University Press.

Horn, H. S., H. H. Shugart, and D. L. Urban. 1989. Simulators as models of forest dynamics, Pages 256–267 *in* J. Roughgarden, R. M. May, and S. A. Levin, eds. Perspectives in Ecological Theory. Princeton, Princeton University Press.

Houston, A. I., and J. M. McNamara. 1999. Models of Adaptive Behaviour. Cambridge, Cambridge University Press.

Howarth, R. W. 1984. The ecological significance of sulfur in the energy dynamics of salt marsh and coastal marine sediments. Biogeochemistry 1:5–27.

———. 2008. Coastal nitrogen pollution: a review of sources and trends globally and regionally. Harmful Algae 8:14–20.

Hubbell, S. P. 1997. A unified theory of biogeography and relative species abundance and its application to tropical rainforests and coral reefs. Coral Reefs 16 (Supplement):9–21.

———. 2001. The Unified Neutral Theory of Biodiversity and Biogeography. Princeton, NJ, Princeton University Press.

———. 2005. Neutral theory in community ecology and the hypothesis of functional equivalence. Functional Ecology 19:166–172.

Hudson, P. J., A. Rizzoli, B. T. Grenfell, J. Heesterbeek, and A. P. Dobson. 2002. The Ecology of Wildlife Diseases. Oxford, Oxford University Press.

Huey, R. B., G. W. Gilchrist, M. L. Carlson, D. Berrigan, and L. Sierra. 2000. Rapid evolution of a geographic cline in size in an introduced fly. Science 287:308–309.

Hughes, C., and R. Eastwood. 2006. Island radiation on a continental scale: exceptional rates of plant diversification after uplift of the Andes. Proceedings of the National Academy of Sciences 103:10334–10339.

Hugie, D. M., and L. M. Dill. 1994. Fish and game: a game theoretic approach to habitat selection by predators and prey. Journal of Fish Biology 45:151–169.

Huisman, J., and F. J. Weissing. 1999. Biodiversity of plankton by species oscillations and chaos. Nature 402:407–410.

Hurlbert, S. H. 1984. Pseudoreplication and the design of ecological field experiments 54:187–211.

Huston, M., and T. Smith. 1987. Plant succession: life history and competition. American Naturalist 130:168–198.

Huston, M. A. 1979. A general hypothesis of species diversity. American Naturalist 113:81–101.

———. 1992. Individual-based forest succession models and the theory of plant competition, Pages 408–420 *in* D. L. DeAngelis, ed. Populations and Communities: An Individual-based Perspective. New York, Chapman and Hall.

———. 1994. Biological Diversity. Cambridge, Cambridge University Press.

Huston, M. A., and D. L. DeAngelis. 1994. Competition and coexistence: the effects of resource transport and supply rates. American Naturalist 144:954–977.

Hutchinson, G. 1957. Concluding remarks. Cold Spring Harbor Symposia on Quantitative Biology 22:415–427.

Hutchinson, G. E. 1959. Homage to Sanda Rosalia or Why are there so many kinds of animals? American Naturalist 93:145–159.

———. 1961. The paradox of the plankton. American Naturalist 95:137–140.

———. 1965. The Ecological Theater and the Evolutionary Play. New Haven, Yale University Press.

———. 1978. An Introduction to Population Biology. New Haven, Yale University Press.

Inchausti, P., and L. R. Ginzburg. 2009. Maternal effects mechanism of population cycling: a formidable competitor to the traditional predator-prey view. Philosophical Transactions of the Royal Society B: Biological Sciences 364:1117–1124.

Inouye, R. S., T. D. Allison, and N. C. Johnson. 1994. Old field succession on a Minnesota sand plain: effects of deer and other factors on invasion by trees. Bulletin of the Torrey Botanical Club 121:266–276.

Inouye, R. S., and D. Tilman. 1995. Convergence and divergence of old-field vegetation after 11 years of nitrogen addition. Ecology 76:1872–1887.

IPCC. 2007. Climate Change 2007: The Physical Science Basis. Contribution of Working Group I to the Fourth Assessment Report of the Intergovernmental Panel on Climate Change [Solomon, S., D. Qin, M. Manning, Z. Chen, M. Marquis, K. B. Averyt, M. Tignor and H. L. Miller (eds.)]. Cambridge, Cambridge University Press.

Jaffe, D., I. McKendry, T. Anderson, and H. Price. 2003. Six "new" episodes of trans-Pacific transport of air pollutants. Atmospheric Environment 37:391–404.

Janzen, D. H. 1967. Why mountain passes are higher in the tropics. American Naturalist 101:233–249.

Jax, K. 2006. Ecological units: definitions and application. Quarterly Review of Biology 81:237–258.

———. 2007. Can we define ecosystems? on the confusion between definition and description of ecological concepts. Acta Biotheoretica 55:341–355.

Jax, K., C. Jones, and S. T. A. Pickett. 1998. The self-identity of ecological units. Oikos 82:253–264.

Jensen, C. X. J., J. M. Jeschke, and L. R. Ginzburg. 2007. A direct, experimental test of resource vs. consumer dependence: comment. Ecology 88:1600–1602.

Jetz, W., and C. Rahbek. 2001. Geometric constraints explain much of the species richness pattern in African birds. Proceedings of the National Academy of Sciences 98:5661–5666.

———. 2002. Geographic range size and determinants avian species richness. Science 279:1548–1551.

Jiang, L., and S. N. Patel. 2008. Community assembly in the presence of disturbance: a microcosm experiment. Ecology 89:1931–1940.

Johnson, E. A. 1979. Succession, an unfinished revolution. Ecology 60:238–240.

Johnson, E. A., and K. Miyanishi. 2008. Testing the assumptions of chronosequences in succession. Ecology Letters 11:419–431.

Johnson, E. A., and K. Miyanishi, eds. 2007. Plant Disturbance Ecology: The Process and the Response. Burlington, MA, Elsevier.

Johnson, R. H. 1910. Determinate Evolution of the Color Pattern of the Lady-beetles. Washington, D.C., Carnegie Institute of Washington.

Jørgensen, S. E. 2007. An integrated ecosystem theory. Annals of the European Academy of Sciences 2006–2007:19–33.

Jost, C., O. Arino, and R. Arditi. 1999. About deterministic extinction in ratio-dependent predator-prey models. Bulletin of Mathematical Biology 61:19–32.

Jost, C., G. Devulder, J. A. Vucetich, R. O. Peterson, and R. Arditi. 2005. The wolves of Isle Royale display scale-invariant satiation and ratio-dependent predation on moose. Journal of Animal Ecology 74:809–816.

Kacelnik, A., and M. Bateson. 1997. Risk-sensitivity: crossroads for theories of decision making. Trends in Cognitive Sciences 1:304–309.

Kalmar, A., and D. J. Currie. 2006. A global model of island biogeography. Global Ecology and Biogeography 15:72–81.

———. 2007. A unified model of avian species richness on islands and continents. Ecology 88:1309–1321.

Keddy, P. A. 1989. Competition. New York, Chapman and Hall.

———. 1992. Assembly and response rules: two goals for predictive community ecology. Journal of Vegetation Science 3:157–164.

Keeling, M. J. 2005. Extensions to mass-action mixing, Pages 107–142 *in* K. Cuddington and B. E. Beisner, eds. Ecological Paradigms Lost: Routes of Theory Change. Burlington, MA, Elsevier.

Keeling, M. J., and P. Rohani. 2008. Modeling Infectious Diseases in Humans and Animals. Princeton, NJ, Princeton University Press.

Keeling, M. J., H. B. Wilson, and S. W. Pacala. 2000. Reinterpreting space, time lags, and functional responses in ecological models. Science 290:1758–1760.

Keeney, R. L., and H. Raiffa. 1993. Decisions with Multiple Objectives: Preference and Value Tradeoffs. Cambridge, Cambridge University Press.

Keever, C. 1950. Causes of succession on oldfields of the Piedmont, North Carolina. Ecological Monographs 20:229–250.

———. 1979. Mechanisms of plant succession on old fields of Lancaster County, Pennsylvania. Bulletin of the Torrey Botanical Club 106:299–308.

———. 1983. A retrospective view of old-field succession after 35 years. American Midland Naturalist 110:397–404.

Keitt, T. H. 1999. Ecological scale: theory and applications. Complexity 4:28–29.

Keller, J. K., M. E. Richmond, and C. R. Smith. 2003. An explanation of patterns of breeding bird species richness and density following clearcutting in northeastern USA forests. Forest Ecology and Management 174:541–564.

Kelly, C. K., and M. G. Bowler. 2005. A new application of storage dynamics: differential sensitivity, diffuse competition, and temporal niches. Ecology 86:1012–1022.

Kempton, R. A., and L. R. Taylor. 1974. Log-series and lognormal parameters as diversity discriminants for the Lepidoptera. Journal of Animal Ecology 43:381–399.

Kendall, B. E., and G. A. Fox. 2003. Unstructured individual variation and demographic stochasticity. Conservation Biology 17:1170–1172.

Kermack, W. P., and A. G. McKendrick. 1927. A contribution to the mathematical theory of epidemics. Proceedings of the Royal Society of London A 115:700–712.

Kerr, B., M. A. Riley, M. W. Feldman, and B. J. M. Bohannan. 2002. Local dispersal promotes biodiversity in a real-life game of rock-paper-scissors. Nature 418:171–174.

Kerr, J. T., M. Perring, and D. J. Currie. 2006. The missing Madagascan mid-domain effect. Ecology Letters 9:149–159.

Kessler, M. 2000. Upslope-directed mass effect in palms along an Andean elevational gradient: a cause for high diversity at mid-elevations? Biotropica 32:756–759.

King, D., and J. Roughgarden. 1982. Multiple switches between vegetative and reproductive growth in annual plants. Theoretical Population Biology 21:194–204.

Kingsland, S. E. 1995. Modeling Nature: Episodes in the History of Population Ecology. Chicago, University of Chicago Press.

———. 2005. The Evolution of American Ecology, 1890–2000. Baltimore, Johns Hopkins University Press.

Kirkpatrick, M., and N. H. Barton. 1997. Evolution of a species' range. American Naturalist 150:1–23.

Kitcher, P. 1989. Explanatory unification and the causal structure of the world, Pages 410–505 *in* P. Kitcher and W. Salmon, eds. Scientific Explanation. Minnesota Studies in the Philosophy of Science, vol. 13. Minneapolis, University of Minnesota Press.

Kitzberger, T., P. M. Brown, E. K. Heyerdahl, T. W. Swetnam, and T. T. Veblen. 2007. Contingent Pacific-Atlantic Ocean influence on multicentury wildfire synchrony over western North America. Proceedings of the National Academy of Sciences 104:543–548.

Kleiber, M. 1932. Body size and metabolism. Hilgardia 6:315–332.

Klironomos, J. N. 2002. Feedback with soil biota contributes to plant rarity and invasiveness in communities. Nature 417:67–70.

Klopfer, P., and R. H. MacArthur. 1960. Niche size and faunal diversity. American Naturalist 94:293–300.

———. 1961. On the causes of tropical species diversity: niche overlap. American Naturalist 95:223–226.

Knapp, A. K., M. D. Sraith, S. L. Collins, N. Zarabatis, M. Peel, S. Emery, J. Wojdak et al. 2004. Generality in ecology: testing North American grassland rules in South African savannas. Frontiers in Ecology and the Environment 2:483–491.

Kneitel, J. M., and J. M. Chase. 2004. Trade-offs and community ecology: linking spatial scales and species coexistence. Ecology Letters 7:69–80.

Kolasa, J. 2005. Complexity, system integration, and susceptibility to change: biodiversity connection. Ecological Complexity 2:431–442.

Kolasa, J., and S. T. A. Pickett. 1989. Ecological systems and the concept of biological organization. Proceedings of the National Academy of Sciences 86:8837–8841.

Kolmogorov, A. N. 1936. Sulla Teoria di Volterra della Lotta per L'Esisttenza. Giornale Instituto Italiana Attuari 7:74–80.

Kooijman, S. A. L. M. 1993. Dynamic Energy Budgets in Biological Systems. Cambridge, Cambridge University Press.

Kotler, B. P., and J. S. Brown. 2007. Community ecology, Pages 397–434 *in* D. W. Stephens, J. S. Brown, and R. C. Ydenberg, eds. Foraging: Behavior and Ecology. Chicago, University of Chicago Press.

Kozlowski, J., and J. Weiner. 1997. Interspecific allometries are by-products of body size optimization. American Naturalist 149:352–380.

Kraft, N. J. B., W. K. Cornwell, C. O. Webb, and D. D. Ackerly. 2007. Trait evolution, community assembly, and the phylogenetic structure of ecological communities. American Naturalist 170:271–283.

Kreft, H., and W. Jetz. 2007. Global patterns and determinants of vascular plant diversity. Proceedings of the National Academy of Sciences of the United States of America 104:5925–5930.

Kreft, H., W. Jetz, J. Mutke, G. Kier, and W. Barthlott. 2008. Global diversity of island floras from a macroecological perspective. Ecology Letters 11:116–127.

Kuhn, T. S. 1962. The Structure of Scientific Revolutions. Chicago, University of Chicago Press.

Kupfer, J., A. Myers, S. McLane, and G. Melton. 2008. Patterns of forest damage in a southern Mississippi landscape caused by Hurricane Katrina. Ecosystems 11:45–60.

Lack, D. 1947. Darwin's Finches: An Essay on the General Biological Theory of Evolution. Cambridge, Cambridge University Press.

———. 1954. The Natural Regulation of Animal Numbers. Oxford, Oxford University Press.

Lafferty, K. D., and A. M. Kuris. 2002. Trophic strategies, animal diversity, and body size. Trends in Ecology and Evolution 17:507–513.

Lake, P. S., N. Bond, and P. Reich. 2007. Linking ecological theory with stream restoration. Freshwater Biology 52:597–615.

Lande, R. 1993. Risks of population extinction from demographic and environmental stochasticity and random catastrophes. American Naturalist 142:911–927.

Lande, R., S. Engen, and B.-E. Saether. 2003. Stochastic Population Dynamics in Ecology and Conservation. Oxford, Oxford University Press.

Landsea, C. W. 1993. A climatology of intense (or major) Atlantic hurricanes. Monthly Weather Review 121:1703–1713.

Landsea, C. W., and W. M. Gray. 1992. The strong association between Western Sahel monsoon rainfall and intense Atlantic hurricanes. Journal of Climate 5:435–453.

Lau, K. M., and J. M. Kim. 2007. How nature foiled the 2006 hurricane forecasts. Eos 88:105–107.

Laudan, L. 1977. Progress and Its Problems: Toward a Theory of Scientific Growth. Berkeley, University of California Press.

Lauenroth, W. K., I. C. Burke, and J. K. Berry. 2003. The status of dynamic quantitative modeling in ecology, Pages 32–48 in C. D. Canham, J. J. Cole, and W. K. Lauenroth, eds. Models in Ecosystem Science. Princeton, NJ, Princeton University Press.

Lauenroth, W. K., C. D. Canham, A. P. Kinzig, K. A. Poiani, W. M. Kemp, S. W. Running, M. L. Pace et al. 1998. Simulation modeling in ecosystem science, Pages 404–415 in P. Groffman and M. Pace, eds. Successes, Limitations, and Frontiers in Ecosystem Science. New York, Springer-Verlag.

Laurance, W. F. 2008. Theory meets reality: how habitat fragmentation research has transcended island biogeographic theory. Biological Conservation 141:1731–1744.

Law, R., and M. A. Leibold. 2005. Assembly dynamics in metacommunities. Metacommunities: Spatial Dynamics and Ecological Communities:263–278.

Law, R., and R. D. Morton. 1996. Permanence and the assembly of ecological communities. Ecology 77:762–775.

Lawlor, T. E. 1986. Comparative biogeography of mammals on islands. Biological Journal of the Linnean Society 28:99–125.

Lawton, J. H. 1999. Are there general laws in ecology? Oikos 84:177–192.

Lees, D. C., and R. K. Colwell. 2007. A strong Madagascan rainforest MDE and no equatorward increase in species richness: re-analysis of 'The missing Madagascan mid-domain effect', by Kerr J. T., Perring M. & Currie D. J. (Ecology Letters 9:149–159, 2006). Ecology Letters 10:E4–E8.

Lees, D. C., C. Kremen, and L. Andriamampianina. 1999. A null model for species richness gradients: bounded range overlap of butterflies and other rainforest endemics in Madagascar. Biological Journal of the Linnean Society 67:529–584.

Legendre, P. 1993. Spatial autocorrelation: trouble or new paradigm? Ecology 74:1659–1673.

Leibold, M. A. 1995. The niche concept revisited: mechanistic models and community context. Ecology 76:1371–1382.

———. 1996. A graphical model of keystone predators in food webs: trophic regulation of abundance, incidence, and diversity patterns in communities. American Naturalist 147:784–812.

———. 1998. Similarity and local co-existence of species in regional biotas. Evolutionary Ecology 12:95–110.

———. 1999. Biodiversity and nutrient enrichment in pond plankton communities. Evolutionary Ecology Research 1:73–95.

Leibold, M. A., E. P. Economo, and P. Peres-Nato. 2010. Letter: phylogenetic signal in metacommunities: separating the roles of environmental filters and historical biogeography. Ecology Letters 13:1290–1299.

Leibold, M. A., M. Holyoak, N. Mouquet, P. Amarasekare, J. M. Chase, M. F. Hoopes, R. D. Holt et al. 2004. The metacommunity concept: a framework for multi-scale community ecology. Ecology Letters 7:601–613.

Leibold, M. A., and M. A. McPeek. 2006. Coexistence of the niche and neutral perspectives in community ecology. Ecology 87:1399–1410.

Leopold, A. 1949. A Sand County Almanac. New York, Oxford University Press.

Lepori, F., and B. Malmqvist. 2008. Deterministic control on community assembly peaks at intermediate levels of disturbance. Oikos 118:471–479.

Leslie, P. H. 1948. Some further notes on the use of matrices in population mathematics. Biometrika 35:213–245.

Lever, C. 1985. Naturalized Mammals of the World. London, Longman.

Levin, L. A., R. J. Etter, M. A. Rex, A. J. Gooday, C. R. Smith, J. Pineda, C. T. Stuart et al. 2001. Environmental influences on regional deep-sea species diversity. Annual Review of Ecology and Systematics 32:51–93.

Levin, S. A. 1970. Community equilibria and stability, and an extension of the competitive exclusion principle. American Naturalist 104:413–423.

———. 1992. The problem of pattern and scale in ecology. Ecology 73:1943–1967.

Levin, S. A., and S. W. Pacala. 1997. Theories of simplification and scaling of spatially distributed processes, Pages 271–296 in D. Tilman and P. Kareiva, eds. Spatial Ecology: The Role of Space in Population Dynamics and Interspecies Interactions. Princeton, Princeton University Press.

Levins, R. 1969. Some demographic and genetic consequences of environmental heterogeneity for biological control. Bulletin of the Entomological Society of America 15:237–240.

———. 1970. Extinction, Pages 75–108 in M. Gerstenhaber, ed. Some Mathematical Questions in Biology. Providence, American Mathematical Society.

———. 1975. Evolution in communities near equilibrium, Pages 16–50 in M. L. Cody and J. M. Diamond, eds. Ecology and Evolution of Communities. Cambridge, Belknap Press of Harvard University Press.

Levins, R., and D. Culver. 1971. Regional coexistence of species and competition between rare species. Proceedings of the National Academy of Sciences 68:1246–1248.

Lewis, D. B., and N. B. Grimm. 2007. Hierarchical regulation of nitrogen export from urban catchments: interactions of storms and landscapes. Ecological Applications 17:2347–2364.

Li, B. L., V. G. Gorshkov, and A. M. Makarieva. 2004. Energy partitioning between different-sized organisms and ecosystem stability. Ecology 85:1811–1813.

Lima, S. L. 1992. Life in a multipredator environment: some considerations for antipredatory vigilance. Annales Zoologici Fennici 29:217–226.

———. 1998. Stress and decision making under the risk of predation: recent developments from behavioral, reproductive and ecological perspectives. Advances in the Study of Behavior 27:215–290.

———. 2002. Putting predators back into behavioral predator-prey interactions. Trends in Ecology and Evolution 17:70–75.

Lima, S. L., and P. A. Bednekoff. 1999. Temporal variation in danger drives antipredator behavior: the predation risk allocation hypothesis. American Naturalist 153:649–707.

Lima, S. L., and L. M. Dill. 1990. Behavioral decisions made under the risk of predation: a review and prospectus. Canadian Journal of Zoology 68:619–640.

Lindeman, R. L. 1942. The trophic-dynamic aspect of ecology. Ecology 23:399–417.

Liu, J., and J. Diamond. 2005. China's environment in a globalizing world: how China and the rest of the world affect each other. Nature 435:1179–1186.

Lockwood, D. R. 2008. When logic fails ecology. Quarterly Review of Biology 83:57–64.

Lockwood, J. L. 1997. An alternative to succession: assembly rules offer guide to restoration efforts. Restoration and Management Notes 15:45–50.

Loeuille, N., and M. A. Leibold. 2008. Evolution in metacommunities: on the relative importance of species sorting and monopolization in structuring communities. American Naturalist 171:788–799.

Lomolino, M. V. 2000a. A call for a new paradigm of island biogeography. Global Ecology and Biogeography 9:1–6.

———. 2000b. A species-based theory of insular zoogeography. Global Ecology and Biogeography 9:39–58.

Lomolino, M. V., and J. H. Brown. 2009. The reticulating phylogeny of island biogeography theory. Quarterly Review of Biology 84:357–390.

Lomolino, M. V., J. H. Brown, and R. Davis. 1989. Island biogeography of montane forest mammals in the American Southwest. Ecology 70:180–194.

Lomolino, M. V., J. H. Brown, and D. F. Sax. 2009. Island biogeography theory: reticulations and reintegration of "a biogeography of the species," Pages 13–51 *in* J. B. Losos and R. E. Ricklefs, eds. The Theory of Island Biogeography Revisited. Princeton, Princeton University Press.

Longino, H. E. 2002. The Fate of Knowledge. Princeton, NJ, Princeton University Press.

Lonsdale, P. 1977. Clustering of suspension-feeding macrobenthos near abyssal hydrothermal vents at oceanic spreading centers. Deep-Sea Research 24:857–863.

Loreau, M. 1998. Ecosystem development explained by competition within and between material cycles. Proceedings of the Royal Society B: Biological Sciences 265:33–38.

Loreau, M., and R. D. Holt. 2004. Spatial flows and the regulation of ecosystems. American Naturalist 163:606–615.

Loreau, M., N. Mouquet, and R. D. Holt. 2003. Meta-ecosystems: a theoretical framework for a spatial ecosystem ecology. Ecology Letters 6:673–679.

Lorenz, E. N. 1964. The problem of deducing the climate from the governing equations. Tellus 16:1–11.

Losos, J. B. 2008. Phylogenetic niche conservatism, phylogenetic signal and the relationship between phylogenetic relatedness and ecological similarity between species. Ecology Letters 11: 995–1007

Losos, J. B., and R. E. Ricklefs. 2009. The Theory of Island Biogeography Revisited. Princeton, Princeton University Press.

Lotka, A. 1925. Elements of Physical Biology. Baltimore, Williams & Wilkins.

Loucks, O. L. 1970. Evolution of diversity, efficiency, and community stability. American Zoologist 10:17–25.

Lovelock, J. E., and L. Margulis. 1974. Atmospheric homeostasis by and for the biosphere: the Gaia hypothesis. Tellus 22:2–9.

Luck, M. A., G. D. Jenerette, J. Wu, and N. B. Grimm. 2001. The urban funnel model and spatially heterogeneous ecological footprint. Ecosystems 4:782–796.

Ludwig, D. 1999. Is it meaningful to estimate a probability of extinction? Ecology 80:298–310.

Ludwig, J. A., R. Bartley, A. A. Hawdon, B. N. Abbott, and D. McJannet. 2007. Patch configuration non-linearly affects sediment loss across scales in a grazed catchment in northeast Australia. Ecosystems 10:839–845.

Lugo, A. E. 2008. Visible and invisible effects of hurricanes on forest ecosystems: an international review. Austral Ecology 33:368–398.

Lundholm, J. T. 2009. Plant species diversity and environmental heterogeneity: spatial scale and competing hypotheses. Journal of Vegetation Science 20:377–391.

Lurquin, P. F. 2003. The Origins of Life and the Universe. New York, Columbia University Press.

Luttbeg, B. T., and T. A. Langen. 2004. Comparing alternative models to empirical data: cognitive models of western scrub-jay foraging behavior. American Naturalist 163:263–276.

Luttbeg, B. T., and A. Sih. 2004. Predator and prey habitat selection games: the effects of how prey balance foraging and predation risk. Israel Journal of Zoology 50:233–254.

Lynch, J. D., and N. V. Johnson. 1974. Turnover and equilibria in insular avifaunas, with special reference to the California Channel Islands. Condor 76:370–384.

Lynch, J. F. 1991. Effects of Hurricane Gilbert on birds in a dry tropical forest in the Yucatan Peninsula. Biotropica 28:577–584.

MacArthur, R. H. 1955. Fluctuations of animal populations, and a measure of community stability. Ecology 36:533–536.

———. 1957. On the relative abundance of bird species. Proceedings of the National Academy of Sciences 43:293–295.

———. 1958. Population ecology of some warblers of northeastern coniferous forests. Ecology 42:710–723.

———. 1960. On the relative abundance of species. American Naturalist 94:25–36.

———. 1962. Some generalized theorems of natural selection. Proceedings of the National Academy of Sciences 48:1893–1897.

———. 1964. Environmental factors affecting bird species diversity. American Naturalist 98:387–397.

———. 1965. Patterns of species diversity. Biological Review 40:510–533.

———. 1969. Patterns of communities in the tropics. Biological Journal of the Linnean Society 1:19–30.

———. 1970a. Graphical analysis of ecological systems, Pages 61–72 in M. Gerstenhaber, ed. Some Mathematical Questions in Biology. Providence, RI, American Mathematics Society.

———. 1970b. Species packing and competitive equilibrium for many species. Theoretical Population Biology 1:1–11.

———. 1972. Geographical Ecology: Patterns in the Distribution of Species. New York, Harper and Row.

MacArthur, R. H., and R. Levins. 1967. The limiting similarity, convergence and divergence of coexisting species. American Naturalist 101:377–385.

MacArthur, R. H., and E. R. Pianka. 1966. On the optimal use of a patchy environment. American Naturalist 100:276–282.

MacArthur, R. H., and E. O. Wilson. 1963. An equilibrium theory of insular zoogeography. Evolution 17:373–387.

———. 1967. The Theory of Island Biogeography. Princeton, NJ, Princeton University Press.

Mack, R. N., D. Simberloff, W. M. Lonsdale, H. Evans, M. Clout, and F. A. Bazzaz. 2000.

Biotic invasions: causes, epidemiology, global consequences and control. Ecological Applications 10:689–710.

MacMahon, J. A. 1980. Ecosystems over time: succession and other types of change, Pages 27–58 *in* R. Waring, ed. Forests: Fresh Perspectives from Ecosystem Analyses. Corvallis, Oregon State University Press.

———. 1981. Successional processes: comparisons among biomes with special reference to probable role of and influences on animals, Pages 277–304 *in* D. C. West, H. H. Shugart, and D. B. Botkin, eds. Forest Succession: Concepts and Applications. New York, Springer-Verlag.

Maddux, R. D. 2004. Self-similarity and the species-area relationship. American Naturalist 163:616–626.

Magill, A. H., J. D. Aber, W. S. Currie, K. J. Nadelhoffer, M. E. Martin, W. H. McDowell, J. M. Melillo et al. 2004. Ecosystem response to 15 years of chronic nitrogen additions at the Harvard Forest LTER, Massachusetts, USA. Forest Ecology and Management 196:7–28.

Malthus, T. R. 1798. An Essay on the Principle of Population. London, Johnson.

Mandelbrot, B. B. 1977. Fractals, Fun, Chance and Dimension. New York, Freeman.

Mangel, M., and C. W. Clark. 1998. Dynamic Modeling in Behavioral Ecology. Princeton, NJ, Princeton University Press.

Mann, M. E., R. S. Bradley, and M. K. Hughes. 1998. Global-scale temperature patterns and climate forcing over the past six centuries. Nature 392:779–787.

Mansfield, E. 1979. Microeconomics: Theory and Applications. New York, W. W. Norton.

Margulis, L., and J. E. Lovelock. 1974. Biological modulation of the Earth's atmosphere. Icarus 21:471–489.

Marquet, P. A. 2002. Of predators, prey, and power laws. Science 295:2229–2230.

Marshall, J. D., J. M. Blair, D. P. C. Peters, G. Okin, A. Rango, and M. Williams. 2008. Predicting and understanding ecosystem responses to climate change at continental scales. Frontiers in Ecology and the Environment 6:273–280.

Matsuda, H., M. Hori, and P. A. Abrams. 1994. Effects of predator-specific defense on community complexity. Evolutionary Ecology 8:628–638.

May, R. M. 1973. Stability and Complexity in Model Ecosystems. Princeton, NJ, Princeton University Press.

———. 1975. Patterns of species abundance and diversity, Pages 81–120 *in* M. L. Cody and J. M. Diamond, eds. Ecology and Evolution of Communities. Cambridge, MA, Belknap Press of Harvard University Press.

———. 1976. Simple mathematical models with very complicated dynamics. Nature 261:459–467.

May, R. M., and R. H. MacArthur. 1972. Niche overlap as a function of environmental variability. Proceedings of the National Academy of Sciences 69:1109–1113.

May, R. M., and C. H. Watts. 1992. The dynamics of predator-prey and resource-harvester systems, Pages 431–457 *in* M. J. Crawley, ed. Natural Enemies. Oxford, Blackwell.

Maynard Smith, J. 1982. Evolution and the Theory of Games. Cambridge, Cambridge University Press.

Mayr, E. 1965. Avifauna: turnover on islands. Science 150:1587–1588.

———. 1982. The Growth of Biological Thought. Cambridge, MA, Belknap Press.

McCain, C. M. 2009. Vertebrate range sizes indicate that mountains may be "higher" in the tropics. Ecology Letters 12:550–560.

McCallum, H., N. Barlow, and J. Hone. 2001. How should pathogen transmission be modeled? Trends in Ecology and Evolution 16:295–300.

McCann, K. S. 2000. The diversity-stability debate. Nature 405:228–233.

———. 2005. Perspectives on diversity, structure, and stability, Pages 183–200 in K. Cuddington and B. E. Beisner, eds. Ecological Paradigms Lost: Routes of Theory Change. New York, Elsevier Academic Press.

McCann, K. S., J. B. Rasmussen, and J. Umbanhowar. 2005. The dynamics of spatially coupled food webs. Ecology Letters 8:513–523.

McCauley, E., W. A. Nelson, and R. M. Nisbet. 2008. Small-amplitude cycles emerge from stage-structured interactions in *Daphnia*-algal systems. Nature 455:1240–1243.

McCoy, M. W., and J. F. Gillooly. 2008. Predicting natural mortality rates of plants and animals. Ecology Letters 11:710–716.

McGill, B. J. 2010. Towards a unification of unified theories of biodiversity. Ecology Letters 13:627–642.

McGill, B. J., B. A. Maurer, and M. D. Weiser. 2006. Empirical evaluation of neutral theory. Ecology 87:1411–1423.

McGill, B. S., R. S. Etienne, J. S. Gray, D. Alonso, M. J. Anderson, H. K. Benecha, M. Dormelas et al. 2007. Species abundance distributions: moving beyond single prediction theories to integration within an ecological framework. Ecology Letters 10:995–1015.

McIntosh, R. P. 1980. The background and some current problems of theoretical ecology, Pages 1–61 in E. Saarinen, ed. Conceptual Issues in Ecology. Boston, D. Reidel.

———. 1985. The Background of Ecology: Concept and Theory. Cambridge, Cambridge University Press.

———. 1987. Pluralism in ecology. Annual Review of Ecology and Systematics 18:321–341.

McKane, A. J., and B. Drossel. 2006. Models of food-web evolution, Pages 223–243 in M. Pascual and J. A. Dunne, eds. Ecological Networks: Linking Structure to Dynamics in Food Webs. Oxford, Oxford University Press.

McNamara, J. M. 1982. Optimal patch use in a stochastic environment. Theoretical Population Biology 21:269–288.

McNamara, J. M., and A. I. Houston. 1986. The common currency for behavioral decisions. American Naturalist 127:358–378.

McRae, B. H., B. G. Dickson, T. H. Keitt, and V. B. Shah. 2008. Using circuit theory to model connectivity in ecology, evolution, and conservation. Ecology 89:2712–2724.

Meiners, S. J., M. L. Cadenasso, and S. T. A. Pickett. 2003. Exotic plant invasions in successional systems: the utility of a long-term approach. Newton Square, PA.

Meiners, S. J., S. N. Handel, and S. T. A. Pickett. 2000. Tree seedling establishment under insect herbivory: edge effects and inter-annual variation. Plant Ecology 151:161–170.

Meiners, S. J., and K. LoGiudice. 2003. Temporal consistency in the spatial pattern of seed predation across a forest-old field edge. Plant Ecology 168:45–55.

Melbourne, B. A., and A. Hastings. 2008. Extinction risk depends strongly on factors contributing to stochasticity. Nature 454:100–103.

Merriam, C. H., and L. Steineger. 1890. Results of a biological survey of the San Francisco Mountain region and the desert of the Little Colorado, Arizona, in North American Fauna Report. Washington, D.C., US Department of Agriculture, Division of Ornithology and Mammalia.

Merrifield, M. A., Y. L. Firing, T. Aarup, W. Argricole, G. Brundit, D. Chang-Seng, R. Farre et al. 2005. Tide gauge observations of the Indian Ocean tsunami, December 26, 2004, Pages L09603, Geophysical Research Letters.

Michener, W. K., E. R. Blood, K. L. Bildstein, M. M. Brinson, and L. R. Gardner. 1997. Climate change, hurricanes and tropical storms, and rising sea level in coastal wetlands. Ecological Applications 7:770–801.

Milchunas, D. G., O. E. Sala, and W. K. Lauenroth. 1988. A generalized model of the effects of grazing by large herbivores on grassland community structure. American Naturalist 132:87–106.

Miles, J. 1979. Vegetation Dynamics. New York, Wiley.

Millennium Ecosystem Assessment. 2005. Ecosystems and Human Well-being: Synthesis. Washington, D.C., Island Press.

Miller, D. A., J. B. Grand, T. E. Fondell, and M. Anthony. 2006. Predator functional response and prey survival: direct and indirect interactions affecting a marked prey population. Journal of Animal Ecology 75:101–110.

Miller, D. H. 1978. The factor of scale: ecosystem, landscape mosaic, and region, Pages 63–88 *in* K. A. Hammond and G. Macinio, eds. Sourcebook on the Environment: A Guide to the Literature. Chicago, University of Chicago Press.

Milne, B. T. 1998. Motivation and benefits of complex systems approaches in ecology. Ecosystems 1:449–456.

Mischel, W., Y. Shoda, and M. L. Rodriquez. 1989. Delay of gratification in children. Science 244:933–938.

Mitchell, S. D. 1997. Pragmatic laws. Philosophy of Science 64:S468–S479.

Mitman, G. 1992. The State of Nature. Chicago, University of Chicago Press.

Mittelbach, G. G. 1981. Foraging efficiency and body size: a study of optimal diet and habitat use by bluegills. Ecology 62:1370–1386.

Mittelbach, G. G., S. M. Scheiner, and C. F. Steiner. 2003. What is the observed relationship between species richness and productivity? Reply. Ecology 84:3390–3395.

Mittelbach, G. G., C. F. Steiner, S. M. Scheiner, K. L. Gross, H. L. Reynolds, R. B. Waide, M. R. Willig et al. 2001. What is the observed relationship between species richness and productivity? Ecology 82:2381–2396.

Möbius, K. 1883. The oyster and oyster-culture, Pages 683–751. Government Printing Office, Washington, D.C., United States Commission of Fish and Fisheries, Part VIII. Report of the Commissioner for 1880. Appendix H. The oyster.

Mönkkönen, M., and P. Viro. 1997. Taxonomic diversity of the terrestrial bird and mammal fauna in temperate and boreal biomes of the northern hemisphere. Journal of Biogeography 24:603–612.

Morgan, J. A., D. G. Milchunas, D. R. LeCain, M. West, and A. R. Mosier. 2007. Carbon dioxide enrichment alters plant community structure and accelerates shrub growth in the shortgrass steppe. Proceedings of the National Academy of Sciences 104:14724–14729.

Morin, P. J. 1999. Community Ecology. Oxford, Blackwell Science.

Morrison, M. 2000. Unifying Scientific Theories: Physical Concepts and Mathematical Structures. Cambridge, Cambridge University Press.

Mouquet, N., M. F. Hoopes, and P. Amarasekare. 2005. The world is patchy and heterogeneous! Trade-off and source-sink dynamics in competitive metacommunities. Metacommunities: Spatial Dynamics and Ecological Communities:237–262.

Mouquet, N., and M. Loreau. 2002. Coexistence in metacommunities: the regional similarity hypothesis. American Naturalist 159:420–426.

———. 2003. Community patterns in source–sink metacommunities. American Naturalist 162:544–557.

Muller, C. H. 1952. Plant succession in arctic heath and tundra in northern Scandinavia. Bulletin of the Torrey Botanical Club 79:296–309.

Murdoch, W. W., C. J. Briggs, and R. M. Nisbet. 2003. Consumer-Resource Dynamics. Princeton, NJ, Princeton University Press.

Murdoch, W. W., and A. Oaten. 1975. Predation and population stability. Advances in Ecological Research 9:1–131.

Murray, B. G., Jr. 2000. Universal laws and predictive theory in ecology and evolution. Oikos 89:403–408.

Myster, R. W. ed. 2008. Post-agricultural Succession in the Neotropics. New York, Springer.

Myster, R. W., and S. T. A. Pickett. 1988. Individualistic patterns of annuals and biennials in early successional oldfields. Vegetatio 78:53–60.

National Research Council. 2008. The Role of Theory in Advancing 21st Century Biology: Catalyzing Transformative Research. Committee on Defining and Advancing the Conceptual Basis of Biological Sciences in the 21st Century. Washington, D.C., National Academies Press.

Nee, S., R. M. May, and M. P. Hassell. 1997. Two-species metapopulation models, Pages 123–147 *in* I. Hanski and M. E. Gilpin, eds. Metapopulation Biology. London, Academic Press.

Newman, J. 2007. Herbivory, Pages 175–220 *in* D. W. Stephens, J. S. Brown, and R. C. Ydenberg, eds. Foraging: Behavior and Ecology. Chicago, University of Chicago Press.

Nicholson, A. J., and V. A. Bailey. 1935. The balance of animal populations, Part I. Proceedings of the Zoological Society, London 3:551–598.

Niering, W. A. 1987. Vegetation dynamics (succession and climax) in relation to plant community management. Conservation Biology 1:287–295.

Nogués-Bravo, D., and M. B. Araujo. 2008. Scale effects and human impact on the elevational species richness gradient. Nature 453:216–219.

Nonacs, P. 2001. State-dependent behavior and the marginal value theorem. Behavioral Ecology 12:71–83.

Novikoff, A. B. 1945. The concept of integrative levels and biology. Science 101:209–215.

Nunney, L. 1985. The effect of long time delays in predator-prey systems. Theoretical Population Biology 27:202–221.

O'Brien, E., R. Field, and R. J. Whittaker. 2000. Climatic gradients in woody plant (tree and shrub) diversity: water-energy dynamics, residual variation, and topography. Oikos 89:588–600.

O'Brien, E. M. 1998. Water-energy dynamics, climate, and prediction of woody plant species richness: an interim general model. Journal of Biogeography 25:379–398.

O'Hara, R. B. 2005. The anarchist's guide to ecological theory. Or, we don't need no stinkin' laws. Oikos 110:390–393.

O'Neill, R. V., D. L. DeAngelis, J. B. Waide, and T. F. H. Allen. 1986. A Hierarchical Concept of Ecosystems. Princeton, NJ, Princeton University Press.

Oaten, A. 1977. Optimal foraging in patches: a case for stochasticity. Theoretical Population Biology 12:263–285.

Odenbaugh, J. 2003. Complex systems, trade-offs and mathematical modeling: a response to Sober and Orzack. Philosophy of Science 70:1496–1507.

———. 2006. The strategy of "The strategy of modeling building in population biology." Biology and Philosophy 21:607–621.

Odum, E. P. 1969. The strategy of ecosystem development. Science 164:262–270.

———. 1971. Fundamentals of Ecology. Philadelphia, PA, W. B. Saunders.

Odum, E. P., and H. T. Odum. 1953. The Fundamentals of Ecology. Philadelphia, W. B. Saunders.

Odum, H. T. 1983. Systems Ecology. New York, John Wiley.

Oksanen, J. 1996. Is the humped relationship between species richness and biomass an artefact due to plot size? Journal of Ecology 84:293–295.

Oksanen, L., S. D. Fretwell, J. Arrunda, and P. Niemela. 1981. Exploitation ecosystems in gradients of primary productivity. American Naturalist 118:240–261.

Olff, H., F. W. M. Vera, J. Bokdam, E. S. Bakker, J. M. Gleichman, K. de Maeyer, and R. Smit. 1999. Shifting mosaics in grazed woodlands driven by the alternation of plant facilitation and competition. Plant Biology 1:127–137.

Orians, G. H. 1962. Natural selection and ecological theory. American Naturalist 46:257–263.

Orrock, J. L., and R. J. Fletcher, Jr. 2005. Changes in community size affect the outcome of competition. American Naturalist 166:107–111.

Orrock, J. L., J. H. Grabowski, J. H. Pantel, S. D. Peacor, B. L. Peckarsky, A. Sih, and E. E. Werner. 2008. Consumptive and nonconsumptive effects of predators on metacommunities of competing prey. Ecology 89:2426–2435.

Ostertag, R., W. L. Silver, and A. E. Lugo. 2005. Factors affecting mortality and resistance to damage following hurricanes in a rehabilitated subtropical moist forest. Biotropica 37:16–24.

Ostfeld, R. S., and F. Keesing. 2000. Biodiversity and disease risk: the case of Lyme disease. Conservation Biology 14:722–728.

Ostfeld, R. S., R. H. Manson, and C. D. Canham. 1999. Interactions between meadow voles and white-footed mice at forest-oldfield edges: competition and net effects on tree invasion of oldfields, Pages 229–247 in G. W. Barrett and J. D. Petes, eds. Landscape Ecology of Small Mammals. New York, Springer.

Ostling, A., J. Harte, J. L. Green, and A. P. Kinzig. 2004. Self-similarity, the power law form of the species-area relationship, and a probability rule: a reply to Maddux. American Naturalist 163:627–633.

Otto, S., and T. Day. 2007. A Biologist's Guide to Mathematical Modeling. Princeton, NJ, Princeton University Press.

Overpeck, J. T., P. J. Bartlein, and T. Webb, III. 1991. Potential magnitude of future vegetation change in eastern North America: comparisons with the past. Science 254:692–695.

Owen-Smith, N. 2002. Adaptive Herbivore Ecology: From Resources to Populations in Variable Environments. Cambridge, Cambridge University Press.

———. 2005. Incorporating fundamental laws of biology and physics into population ecology: the metaphysiological approach. Oikos 111:611–615.

Pacala, S. W., C. D. Canham, and J. A. Silander, Jr. 1993. Forest models defined by field measurements: the design of a northeastern forest simulator. Canadian Journal of Forest Research 23:1980–1988.

Pagel, M. D., R. M. May, and A. R. Collie. 1991. Ecological aspects of the geographical distribution and diversity of mammalian species. American Naturalist 137:791–815.

Palmer, M., E. Bernhardt, E. Chornesky, S. Collins, A. Dobson, C. Duke, B. Gold et al. 2004. Ecology for a crowded planet. Science 304:1251–1252.

Parker, V. T. 2004. Community of the individual: implications for the community concept. Oikos 104:27–34.

Parmesan, C., and G. Yohe. 2003. A globally coherent fingerprint of climate change impacts across natural systems. Nature 421:37–42.

Pastor, J., and W. M. Post. 1986. Influence of climate, soil moisture, and succession on forest carbon and nitrogen cycles. Biogeochemistry 2:3–27.

Patten, B. C. 1981. Environs: the superniches of ecosystems. American Zoologist 21:845–852.

Peck, S. L. 2004. Simulation as experiment: a philosophical reassessment for biological modeling. Trends in Ecology and Evolution 19:530–534.

Peet, R. K. 1992. Community structure and persistence, Pages 103–151 in D. C. Glenn-Lewin, R. K. Peet, and T. T. Veblen, eds. Plant Succession: Theory and Prediction. New York, Chapman and Hall.

Peters, D. P. C., B. T. Bestelmeyer, J. E. Herrick, H. C. Monger, E. Fredrickson, and K. M. Havstad. 2006. Disentangling complex landscapes: new insights to forecasting arid and semiarid system dynamics. BioScience 56:491–501.

Peters, D. P. C., B. T. Bestelmeyer, A. K. Knapp, J. E. Herrick, H. C. Monger, and K. M. Havstad. 2009. Approaches to predicting broad-scale regime shifts using changing pattern-process relationships across scales, Pages 47–72 in S. Miao, S. Carstenn, and M. Nungesser, eds. Real World Ecology: Large-scale and Long-term Case Studies and Methods. New York, Springer.

Peters, D. P. C., B. T. Bestelmeyer, and M. G. Turner. 2007. Cross-scale interactions and changing pattern-process relationships: consequences for system dynamics. Ecosystems 10:790–796.

Peters, D. P. C., P. M. Groffman, K. J. Nadelhoffer, N. B. Grimm, S. L. Collins, W. K. Michener, and M. A. Huston. 2008. Living in a connected world: a framework for continental-scale environmental science. Frontiers in Ecology and the Environment 5:229–237.

Peters, D. P. C., R. A. Pielke, Sr, B. T. Bestelmeyer, C. D. Allen, S. Munson-McGee, and K. M. Havstad. 2004. Cross scale interactions, nonlinearities, and forecasting catastrophic events. Proceedings of the National Academy of Sciences 101:15230–15135.

Peters, R. H. 1991. A Critique for Ecology. Cambridge, Cambridge University Press.

Peterson, A. T., M. A. Ortega-Huerta, J. Bartley, V. Sánchez-Cordero, J. Soberón, R. H. Buddemeier, and D. R. B. Stockwell. 2002. Future projections for Mexican faunas under global climatic change scenarios. Nature:626–629.

Peterson, G. D. 2005. Ecological management: control, uncertainty, and understanding, Pages 371–396 in K. Cuddington and B. E. Beisner, eds. Ecological Paradigms Lost: Routes of Theory Change. Burlington, MA, Elsevier Academic Press.

Peterson, G. D., S. R. Carpenter, and W. A. Brock. 2003. Uncertainty and management of multi-state ecosystems: an apparently rational route to collapse. Ecology 84:1403–1411.

Petit, J. R., J. Jouzel, D. Raynaud, N. I. Barkov, and J. M. Barnola. 1999. Climate and atmospheric history of the past 420000 years from the Vostok ice core, Antarctica. Nature 399:429–436.

Petraitis, P. S., R. E. Latham, and R. A. Niesenbaum. 1989. The maintenance of species diversity by disturbance. Quarterly Review of Biology 64:393–418.

Pianka, E. R. 1966. Latitudinal gradients in species diversity: a review of concepts. American Naturalist 100:33–46.

Pianka, E. R., and H. S. Horn. 2005. Ecology's legacy from Robert MacArthur, Pages 213–232 in K. Cuddington and B. E. Biesner, eds. Ecological Paradigms Lost: Routes of Theory Change. New York, Elsevier Academic Press.

Pickett, S. T. A. 1976. Succession: an evolutionary interpretation. American Naturalist 110:107–119.

———. 1989. Space-for-time substitution as an alternative to long-term studies, Pages 110–135 *in* G. E. Likens, ed. Long-term Studies in Ecology: Approaches and Alternatives. New York, Springer-Verlag.

Pickett, S. T. A., and M. L. Cadenasso. 2002. Ecosystem as a multidimensional concept: meaning, model and metaphor. Ecosystems 5:1–10.

———. 2005. Vegetation succession, Pages 172–198 *in* E. van der Maarel, ed. Vegetation Ecology. Blackwell Publishing, Malden, MA.

Pickett, S. T. A., M. L. Cadenasso, and C. G. Jones. 2001a. Generation of heterogeneity by organisms: creation, maintenance, and transformation, Pages 33–52 *in* M. L. Hutchings, E. A. John, and A. J. A. Stewart, eds. Ecological Consequences of Habitat Heterogeneity. London, Blackwell.

Pickett, S. T. A., M. L. Cadenasso, and S. J. Meiners. 2008. Ever since Clements: from succession to vegetation dynamics and understanding to intervention. Applied Vegetation Science 12:9–21.

Pickett, S. T. A., M. L. Cadenasso, J. M. Grove, C. H. Nilon, R. V. Pouyat, W. C. Zipperer, and R. Costanza. 2001b. Urban ecological systems: linking terrestrial ecological, physical and socioeconomic components of metropolitan areas. Annual Review of Ecology and Systematics 32:127–157.

Pickett, S. T. A., S. L. Collins, and J. J. Armesto. 1987a. A hierarchical consideration of causes and mechanisms of succession. Vegetatio 69:109–114.

———. 1987b. Models, mechanisms and pathways of succession. Botanical Review 53:335–371.

Pickett, S. T. A., J. Kolasa, J. J. Armesto, and S. L. Collins. 1989. The ecological concept of disturbance and its expression at various hierarchical levels. Oikos 54:129–136.

Pickett, S. T. A., J. Kolasa, and C. G. Jones. 2007. Ecological Understanding: The Nature of Theory and the Theory of Nature. 2nd ed. New York, Elsevier.

Pickett, S. T. A., and M. J. McDonnell. 1989. Changing perspectives in community dynamics: a theory of successional forces. Trends in Ecology and Evolution 4:241–245.

Pickett, S. T. A., and K. H. Rogers. 1997. Patch dynamics: the transformation of landscape structure and function, Pages 101–127 *in* J. A. Bissonette, ed. Wildlife and Landscape Ecology: Effects of Pattern and Scale. New York, Springer-Verlag.

Pickett, S. T. A., and P. S. White. 1985. The Ecology of Natural Disturbance and Patch Dynamics. San Diego, Academic Press.

Pielke, R. A., Jr, and C. N. Landsea. 1999. La Niña, El Niño, and Atlantic hurricane damages in the United States. Bulletin of the American Meteorological Society 80:2027–2033.

Pielke, R. A., Sr, G. Marland, R. A. Betts, T. N. Chase, J. L. Eastman, J. O. Niles, D. D. S. Niyogi et al. 2002. The influence of land-use change and landscape dynamics on the climate system: relevance to climate change policy beyond the radiative effect of greenhouse gases. Philosophical Transactions of the Royal Society A 360:1705–1719.

Pielou, E. C. 1969. Introduction to Mathematical Ecology. New York, Wiley.

———. 1975. Ecological Diversity. New York, Wiley.

———. 1977. The latitudinal spans of seaweed species and their patterns of overlap. Journal of Biogeography 4:299–311.

Pimm, S. L. 1980. Food web design and the effects of species deletion. Oikos 35:139–149.

———. 1984. The complexity and stability of ecosystems. Nature 307:321–326.

Pineda, J., and H. Caswell. 1998. Bathymetric species-diversity patterns and boundary constraints on vertical range distributions. Deep-Sea Research II 45:83–101.

Poe, E. A. 1844. "The premature burial," The Philadelphia Dollar Newspaper. Philadelphia.

Porter, W. P., and D. M. Gates. 1969. Thermodynamic equilibria of animals with environment. Ecological Monographs 39:227–244.

Post, J. R., M. Sullivan, S. Cox, N. P. Lester, C. J. Walters, E. A. Parkinson, A. J. Paul et al. 2002. Canada's recreational fisheries: the invisible collapse? Fisheries 27:6–17.

Potter, C. S., S. Wang, N. T. Nikolov, A. D. McGuire, J. Liu, A. W. King, J. S. Kimball et al. 2001. Comparison of boreal ecosystem model sensitivity to variability in climate and forest site parameters. Journal of Geophysical Research 106:671–687.

Power, M. E., W. J. Matthews, and A. J. Stewart. 1985. Grazing minnows, piscivorous bass, and stream algae: dynamics of a strong interaction. Ecology 66:1448–1456.

Preisser, E. L., D. I. Bolnick, and M. F. Benard. 2005. Scared to death? The effects of intimidation and consumption in predator-prey interactions. Ecology 86:501–509.

Pressey, R. L., M. Cabeza, M. E. Watts, R. M. Cowling, and K. A. Wilson. 2007. Conservation planning in a changing world. Trends in Ecology and Evolution 22:583–592.

Preston, F. W. 1948. The commonness and rarity of species. Ecology 29:254–283.

———. 1962a. The canonical distribution of commonness and rarity: part I. Ecology 43:182–215.

———. 1962b. The canonical distribution of commonness and rarity: part II. Ecology 43:410–432.

Pueyo, S., F. He, and T. Zillio. 2007. The maximum entropy formalism and the idiosyncratic theory of biodiversity. Ecology Letters 10:1017–1028.

Pulliam, H. R. 2000. On the relationship between niche and distribution. Ecology Letters 3:349–361.

Quetier, F., A. Thebault, and S. Lavorel. 2007. Plant traits in a state and transition framework as markers of ecosystem response to land-use change. Ecological Monographs 77:33–52.

Quine, W. V. O. 1960. Word and Object. Cambridge, MA, MIT Press.

Raffel, T. R., L. B. Martin, and J. R. Rohr. 2008. Parasites as predators: unifying natural enemy ecology. Trends in Ecology and Evolution 23:610–618.

Rahbek, C. 1995. The elevational gradient of species richness: a uniform pattern? Ecography 19:200–205.

———. 1997. The relationship between area, elevation and regional species richness in Neotropical birds. American Naturalist 149:875–902.

———. 2005. The role of spatial scale in the perception of large-scale species-richness patterns. Ecology Letters 8:224–239.

Rahbek, C., N. Gotelli, R. K. Colwell, G. L. Entsminger, T. F. L. V. B. Rangel, and G. R. Graves. 2007. Predicting continental-scale patterns of bird species richness with spatially explicit models. Proceedings of the Royal Society B: Biological Sciences 274:165–174.

Rahbek, C., and G. R. Graves. 2000. Detection of macro-ecological patterns in South American hummingbirds is affected by spatial scale. Proceedings of the Royal Society B: Biological Sciences 267:2259–2265.

———. 2001. Multiscale assessment of patterns of avian species richness. Proceedings of the National Academy of Sciences 98:4534–4539.

Ramalay, F. 2008. The growth of a science. University of Colorado Studies 26:3–14.

Ramos-Jiliberto, R. 2005. Resource-consumer models and the biomass conversion principle. Environmental Modelling and Software 20:85–91.

Randolph, S. E., C. Chemini, C. Furlanello, C. Genchi, R. S. Halls, P. J. Hudson, L. D. Jones et al. 2002. The ecology of tick-borne infections in wildlife reservoirs, Pages 119–138 in P. J. Hudson, A. Rizzoli, B. T. Grenfell, H. Heesterbeek, and A. P. Dobson, eds. The Ecology of Wildlife Diseases. Oxford, Oxford University Press.

Rangel, T. F., and J. A. F. Diniz-Filho. 2005a. Neutral community dynamics, the mid-domain effect and spatial patterns in species richness. Ecology Letters 8:783–790.

Rangel, T. F., J. A. F. Diniz-Filho, and L. M. Bini. 2010. SAM: a comprehensive application for Spatial Analysis in Macroecology. Ecography 33:46–50.

Rangel, T. F. L. V. B., and J. A. F. Diniz-Filho. 2005b. An evolutionary tolerance model explaining spatial patterns in species richness under environmental gradients and geometric constraints. Ecography 28:253–263.

Rangel, T. F. L. V. B., J. A. F. Diniz-Filho, and R. K. Colwell. 2007. Species richness and evolutionary niche dynamics: a spatial pattern-oriented simulation experiment. American Naturalist 170:602–616.

Rapoport, E. H. 1975. Areografia: Estrategias Geograficas de las Especies. Mexico City, Fondo de Cultura Economica.

———. 1982. Areography. Geographical Strategies of Species. 1st English ed. Oxford, Pergamon Press.

Real, L., and T. Caraco. 1986. Risk and foraging in stochastic environments. Annual Review of Ecology and Systematics 17:371–390.

Real, L. A., and J. H. Brown. 1991. Foundations of Ecology. Chicago, University of Chicago Press.

Redfield, A. C. 1958. The biological control of chemical factors in the environment. American Scientist 46:205–221.

Rehage, J. S., B. K. Barnett, and A. Sih. 2005. Behavioral responses to a novel predator and competitor of invasive mosquitofish and their non-invasive relatives (Gambusia sp.). Behavioral Ecology and Sociobiology 57:256–266.

Reiners, W. A. 1986. Complementary models for ecosystems. American Naturalist 127:59–73.

Reiners, W. A., and J. A. Lockwood. 2009. Philosophical Foundations for the Practices of Ecology. Cambridge, Cambridge University Press.

Relyea, R. A. 2001. Morphological and behavioural plasticity of larval anurans in response to different predators. Ecology 82:523–540.

Reynolds, H. L., A. Packer, J. D. Bever, and K. Clay. 2003. Grassroots ecology: plant-microbe-soil interactions as drivers of plant community structure and dynamics. Ecology 84:2281–2291.

Rial, J. A., R. A. Pielke, Sr, M. Beniston, M. Claussen, J. Canadell, P. Cox, H. Held et al. 2004. Nonlinearities, feedbacks, and critical thresholds within the Earth's climate system. Climatic Change 65:11–38.

Ricklefs, R. E. 1987. Community diversity: relative roles of local and regional processes. Science 235:167–171.

Ricklefs, R. E., and D. Schluter. 1993. Species Diversity in Ecological Communities: Historical and Geographical Perspectives. Chicago, University of Chicago Press.

Rietkerk, M., S. C. Dekker, P. C. de Ruiter, and J. van de Koppel. 2004. Self-organized patchiness and catastrophic shifts in ecosystems. Science 305:1926–1929.

Robinson, B. W., and D. S. Wilson. 1998. Optimal foraging, specialization, and a solution to Liem's Paradox. American Naturalist 151:223–235.

Rodriguez, M. A., J. A. Belmontes, and B. A. Hawkins. 2005. Energy, water and large-scale patterns of reptile and amphibian species richness in Europe. Acta Oecologica-International Journal of Ecology 28:65–70.

Rohde, K., M. Heap, and D. Heap. 1993. Rapoport's rule does not apply to marine teleosts and cannot explain latitudinal gradients in species richness. American Naturalist 142:1–16.

Root, R. B. 1967. The niche exploitation pattern of the blue-grey gnatcatcher. Ecological Monographs 37:317–350.

Root, T. L., J. T. Price, K. R. Hall, S. H. Schneider, C. Rosenzweig, and J. A. Pounds. 2003. Fingerprints of global warming on wild animals and plants. Nature 421:57–60.

Rosenzweig, M. L. 1971. Paradox of enrichment: destabilization of exploitation ecosystems in ecological time. Science 171:385–387.

———. 1995. Species Diversity in Space and Time. Cambridge, Cambridge University Press.

———. 2001. Loss of speciation rate will impoverish future diversity. Proceedings of the National Academy of Sciences 98:5404–5410.

Rosenzweig, M. L., and Z. Abramsky. 1993. How are diversity and productivity related?, Pages 52–65 in R. E. Ricklefs and D. Schluter, eds. Species Diversity in Ecological Communities. Chicago, University of Chicago Press.

Rosenzweig, M. L., and R. H. MacArthur. 1963. Graphical representation of stability conditions of predator-prey interactions. American Naturalist 97:209–223.

Roughgarden, J. 1998. Primer of Ecological Theory. Upper Saddle River, NJ, Prentice Hall.

———. 2009. Is there a general theory of community ecology? Biology and Philosophy 24:521–529.

Royama, T. 1992. Analytical Population Dynamics. Berlin, Springer.

Running, S. W. 2008. Ecosystem disturbance, carbon, and climate. Science 321:652–653.

Ruscoe, W. A., J. S. Elkinton, D. Choquenot, and R. B. Allen. 2005. Predation of beech seedide by mice: effects of numerical and functional responses. Journal of Animal Ecology 74:1005–1019.

Ryder, J. J., M. R. Miller, A. White, R. J. Knell, and M. Boot. 2007. Host-parasite population dynamics under combined frequency- and density-dependent transmission. Oikos 11:2017–2026.

Sala, O. E., W. K. Lauenroth, S. J. McNaughton, G. Rusch, and X. Zhang. 1996. Biodiversity and ecosystem functioning in grasslands, Pages 129–149 in H. A. Mooney, J. H. Cushman, E. Medine, O. E. Sala, and E. D. Schulze, eds. Functional Roles of Biodiversity: A Global Perspective. New York, John Wiley & Sons.

Salo, P., E. Korpimaki, P. B. Banks, M. Nordstrom, and C. R. Dickman. 2007. Alien predators are more dangerous than native predators to prey populations. Proceedings of the Royal Society B: Biological Sciences 274:1237–1243.

Sandel, B. S., and M. J. McKone. 2006. Reconsidering null models of diversity: do geometric constraints on species ranges necessarily cause a mid-domain effect? Diversity and Distributions 12:467–474.

Sarnelle, O. 2003. Nonlinear effects of an aquatic consumer: causes and consequences. American Naturalist 161:478–496.

Saunders, M. A., R. E. Chandler, C. L. Merchant, and F. P. Roberts. 2000. Atlantic hurricanes and NW Pacific typhoons: ENSO spatial impacts on occurrence and landfall. Geophysical Research Letters 27:1147–1150.

Savage, V. M., J. F. Gillooly, J. H. Brown, G. B. West, and E. L. Charnov. 2004. Effects of body size and temperature on population growth. American Naturalist 163:429–441.

Sax, D. F., and S. D. Gaines. 2003. Species diversity: from global decreases to local increases. Trends in Ecology and Evolution 18:561–566.

———. 2006. The biogeography of naturalized species and the species-area relationship, Pages 449–480 in M. W. Cadotte, S. M. McMahon, and T. Fukami, eds. Conceptual Ecology and Invasions Biology: Reciprocal Approaches to Nature. Dordrecht, Springer.

———. 2008. Species invasions and extinction: the future of native biodiversity on islands. Proceedings of the National Academy of Sciences 105:11490–11497.

Sax, D. F., S. D. Gaines, and J. H. Brown. 2002. Species invasions exceed extinctions on islands worldwide: a comparative study of plants and birds. American Naturalist 160:766–783.

Sax, D. F., B. P. Kinlan, and K. F. Smith. 2005. A conceptual framework for comparing species assemblages in native and exotic habitats. Oikos 108:457–464.

Sax, D. F., J. J. Stachowicz, J. H. Brown, J. F. Bruno, M. N. Dawson, S. D. Gaines, R. K. Grosberg et al. 2007. Ecological and evolutionary insights from species invasions. Trends in Ecology and Evolution 22:465–471.

Schaffer, W. N. 1983. On the application of optimal control theory to the general life history problem. American Naturalist 121:418–431.

Scheffer, M., S. R. Carpenter, J. A. Foley, C. Folke, and B. Walker. 2001. Catastrophic shifts in ecosystems. Nature 413:591–596.

Scheiner, S. M. 2003. Six types of species-area curves. Global Ecology and Biogeography 12:441–447.

———. 2004a. Experiments, observations, and other kinds of evidence, Pages 51–71 in M. L. Taper and S. R. Lele, eds. The Nature of Scientific Evidence. Chicago, University of Chicago Press.

———. 2004b. A mélange of curves: further dialogue about species-area relationships. Global Ecology and Biogeography 13:479–484.

———. 2010. Towards a conceptual framework for biology. Quarterly Review of Biology 85:293–318.

Scheiner, S. M., and J. M. Rey-Benayas. 1994. Global patterns of plant diversity. Evolutionary Ecology 8:331–347.

Scheiner, S. M., and M. R. Willig. 2005. Developing unified theories in ecology as exemplified with diversity gradients. American Naturalist 166:458–469.

———. 2008. A general theory of ecology. Theoretical Ecology 1:21–28.

Schenk, D., L.-F. Bersier, and S. Bacher. 2005. An experimental test of the nature of predation: neither prey- nor ratio-dependent. Journal of Animal Ecology 74:86–91.

Schindler, D. W. 2006. Recent advances in the understanding and management of eutrophication. Limnology and Oceanography 51:356–363.

Schindler, D. W., and R. E. Hecky. 2009. Eutrophication: more nitrogen data needed. Science 324:721–722.

Schindler, D. W., R. E. Hecky, D. L. Findlay, M. P. Stainton, B. R. Parker, M. J. Paterson, K. G. Beaty et al. 2008. Eutrophication of lakes cannot be controlled by reducing nitrogen input: results of a 37-year whole-ecosystem experiment. Proceedings of the National Academy of Sciences 105:11254–11258.

Schlaepfer, M. A., M. C. Runge, and P. W. Sherman. 2002. Ecological and evolutionary traps. Trends in Ecology and Evolution 17:474–480.

Schmida, A., and M. V. Wilson. 1985. Biological determinants of species diversity. Journal of Biogeography 12:1–20.

Schmitz, O. J. 2004. Trophic cascades: the primacy of trait-mediated indirect interactions. Ecology Letters 7:153–163.

Schmitz, O. J., J. H. Grabowski, B. L. Peckarsky, E. L. Preisser, G. C. Trussell, and J. R. Vonesh. 2008. From individuals to ecosystem function: toward an integration of evolutionary and ecosystem ecology. Ecology 89:2436–2445.

Schneider, D. C. 2001. The rise of the concept of scale in ecology. BioScience 51:545–553.

Schneinder, S. H., and R. Londer. 1984. The Coevolution of Climate and Life. New York, Random House Books.

Schoener, T. W. 1971. Theory of feeding strategies. Annual Review of Ecology and Systematics 2:369–404.

———. 1974. Resource partitioning in ecological communities. Science 185:27–39.

———. 1986. Overview: kinds of ecological communities: ecology becomes pluralistic, Pages 467–479 in J. M. Diamond and T. J. Case, eds. Community Ecology. New York, Harper & Row.

———. 1989. The ecological niche, Pages 79–114 in J. M. Cherrett, ed. Ecological Concepts: The Contribution of Ecology to an Understanding of the Natural World. Oxford, Blackwell Scientific.

———. 2009. Ecological niche, Pages 3–13 in S. A. Levin, ed. The Encyclopedia of Ecology. Princeton, NJ, Princeton University Press.

Schoennagel, T., T. T. Veblen, and W. H. Romme. 2004. The interaction of fire, fuels, and climate across rocky mountain forests. BioScience 54:661–676.

Schooley, R. L., and L. C. Branch. 2007. Spatial heterogeneity in habitat quality and cross-scale interactions in metapopulations. Ecosystems 10:846–853.

Schopf, J. W. 1983. Evolution of earth's earliest ecosystems: recent progress and unsolved problems, Pages 361–364 in J. W. Schopf, ed. Earth's Earliest Biosphere. Princeton, NJ, Princeton University Press.

Schweiger, E. W., J. E. Diffendorfer, R. D. Holt, R. Pierotti, and M. S. Gaines. 2000. The interaction of habitat fragmentation, plant, and small mammal succession in an old field. Ecological Monographs 70:383–400.

Seager, R., M. Ting, I. Held, Y. Kushnir, J. Lu, G. Vecchi, H.-P. Huang et al. 2007. Model projections of an imminent transition to a more arid climate in southwestern North America. Science 316:1181–1184.

Seastedt, T. R., R. J. Hobbs, and K. N. Suding. 2008. Management of novel ecosystems: are novel approaches required? Frontiers in Ecology and the Environment 6:547–553.

Seger, J. 1992. Evolution of exploiter-victim relationships, Pages 3–25 in M. J. Crawley, ed. Natural Enemies: The Population Biology of Predators, Parasites, and Diseases. Oxford, Oxford University Press.

Shaw, M. R., E. S. Zavaleta, N. R. Chiariello, E. E. Cleland, H. A. Mooney, and C. B. Field. 2002. Grassland responses to global environmental changes suppressed by elevated CO_2. Science 298:1987–1990.

Sherman, R. E., T. J. Fahey, and P. Martinez. 2001. Hurricane impacts on a mangrove forest in the Dominican Republic: damage patterns and early recovery. Biotropica 33:393–408.

Shugart, H. H. 1984. A Theory of Forest Dynamics: The Ecological Implications of Forest Succession Models. New York, Springer-Verlag.

Shurin, J. 2000. Dispersal limitation, invasion resistance, and the structure of pond zooplankton communities. Ecology 81:3074–3086.

Siegenthaler, U., T. F. Stocker, E. Monnin, D. Luthi, J. Schwander, B. Stauffer, D. Raynaud et al. 2005. Stable carbon cycle-climate relationship during the late Pleistocene. Science 310:1313–1317.

Siemann, E., W. E. Rogers, and J. B. Grace. 2007. Effects of nutrient loading and extreme rainfall events on coastal tallgrass prairies: invasion intensity, vegetation responses, and carbon and nitrogen distribution. Global Change Biology 13:2184–2192.

Sih, A. 1980. Optimal behavior: can foragers balance two conflicting demands? Science 210:1041–1042.

———. 1987. Predators and prey lifestyles: an evolutionary and ecological overview, Pages 203–224 *in* W. C. Kerfoot, and A. Sih, eds. Predation: Direct and Indirect Impacts on Aquatic Communities. Hanover, NH, University Press of New England.

———. 1992. Prey uncertainty and the balancing of antipredator and feeding needs. American Naturalist 139:1052–1069.

———. 1998. Game theory and predator-prey response races, Pages 221–238 *in* L. A. Dugatkin and H. K. Reeve, eds. Game Theory and Animal Behavior. Oxford, Oxford University Press.

Sih, A., D. I. Bolnick, B. Luttbeg, J. L. Orrock, S. D. Peacor, L. M. Pintor, E. Preisser et al. 2010. Predator-prey naïveté, antipredator behavior, and the ecology of predator invasions. Oikos 119.

Sih, A., and B. Christensen. 2001. Optimal diet theory: when does it work, and when and why does it fail? Animal Behaviour 61:379–390.

Sih, A., and R. D. Moore. 1990. Interacting effects of predator and prey behavior in determining diets *in* R. N. Hughes, ed. Behavioural Mechanisms of Food Selection. NATO ASI Series. New York, Springer-Verlag.

Silvert, W. 1995. Is the logistic equation a Lotka-Volterra model? Ecological Modelling 77:95–96.

Simberloff, D. 1974. Equilibrium theory of island biogeography and ecology. Annual Review of Ecology and Systematics 15:161–182.

———. 1978. Using island biogeographic distributions to determine if colonization is stochastic. American Naturalist 112:713–726.

———. 1983. Competition theory, hypothesis-testing, and other community ecology buzzwords. American Naturalist 122:626–635.

———. 2004. Community ecology: is it time to move on? American Naturalist 163:787–799.

Simberloff, D., and E. O. Wilson. 1970. Experimental zoogeography of islands: a two year record of colonization. Ecology 51:934–937.

Simberloff, D. S., and W. Boecklen. 1981. Santa Rosalia reconsidered: size ratios and competition. Evolution 35:1206–1228.

Simberloff, D. S., and E. O. Wilson. 1969. Experimental zoogeography of islands: the colonization of empty Islands. Ecology 50:278–296.

Sitch, S., C. Huntingford, N. Gedney, P. E. Levy, M. Lomas, S. L. Piao, R. Betts et al. 2008. Evaluation of the terrestrial carbon cycle, future plant geography and climate-carbon cycle feedbacks using five Dynamic Global Vegetation Models (DGVMs). Global Change Biology 14:2015–2039.

Šizling, A. L., D. Storch, E. Šizlingová, J. Reif, and K. Gaston. 2009. Species abundance distribution results from a spatial analogy of central limit theorem. Proceedings of the National Academy of Sciences 106:6691–6695.

Skalski, G. T., and J. F. Gilliam. 2001. Functional responses with predator interference: viable alternatives to the Holling Type II model. Ecology 82:3083–3092.

Skellam, J. G. 1951. Random dispersal in theoretical populations. Biometrika 38:196–218.

Smith, V. G. 1928. Animal communities of a deciduous forest succession. Ecology 9:479–500.

Smith, V. H. 2007. Host resource supplies influence the dynamics and outcome of infectious disease. Integrative and Comparative Biology 47:310–316.

Smith, V. H., and R. D. Holt. 1996. Resource competition and within-host disease dynamics. Trends in Ecology and Evolution 11:386–389.

Snyder, R. E., and P. Chesson. 2004. How the spatial scales of dispersal, competition, and environmental heterogeneity interact to affect coexistence. American Naturalist 164:633–650.

Sober, E. 1997. Two outbreaks of lawlessness in recent philosophy of biology. Philosophy of Science 64:S458–S467.

Solé, R. V., D. Alonso, and J. Saldaña. 2004. Habitat fragmentation and biodiversity collapse in neutral communities. Ecological Complexity 1:65–75.

Sousa, W. P. 1984. Intertidal mosaics: propagule availability, and spatially variable patterns of succession. Ecology 65:1918–1935.

Southwood, T. R. E. 1977. Habitat, the template for ecological strategies. Journal of Animal Ecology 46:337–350.

Sprengel, C. 1839. Die Lehre vom Dünger oder Beschreibung aller bei der Landwirthschaft Gebräuchlicher Vegetabilischer, Animalischer und Mineralischer Düngermaterialien, nebst Erklärung ihrer Wirkungsart. Leipzig.

Srivastava, D. S., and J. H. Lawton. 1998. Why more productive sites have more species: an experimental test of theory using tree-hole communities. American Naturalist 152:510–529.

Stachowicz, J. J., and D. Tilman. 2005. Species invasions and the relationships between species diversity, community saturation, and ecosystem functioning, Pages 41–46 in D. F. Sax, S. D. Gaines, and J. J. Stachowicz, eds. Species Invasions: Insights into Ecology, Evolution, and Biogeography. Sunderland, MA, Sinauer Associates.

Stephens, D. W. 2002. Discrimination, discounting and impulsivity: a role for an informational constraint. Philosophical Transactions of the Royal Society B: Biological Sciences 357:1527–1537.

———. 2007. Models of information use, Pages 31–58 in D. W. Stephens, J. S. Brown, and R. C. Ydenberg, eds. Foraging. Behavior and Ecology. Chicago, University of Chicago Press.

Stephens, D. W., J. S. Brown, and R. C. Ydenberg. 2007. Foraging. Behavior and Ecology. Chicago, University of Chicago Press.

Stephens, D. W., and J. R. Krebs. 1986. Foraging Theory. Princeton, NJ, Princeton University Press.

Sterner, R. W., and J. J. Elser. 2002. Ecological Stoichiometry: The Biology of Elements from Molecules to the Biosphere. Princeton, NJ, Princeton University Press.

Stevens, G. C. 1989. The latitudinal gradient in geographical range: how so many species coexist in the tropics. American Naturalist 133:240–256.

———. 1992. The elevational gradient in altitudinal range: an extension of Rapoport's latitudinal rule to altitude. American Naturalist 140:893–911.

———. 1996. Extending Rapoport's rule to Pacific marine fishes. Journal of Biogeography 23:149–154.

Stevens, M. H. H., and W. P. Carson. 1999. The significance of assemblage level thinning for species richness. Journal of Ecology 87:490–502.

Stevens, R. D., and M. R. Willig. 2002. Geographical ecology at the community level: perspectives on the diversity of New World bats. Ecology 83:545–560.

Stevenson, R. J., C. G. Peterson, D. B. Kirschtel, C. C. King, and N. C. Tuchman. 1991. Density-dependent growth, ecological strategies, and the effects of nutrients and shading on benthic diatom succession in streams. Journal of Phycology 27:59–69.

Stiles, A., and S. M. Scheiner. 2010. Effects of fragmentation on remnant plant species richness in an urban landscape. Journal of Biogeography 37:1721–1729.

Stockwell, D. 2007. Niche Modeling: Predictions from Statistical Distributions. Boca Raton, Chapman and Hall.

Stone, L. 2004. Population ecology: a three-player solution. Nature 430:299–300.

Storch, D., R. G. Davies, S. Zajicek, C. D. L. Orme, V. Olson, G. H. Thomas, T. S. Ding et al.

2006. Energy, range dynamics and global species richness patterns: reconciling mid-domain effects and environmental determinants of avian diversity. Ecology Letters 9:1308–1320.

Strong, D. R. 1984. Exorcising the ghost of competition past: phytophagous insects, Pages 28–41 *in* D. R. Strong, D. Simberloff, L. G. Abele, and A. B. Thistle, eds. Ecological Communities: Conceptual Issues and the Evidence. Princeton, NJ, Princeton University Press.

Strong, D. R., Jr., D. Simberloff, L. G. Abele, and A. B. Thistle. 1984. Ecological Communities: Conceptual Issues and the Evidence. Princeton, NJ, Princeton University Press.

Suppe, F. 1977. The Structure of Scientific Theories. Urbana, University of Illinois Press.

Svanback, R., and D. I. Bolnick. 2005. Intraspecific competition affects the strength of individual specialization: an optimal diet theory method. Evolutionary Ecology Research 7:993–1012.

———. 2007. Intraspecific competition drives increased resource use diversity within a natural population. Proceedings of the Royal Society B: Biological Sciences 274:839–844.

Swinton, J., M. E. J. Woolhouse, M. E. Begon, A. P. Dobson, E. Ferroglio, B. T. Grenfell, V. Guberti et al. 2002. Microparasite transmission and persistence, Pages 83–101 *in* P. J. Hudson, A. Rizzoli, B. T. Grenfell, H. Heesterbeek, and A. P. Dobson, eds. The Ecology of Wildlife Diseases. Oxford, Oxford University Press.

Tansley, A. G. 1935. The use and abuse of vegetational concepts and terms. Ecology 16:284–307.

Taylor, P., and S. Gaines. 1999. Can Rapoport's rule be rescued? Modeling causes of the latitudinal gradient in species richness. Ecology 80:2474–2482.

Taylor, R. J. 1984. Predation. London, Chapman & Hall.

Templeton, A. R. 2006. Population Genetics and Microevolutionary Theory. New York, John Wiley & Sons.

Terborgh, J. 1973a. Chance, habitat and dispersal in distribution of birds in West-Indies. Evolution 27:338–349.

———. 1973b. On the notion of favorableness in plant ecology. American Naturalist 107:481–501.

———. 1974. Preservation of natural diversity: the problem of extinction prone species. Bioscience 24:715–722.

Theobold, D. M. 2005. Landscape patterns of exurban growth in the USA from 1980 to 2020, Ecology and Society 10:32, http://www.ecologyandsociety.org/vol10/iss1/art32.

Thompson, J. N. 2005. The Geographic Mosaic of Coevolution. Chicago, University of Chicago Press.

Tilman, D. 1976. Ecological competition between algae: experimental confirmation of resource-based competition theory. Science 192:463–465.

———. 1982. Resource Competition and Community Structure. Princeton, NJ, Princeton University Press.

———. 1988. Plant Strategies and the Dynamics and Structure of Plant Communities. Princeton, NJ, Princeton University Press.

———. 1991. Constraints and tradeoffs: toward a predictive theory of competition and succession. Oikos 58:3–15.

———. 1994. Competition and biodiversity in spatially structured habitats. Ecology 75:2–16.

———. 2004. Niche tradeoffs, neutrality, and community structure: a stochastic theory of resource competition, invasion, and community assembly. Proceedings of the National Academy of Sciences 101:10854–10861.

Tilman, D., and J. A. Downing. 1994. Biodiversity and stability in grasslands. Nature 367:363–365.

Tilman, D., and P. M. Kareiva. 1997. Spatial Ecology: The Role of Space in Population Dynamics and Interspecific Interactions. Princeton, NJ, Princeton University Press.

Tilman, D., J. Knops, D. Wedin, P. Reich, M. Ritchie, and E. Siemann. 1997. The influence of functional diversity and composition on ecosystem processes. Science 277:1300–1302.

Tilman, D., C. L. Lehman, and C. E. Bristow. 1998. Diversity-stability relationships: statistical inevitability or ecological consequence? American Naturalist 151:277–282.

Tilman, D., R. M. May, C. L. Lehman, and M. A. Nowak. 1994. Habitat destruction and the extinction debt. Nature 371:65–66.

Tilman, D., and S. Pacala. 1993. The maintenance of species richness in plant communities, Pages 13–25 *in* R. E. Ricklefs and D. Schluter, eds. Species Diversity in Ecological Communities. Chicago, University of Chicago Press.

Tjørve, E. 2003. Shapes and functions of species-area curves: a review of possible models. Journal of Biogeography 30:827–835.

Toft, C. A., and T. W. Schoener. 1983. Abundance and diversity of orb spiders on 106 Bahamian islands: biogeography at an intermediate trophic level. Oikos 41:411–426.

Tokeshi, M. 1993. Species abundance patterns and community structure. Advances in Ecological Research 24:111–186.

Tollrian, R., and C. D. Harvell. eds. 1999. The Ecology and Evolution of Inducible Defenses. Princeton, NJ, Princeton University Press.

Trenberth, K. E. 2005. Uncertainty in hurricanes and global warming. Science 308:1753–1754.

Trexler, J. C., W. F. Loftus, and S. Perry. 2005. Disturbance frequency and community structure in a twenty-five year intervention study. Oecologia 145:140–152.

Turchin, P. 2001. Does population ecology have general laws? Oikos 94:17–26.

———. 2003. Complex Population Dynamics: A Theoretical/Empirical Synthesis. Princeton, NJ, Princeton University Press.

Turner, J. R. G., J. J. Lennon, and J. A. Lawrenson. 1988. British bird species distributions and the energy theory. Nature 335:539–541.

Turner, M. G., V. H. Dale, and R. H. Gardner. 1989. Predicting across scales: theory development and testing. Landscape Ecology 3:245–252.

Turner, M. G., R. H. Gardner, and R. V. O'Neill. 2001. Landscape Ecology in Theory and Practice. New York, Springer-Verlag.

Ugland, K. I., and J. S. Gray. 1982. Lognormal distributions and the concept of community equilibrium. Oikos 39:171–178.

Ulanowicz, R. E. 1997. Ecology, the Ascendent Perspective. New York, Columbia University.

Urban, D. L., R. V. O'Neill, and H. H. Shugart. 1987. Landscape ecology: a hierarchical perspective can help scientists understand spatial patterns. BioScience 37:119–127.

Urban, F. E., J. E. Cole, and J. T. Overpeck. 2000. Influence of mean climate change on climate variability from a 155-year tropical Pacific coral record. Nature 407:989–993.

Urban, M. C., M. A. Leibold, P. Amarasekare, L. De Meester, R. Gomulkiewicz, M. E. Hochberg, C. A. Klausmeier et al. 2008. The evolutionary ecology of metacommunities. Trends in Ecology and Evolution 23:311–317.

Uriarte, M., R. Condit, C. D. Canham, and S. P. Hubbell. 2004. A spatially explicit model of sampling growth in a tropical forest: does the identity of the neighbours matter? Journal of Ecology 348–360.

Usher, M. B. 1979. Markovian approaches to ecological succession. Journal of Animal Ecology 48:413–426.

———. 1992. Statistical models of succession, Pages 215–248 in D. C. Glenn-Lewin, R. K. Peet, and T. T. Veblen, eds. Plant Succession: Theory and Prediction. New York, Chapman and Hall.

Vandermeer, J. H. 1972. Niche theory. Annual Review of Ecology and Systematics 3:107–132.

VanderMeulen, M. A., A. J. Hudson, and S. M. Scheiner. 2001. Three evolutionary hypotheses for the hump-shaped productivity-diversity curve. Evolutionary Ecology Research 3:379–392.

van der Ploeg, R. R., W. Böhm, and M. B. Kirkham. 1999. On the origin of the theory of mineral nutrition of plants and the law of the minimum. Soil Science Society of America Journal 63:1055–1062.

van Fraassen, B. 1980. The Scientific Image. Oxford, Oxford University Press.

Vasseur, D. A., and K. S. McCann. 2005. A mechanistic approach for modeling temperature-dependent consumer-resource dynamics. American Naturalist 166:184–198.

Veblen, T. T. 1992. Regeneration dynamics, Pages 152–187 in D. C. Glenn-Lewin, R. K. Peet, and T. T. Veblen, eds. Plant Succession: Theory and Prediction. New York, Chapman and Hall.

Velhurst, P. F. 1838. Notice sur la loi que la population poursuit dans son accroissement. Correspondance mathématique et physique 10:113–121.

Vellend, M., L. J. Harmon, J. L. Lockwood, M. M. Mayfield, A. R. Hughes, J. P. Wares, and D. F. Sax. 2007. Effects of exotic species on evolutionary diversification. Trends in Ecology and Evolution 22:481–488.

Villalobos, F., and H. Arita. 2009. The diversity field of New World leaf-nosed bats (Phyllostomidae). Global Ecology and Biogeography 19:200–211.

Vincent, T. L. S., D. Scheel, J. S. Brown, and T. L. Vincent. 1996. Trade-offs and coexistence in consumer-resource models: it all depends on what and where you eat. American Naturalist 148:1038.

Vitousek, P. M. 2004. Nutrient Cycling and Limitation: Hawai'i as a Model System. Princeton, Princeton University Press.

Vitousek, P. M., J. D. Aber, R. W. Howarth, G. E. Likens, P. A. Matson, D. W. Schindler, W. H. Schlesinger et al. 1997a. Human alteration of the global nitrogen cycle: sources and consequences. Ecological Applications 7:737–750.

Vitousek, P. M., P. R. Ehrlich, A. H. Ehrlich, and P. A. Matson. 1986. Human appropriation of the products of photosynthesis. BioScience 36:368–373.

Vitousek, P. M., and J. M. Melillo. 1979. Nitrate losses from disturbed forests: patterns and mechanisms. Forest Science 25:605–619.

Vitousek, P. M., H. A. Mooney, J. Lubchenco, and J. M. Melillo. 1997b. Human domination of Earth's ecosystems. Science 277:494–499.

Vitousek, P. M., and W. A. Reiners. 1975. Ecosystem succession and nutrient retention: a hypothesis. BioScience 25:376–381.

Volkov, I., J. R. Banavar, F. He, S. P. Hubbell, and A. Maritan. 2005. Density dependence explains tree species abundance and diversity in tropical forests. Nature 438:658–661.

Volkov, I., J. R. Banavar, S. P. Hubbell, and A. Maritan. 2003. Neutral theory and relative species abundance in ecology. Nature 424:1035–1037.

———. 2007. Patterns of relative species abundance in rainforests and coral reefs. Nature 450:45–49.

Volterra, V. 1926. Fluctuations in the abundance of a species considered mathematically. Nature 118:558–560.

———. 1931. Lecons sur la Mathematique de la Lutte pour la Via. Paris, Marcel Brelot.

von Humboldt, A. 1808. Ansichten der Natur mit wissenschaftlichen Erlauterungen. Tübingen, Germany, J. G. Cotta.

Vucetich, J. A., R. O. Peterson, and C. L. Schaefer. 2002. The effect of prey and predator densities on wolf predation. Ecology 83:3003–3013.

Wächtershäuser, G. 1990. Evolution of the first metabolic cycles. Proceedings of the National Academy of Sciences 87:200–204.

Waide, R. B., M. R. Willig, C. F. Steiner, G. Mittelbach, L. Gough, S. I. Dodson, G. P. Juday et al. 1999. The relationship between productivity and species richness. Annual Review of Ecology and Systematics 30:257–300.

Waite, T. A., and K. L. Field. 2007. Foraging with others: games social foragers play, Pages 331–362 in D. W. Stephens, J. S. Brown, and R. C. Ydenberg, eds. Foraging: Behavior and Ecology. Chicago, University of Chicago Press.

Walker, L. R. 1993. Nitrogen fixers and species replacements in primary succession, Pages 249–272 in J. Miles and D. W. H. Walton, eds. Primary Succession on Land. Oxford, Blackwell Scientific Publication.

———. ed. 1999. Ecosystems of Disturbed Ground. New York, Elsevier.

Walker, L. R., and R. del Moral. 2003. Primary Succession and Ecosystem Rehabilitation. New York, Cambridge University Press.

Walker, L. R., J. Walker, and R. J. Hobbs. eds. 2007. Linking Restoration and Succession in Theory and Practice. New York, Springer.

Wallace, A. R. 1878. Tropical Nature and Other Essays. London, Macmillan & Co.

Walther, G.-R., E. Post, P. Convey, A. Menzel, C. Parmesan, T. J. C. Beebee, J.-M. Fromentin et al. 2002. Ecological responses to recent climate change. Nature 416:389–395.

Watt, A. S. 1947. Pattern and process in the plant community. Journal of Ecology 35:1–22.

Webster, P. J., G. J. Holland, J. A. Curry, and H. R. Chang. 2005. Changes in tropical cyclone number, duration, and intensity in a warming environment. Science 309:1844–1846.

Werner, E. E. 1977. Species packing and niche complementary in 3 sunfishes. American Naturalist 111:553–578.

Werner, E. E., and B. R. Anholt. 1993. Ecological consequences of the tradeoff between growth and mortality rates mediated by foraging activity. American Naturalist 142:242–272.

Werner, E. E., and J. F. Gilliam. 1984. The ontogenetic niche and species interactions in size-structured populations. Annual Review of Ecology and Systematics 15:393–425.

Werner, E. E., J. F. Gilliam, D. L. Hall, and G. G. Mittelbach. 1983. An experimental test of the effects of predation risk on habitat use in fish. Ecology 64:1540–1548.

Werner, E. E., and D. L. Hall. 1974. Optimal foraging and size selection of prey by bluegill sunfish (Lepomis macrochirus). Ecology 5:1042–1052.

West, G. B., J. H. Brown, and B. J. Enquist. 1997. A general model for the origin of allometric scaling laws in biology. Science 276:122–126.

Whelan, C. J., and K. A. Schmidt. 2007. Food acquisition, processing, and digestion, Pages 141–172 in D. W. Stephens, J. S. Brown, and R. C. Ydenberg, eds. Foraging: Behavior and Ecology. Chicago, University of Chicago Press.

Whewell, W. 1858. Novum Organon Renovatum. London, J. W. Parker and son.

White, P. S., and A. Jentsch. 2001. The search for generality in studies of disturbance and ecosystem dynamics. Progress in Botany 62 399–449.

White, P. S., and S. T. A. Pickett. 1985. Natural disturbance and patch dynamics: an introduction, Pages 3–13 *in* S. T. A. Pickett and P. S. White, eds. The Ecology of Natural Disturbance and Patch Dynamics. Orlando, FL, Academic Press.

Whitfield, J. 2004. Ecology's big, hot idea. PLoS Biology 2:2023–2027.

Whitham, T. G., W. P. Young, G. D. Martinsen, C. A. Gehring, J. A. Schweitzer, S. M. Shuster, G. M. Wimp et al. 2003. Community and ecosystem genetics: a consequence of the extended phenotype. Ecology 84:559–573.

Whittaker, R. H. 1965. Dominance and diversity in land plant communities. Science 147:250–260.

Whittaker, R. H., S. A. Levin, and R. B. Root. 1973. Niche, habitat, and ecotope. American Naturalist 107:321–338.

Whittaker, R. J. 1998. Island Biogeography: Ecology, Evolution, and Conservation. Oxford, Oxford University Press.

———. 2000. Scale, succession and complexity in island biogeography: are we asking the right questions? Global Ecology and Biogeography 9:75–85.

———. 2006. Island species-energy theory. Journal of Biogeography 33:11–12.

Whittaker, R. J., M. B. Araujo, P. Jelpson, R. J. Ladle, J. E. M. Watson, and K. J. Willis. 2005. Conservation biogeography: assessment and prospect. Diversity and Distributions 11:3–23.

Whittaker, R. J., and M. B. Bush. 1993. Dispersal and establishment of tropical forest assemblages, Krakatoa, Indonesia, Pages 147–160 *in* J. Miles and D. W. H. Walton, eds. Primary Succession on Land. Oxford, Blackwell Scientific Publications.

Whittaker, R. J., and J. M. Fernandez-Palacios. 2007. Island Biogeography: Ecology, Evolution, and Conservation. 2nd ed. New York, Oxford University Press.

Whittaker, R. J., and E. Heegaard. 2003. What is the observed relationship between species richness and productivity? Comment. Ecology 84:3384–3390.

Whittaker, R. J., K. A. Triantis, and R. J. Ladle. 2008. A general dynamic theory of oceanic island biogeography. Journal of Biogeography 35:977–994.

Whittaker, R. J., K. J. Willis, and R. Field. 2001. Scale and species richness: towards a general, hierarchical theory of species diversity. Journal of Biogeography 28:453–470.

Wiens, J., and M. Donoghue. 2004. Historical biogeography, ecology and species richness. Trends in Ecology and Evolution 19:639–644.

Wiens, J. A. 1984. On understanding a non-equilibrium world: myth and reality in community patterns and processes, Pages 439–457 Ecological Communities: Conceptual Issues and the Evidence. Princeton, NJ, Princeton University Press.

———. 1989. Spatial scaling in ecology. Functional Ecology 3:385–397.

Wigley, T. M. L., and S. D. Schimel. eds. 2000. The Carbon Cycle. Cambridge, Cambridge University Press.

Williams, G. C. 1966. Adaptation and Natural Selection. Princeton, NJ, Princeton University Press.

Williams, J. W., and S. T. Jackson. 2007. Novel climates, no-analog communities, and ecological surprises. Frontiers in Ecology and the Environment 9:475–482.

Williams, J. W., S. T. Jackson, and J. E. Kutzbach. 2007a. Projected distributions of novel and disappearing climates by 2100 AD. Proceedings of the National Academy of Sciences 104:5738–5742.

Williams, R. J., U. Brose, and N. D. Martinez. 2007b. Homage to Yodzis and Innes 1992: Scaling up feeding-based population dynamics to complex ecological networks, Pages 37–52 *in*

N. Rooney, K. S. McCann, and D. L. G. Noakes, eds. From Energetics to Ecosystems: The Dynamics and Structure of Ecological Systems. Dordrecht, Netherlands, Springer.

Williamson, C. E., W. Dodds, T. K. Kratz, and M. A. Palmer. 2008. Lakes and streams as sentinels of environmental change in terrestrial and atmospheric processes. Frontiers in Ecology and the Environment 6:247–254.

Williamson, M., and K. J. Gaston. 2005. The lognormal distribution is not an appropriate null hypothesis for the species-abundance distribution. Journal of Animal Ecology 74:409–422.

Willig, M. R., C. P. Bloch, N. Brokaw, C. Higgens, J. Thompson, and C. R. Zimmerman. 2007. Cross-scale responses of biodiversity to hurricane and anthropogenic disturbance in a tropical forest. Ecosystems 10:824–838.

Willig, M. R., D. M. Kaufman, and R. D. Stevens. 2003. Latitudinal gradients of biodiversity: pattern, process, scale, and synthesis. Annual Review of Ecology, Evolution, and Systematics 34:273–309.

Willig, M. R., and S. K. Lyons. 1998. An analytical model of latitudinal gradients of species richness with an empirical test for marsupials and bats in the New World. Oikos 81:93–98.

Wilmshurst, J. F., J. M. Fryxell, and C. M. Bergman. 2000. The allometry of patch selection in ruminants. Proceedings of the Royal Society B: Biological Sciences 267:345–349.

Wilmshurst, J. F., J. M. Fryxell, and P. E. Colucci. 1999. What constrains daily intake in Thomson's gazelles? Ecology 80:2338–2347.

Wilsey, B. J., D. R. Chalcraft, C. M. Bowles, and M. R. Willig. 2005. Two dimensional nature of spatial variation in species diversity of grassland communities. Ecology 86:1178–1184.

Wilson, D. S. 1988. Holism and reductionism in evolutionary ecology. Oikos 53:269–273.

———. 1997. Biological communities as functionally organized units. Ecology 78:2018–2024.

Wilson, E. O. 1961. The nature of the taxon cycle in the Melanesian ant fauna. American Naturalist 95:169–193.

———. 1969. The species equilibrium, Pages 38–47 in G. M. Woodwell and H. H. Smith, eds. Diversity and Stability in Ecological Systems. Upton, NY, Brookhaven National Laboratory.

———. 1993. Naturalist. New York, HarperCollins.

Wilson, E. O., and G. E. Hutchinson. 1982. Robert Helmer MacArthur. Biographical Memoirs 58:319–327.

Wilson, E. O., and E. O. Willis. 1975. Applied Biogeography. Cambridge, Belknap Press of Harvard University.

Wilson, K. A., and T. R. Hrabik. 2006. Ecological change and exotic invaders, Pages 151–167 in J. J. Magnuson, T. K. Kratz, and B. J. Benson, eds. Long-Term Dynamics of Lakes on the Landscape. Oxford, Oxford University Press.

Wilson, W. G., and P. A. Abrams. 2005. Coexistence of cycling consumer species having localized interactions; Armstrong and McGehee in space. American Naturalist 195:193–205.

Wilson, W. G., C. W. Osenberg, R. J. Schmitt, and R. M. Nisbet. 1999. Complementary foraging behavior allows coexistence of two grazers. Ecology 80:2358–2372.

Wimsatt, W. C. 2007. Re-engineering Philosophy for Limited Beings. Cambridge, MA, Harvard University Press.

Winsberg, E. 1999. Sanctioning models: the epistemology of a simulation. Science in Context 12:247–260.

———. 2001. Simulations, models, and theories: complex physical systems and their representations. Philosophy of Science 68:S442–S454.

———. 2003. Simulated experiments: methodology for a virtual world. Philosophy of Science 70:105–125.

Wollkind, D. J. 1976. Exploitation in three trophic levels: an extension allowing intraspecies carnivore interaction. American Naturalist 110:431–447.

Woodward, F. I., and C. K. Kelly. 2008. Responses of global plant diversity capacity to changes in carbon dioxide concentration and climate. Ecology Letters 11:1229–1237.

Wootton, J. T. 1998. Effects of disturbance on species diversity: a multitrophic perspective. American Naturalist 152:803–825.

Wright, D. H. 1983. Species-energy theory: an extension of species-area theory. Oikos 41:496–506.

Wu, J. 2007. Scale and scaling: a cross-disciplinary perspective, Pages 115–142 in J. Wu and R. J. Hobbs, eds. Key Topics in Landscape Ecology. Cambridge, Cambridge University Press.

Wyckoff, P. H., and J. S. Clark. 2002. The relationship between growth and mortality for seven co-occurring tree species in the southern Appalachian Mountains. Journal of Ecology 90:604–615.

Wylie, J. L., and D. J. Currie. 1993. Species energy theory and patterns of species richness: I. Patterns of bird, angiosperm, and mammal richness on islands. Biological Conservation 63:137–144.

Wynne-Edwards, V. C. 1962. Animal Dispersion in Relation to Social Behaviour. Edinburgh, Oliver and Boyd.

Yao, J., D. P. C. Peters, K. M. Havstad, R. P. Gibbens, and J. E. Herrick. 2006. Multi-scale factors and long-term responses of Chihuahuan Desert grasses to drought. Landscape Ecology 21:1217–1231.

Yi, L., W. You, and Y. Song. 2006. Soil animal communities at five succession stages in the litter of evergreen broad-leaved forest in Tiantong, China. Acta Ecologica Sinica 1:142–150.

Yodzis, P. 1989. Introduction to Theoretical Ecology. New York, Harper and Row.

Yodzis, P., and S. Innes. 1992. Body size and consumer-resource dynamics. American Naturalist 139:1151–1175.

Young, D. R., J. H. Porter, C. M. Bachmann, G. Shao, R. A. Fusina, J. H. Bowles, D. Korwan et al. 2007. Cross-scale patterns in shrub thicket dynamics in the Virginia Barrier Complex. Ecosystems:854–863.

Yu, D. W., and H. B. Wilson. 2001. The competition-colonization trade-off is dead; long live the competition-colonization trade-off. American Naturalist 158:49–63.

Zak, D. R., C. B. Blackwood, and M. P. Waldrop. 2006. A molecular dawn for biogeochemistry. Trends in Ecology and Evolution 21:288–295.

Zapata, F. A., K. J. Gaston, and S. L. Chown. 2003. Mid-domain models of species richness gradients: assumptions, methods and evidence. Journal of Animal Ecology 72:677–690.

———. 2005. The mid-domain effect revisited. American Naturalist 166:E144–E148.

Zipf, G. K. 1965. Human Behavior and the Principle of Least Effort. New York, Hafner.

INDEX